Lecture Notes in Earth Sciences 103

Springer
Berlin
Heidelberg
New York
Hong Kong
London
Milan
Paris
Tokyo

Michael Kühn

Reactive Flow Modeling of Hydrothermal Systems

With 68 Colour Illustrations

Springer

Author

Dr. Michael Kühn
CSRIO – ARRC / Exploration and Mining
26 Dick Perry Avenue, Technology Park
Kensington, Perth, WA 6151
Australia

Cataloging-in-Publication Data applied for

Bibliographic information published by Die Deutsche Bibliothek.
Die Deutsche Bibliothek lists this publication in the Deutsche Nationalbibliographie; detailed bibliographic data is available in the Internet at <http://dnb.ddb.de>.

"For all Lecture Notes in Earth Sciences published till now please see final pages of the book"

ISSN 0930-0317
ISBN 3-540-20338-9 Springer-Verlag Berlin Heidelberg New York

Springer-Verlag is a part of Springer Science+Business Media

springeronline.com

Cover design: Erich Kirchner, Heidelberg
Typesetting: Camera ready by author
Printed on acid-free paper 32/3142/du - 5 4 3 2 1 0

Preface

This book introduces the topic of geochemical reaction modeling of the fluids in subsurface and hydrothermal systems. It is designed for readers first entering into this world as well as for sophisticated researchers who want to improve their knowledge especially of the interaction between chemical reactions and other processes like fluid flow and heat transfer in hydrothermal systems. Furthermore, it will be of interest for characterization, delineation, and exploration of geothermal reservoirs as well as for mineral geology, to explain or to reassess existing deposits and to search for new ones.

The intention behind this manuscript has been to serve as a textbook for graduate students in aqueous, environmental, and groundwater geochemistry, despite the fact that its focus is on the special topic of geochemistry in hydrothermal systems, and to provide new insights for experienced researches with respect to the topic of reactive transport. The overall purpose of the book is to give the reader an understanding of the processes that control the chemical composition of waters in hydrothermal systems and to point out the interfaces between chemistry, geothermics, and hydrogeology.

The first chapter displays the significance of geochemical models of hydrothermal systems and their use in reservoir exploration and exploitation. It outlines the objectives of this book and the conducted research work.

The second chapter consists of two parts. The first part explains main concepts of different types of geothermal systems. Particular attention is given to chemical, physical, and geometric features. In the second part different types of water existing in geothermal reservoirs worldwide are reviewed. Their compositions are discussed and related to the basic processes that dominate their chemistry.

In the third chapter geochemical modeling theories are presented in a sequence of increasing complexity from geochemical equilibrium models to kinetic, reaction path, and finally coupled transport and reaction models. The state of the art of hydrothermal reactive transport simulation is delineated. Available numerical codes are presented, which are capable to simulate the processes fluid flow, heat transfer, transport, and chemical reactions, necessary for a comprehensive study of hydrothermal systems. Uncertainty, usefulness, and limitations of hydrogeochemical models are discussed in general.

In the fourth chapter specific features of coupled fluid flow and chemical reaction are investigated in more detail. Distinct types of reactive environments are described in combination with permeability-porosity relationships resulting in specific flow induced reaction patterns. The specific phenomena of reactive infiltration instability and free thermal convection are investigated accordingly.

Reactive transport modeling of the history of fossil hydrothermal systems is presented in the fifth chapter. Firstly, a brief overview is given about numerical simulations done to investigate genesis of ore deposits as well as progress of diagenetic processes. Secondly, a detailed examination of formation scenarios is presented in order to understand the observed anhydrite cementation at the location Allermöhe (Germany). This is done under special consideration of the recent structure of the Allermöhe site and its geological history.

Aim of numerical investigations of recent hydrothermal systems in chapter six is to set-up or to evaluate conceptual models of geothermal areas. Within the first part of this chapter some of the currently published numerical studies are summarized. The following second part is a detailed case study of the shallow hydrothermal system of Waiwera (New Zealand), investigating the complex interaction of density driven flow, heat transfer, and chemical reactions.

In chapter seven opportunities are discussed of the application of reactive transport modeling for reservoir management purposes. The case study of the long term performance of the geothermal potential Stralsund (Germany) is shown in detail.

For the interested reader a CD-ROM is available from the author, which includes the complete database of the geothermal waters compiled from an extensive literature study (Chap. 2, readable with MS Access®) and all numerical models presented here (Chap. 4 to 7) with a comprehensive selection of the produced results. The models may be investigated with the help of either Processing SHEMAT (Clauser 2003, available at Springer Publishers, Berlin-Heidelberg) or the SHEMAT Viewer (provided with the CD-ROM).

Finally, the reader must note that I take full responsibility for the contents of this book. If some of the theories and concepts taken from the literature are misinterpreted, this was unintentional and does not reflect disregard of the original author's work.

Acknowledgements

My thanks go to the German Federal Ministry for Education, Science, Research, and Technology (BMBF) and the German Federal Ministry for Economic Affairs (BMWi) for the financial support during the past five years. The contents of the work in hand arose in the context of the projects “Hydraulic, thermal and mechanical behavior of geothermally used aquifers” (BMBF, under grant 032 69 95) and “Scenarios of the Emergence of Anhydrite Cementation in Geothermal Reservoirs” (BMWi, under grant 032 70 95).

When a work like this is produced over a long period of time, it is difficult to limit the acknowledgements as many people have contributed in different ways. However, I should first like to thank Prof. Dr.-Ing. Wilfried Schneider who engaged me at the TU Hamburg-Harburg and who encouraged me to write this book and provided moral support and valuable discussions. I am thankful for his confidence in my work and the freedom he allowed me to guide the "Geothermal" projects in my sole discretion.

I would like to thank Prof. Dr.-Ing. Knut Wichmann who affiliated me in the Department of Water Management and Water Supply at the Technical University of Hamburg-Harburg. He provided any support I needed for my research work. Thanks to all my colleagues and friends from the Department, who gave me a helping hand when I needed one and who readily shared their computer with me when I needed more computational power.

I am indebted to a number of people. Special thanks go to Dr. Jörn Bartels who had great stake in my employment at the TU Hamburg-Harburg and I appreciate his advice during the learning of computer programming. His endless input and critical discussions were invaluable for the "fully coupling" of his physical and my geochemical ideas and experiences.

I would like to express my sincere gratitude to Dipl.-Ing. Heinke Stöfen for her continuing support as collegiate assistant, diploma student, and colleague and for the fruitful discussions we had.

I am thankful for adjuvant suggestions and efforts of Prof. Dr. Christoph Clauser. Furthermore, I am grateful to Dr. Hansgeorg Pape, Dipl.-Geol. Joachim Iffland, and Dr. Andreas Günther for their geological and petrophysical input and explanations.

I am much obliged to Prof. Patrick R.L. Browne, Prof. Arnold Watson, and the whole staff of the Geothermal Institute at the University of Auckland for their warm welcome, help and advices during my stays in New Zealand in 1999 and 2001. Moreover I am thankful for the all-embracing support of Stephen Crane from the Auckland Regional Council concerning the Waiwera geothermal field (New Zealand) and the help of Francisco da Costa Monteiro during my field studies at Waiwera.

Furthermore I am grateful to Dr. Jörn Bartels, Dr. Martin Kölling, Dr. Paul Hoskin, Dipl.-Ing. Heinke Stöfen, and Dipl.-Ing. Thomas Nuber for finding time in busy schedules to help with reviewing the manuscript. Various drafts were improved immensely thanks to their vigilant proofreading.

Thanks also due to ITA Jens-Uwe Stoß, who assisted with several graphics, and to Ulrike Witt and Ciprian Scurtu for build up of the geothermal water database.

Throughout all this time however, my greatest supporter and source of encouragement and love has been Silke Gößling. Finally I would like to thank her, my children Lisa and Lukas, and my family and friends for moral support and for allowing me to use my free time to finish this book.

Michael Kühn

Contents

1 General Significance of Geochemical Models of Hydrothermal Systems

Hydrothermal systems are highly heterogeneous, consisting of the host rock formation and the inherent water. Their investigation and the development of geochemical models describing these systems arise from and focus on geothermal energy production and ore deposit exploitation. Both are the two main economical topics.

Between 1999 and 2020, the world's energy consumption will rise by about 50 % mainly due to the increase of population (Energy Information Administration 2001). This will happen especially in rapidly developing parts of the world. Finding the supply to meet this demand will be a Herculean task, yet that is just one part of the energy challenge. Global needs must be satisfied in a sustainable way, hence, the energy must be used with great efficiency. In the face of global warming it is clear that technological leaps, strong policies, and large investments will be required. The conventional sources of energy - oil, gas, and coal - used for energetic transformation developed in a period of many million years. But they are used up irretrievably within a few one hundred years by human exploitation.

Adequate and reliable supplies of affordable energy, obtained in environmentally sustainable ways, are essential to economic prosperity, environmental quality, and political stability around the world. Geothermal energy, as discussed here, is among other technologies, such as wind and solar energy, or biomass, especially suitable, due to its ubiquitous occurrence. However, no technology should be viewed in isolation; each is just one element of the entire system. Due to the fact that the existing energy-related infrastructure is designed for fossil fuels, it seems certain that the world will continue to rely heavily on hydrocarbon combustion in the medium-term. However, since we cannot ignore the long-term impacts of continued hydrocarbon combustion, we must develop alternative energy sources.

At the time of the oil price shocks of the 1970s, it was believed that market development for renewables would evolve smoothly from niches to major energy markets. But as a result of the sharp declines in energy prices in the 1980s and late 1990s, the transition to major markets has proven to be difficult. For example, at current prices, combined-cycle systems fired by natural gas provide electricity at lower costs than most alternative systems could. Deregulation and restructuring of the energy market have, in some respects, increased the difficulties of alternatives and renewables. The markets may perceive renewable systems as being financially risky, because their capital costs are high. On the other hand, solar, wind, and geothermal systems are immune to the risk of fuel cost increase. Renewable energy technologies are often recognized as technically risky, as is any nascent technology unfamiliar to its potential users. Implementing geothermal energy there is especially the drilling, which is of a certain risk to fail. In competition with low-cost fossil fuels, renewable energy technologies face difficulties in achieving market and production scales large enough to drive costs down (Baldwin 2002).

Yet the strong interest in renewable energies, here in geothermics, comes from the fact that conventional energy systems are the principal source of air pollution and green house gases. Furthermore, most countries require the import of fossil fuels, often from politically volatile regions of the world. Geothermal energy can similarly provide electricity, heat, and process heat for industry, but with much less environmental impact. The inherent cleanliness of geothermal technologies minimizes decommissioning costs and long term health and safety concerns. An additional advantage of geothermal energy systems is the fact that in the decades ahead the demand of energy will increase mostly in developing countries where active geothermal areas are often to be found.

Geothermal energy is usually classified as renewable and sustainable. Renewable describes a property of the energy source, whereas sustainable describes how the resource is utilized. The most critical aspect for the classification of geothermal energy as a renewable energy source is the rate of energy recharge. If the recharge of energy during exploitation of geothermal systems takes place by advection of thermal water, on the same time scale as production, the resource can be classified as renewable.

Nevertheless, climate protection, declining resources, and the commandment of a sustainable development for all people require clearly structured changes in the worldwide energy supply within the next decades. Particularly alternative energies must contribute to these changes (Fischedick et al. 2000). Therefore, geothermal energy use is of great interest for Germany especially due to the decision to turn away from using nuclear power.

Based on the global-average temperature gradient of about 25°C km^{-1}, one can calculate that the heat stored in the upper few kilometers of the Earth's crust would be sufficient to supply the world's consumption of energy indefinitely. But in general such calculations have little practical relevance because successful exploitation of geothermal energy requires that it is concentrated well above "background" levels. Nevertheless, the energy potential of geothermal water systems is enormous.

Here, the focus will be on hydrogeothermal reservoir exploitation referring to the geological situation of Germany and the fact that high temperature reservoirs are not available there. By the end of 1999 direct thermal use of geothermal energy in Germany amounted to an installed thermal power of roughly 397 $MW_{thermal}$ (Schellschmidt et al. 2000). At present no electric power at all is produced from geothermal resources in Germany.

Most economically significant ore deposits exist because of the advective transport of solutes and heat by flowing groundwater. Mobilization, transport, and deposition of chemical species are all linked to fluid flow. Many ore deposits are associated with magmatic-hydrothermal systems or metamorphic environments and are therefore an important topic within the chemical investigation of geothermal systems. Although commercial extraction of heat from active hydrothermal systems has been growing steadily over the past few decades, extraction of minerals from fossil hydrothermal systems continues to have large economic significance and provides a practical impetus for research on these systems.

An overall goal for studying ore deposits in hydrothermal fields is to precisely describe the triggering processes within the geological environment with the aim to define the stratigraphic, structural, and petrological constraints that localize the mineral deposit. Basis for that is the knowledge of the chemistry of the fluids involved in the hydrothermal process. The geochemical signature of geothermal waters should enable the delineation of the source region and the geochemical environments through which the fluid has moved. Thus, chemical, quantitative modeling can be used as a predictive tool to guide an exploration program and to provide an understanding of the definite location of mineralization and the specific pattern of ore deposit (size, mineralogy, grade, zoning, etc.). In particular, quantitative modeling enables a large number of "what if" scenarios to be explored for old and for undiscovered new types of mineralization. Once an exploration model has been developed, quantitative modeling allows various scenarios for ore formation to be explored in an economical manner before drilling commences.

The challenge for the exploration industry is to find cost-effective ways of locating high quality resources (large tonnage, high grade, and suitable metallurgical properties). New exploration concepts and tools are required to sustain this enhanced exploration effort in the modern era. Models of resource formation are needed that will promote the effective selection and evaluation of large areas of the Earth's crust.

The aim of this book is to firstly give a review of typical geothermal reservoirs and their inherent waters. On that basis various chemical models of increasing complexity will be applied in order to understand and explain hydrothermal systems. Finally reactive transport models are shown, assumed to be of utmost importance as tools to support the investigation of hydrogeothermal reservoirs concerning geothermal energy production and exploitation of ore deposits. The detailed examination of geothermal systems subdivides into:

(1) Fossil hydrothermal systems (Chap. 5) including
    a. Development of ore deposits
    b. Diagenetic processes
(2) Recent hydrothermal systems (Chap. 6) and
    a. Investigation of their development
    b. Determination of their structure
(3) Reservoir management (Chap. 7) with
    a. Brine rock interaction and reactive tracer
    b. Recovery of minerals from geothermal brines
    c. Long-term prediction and productivity control

1.1 Fossil and Recent Hydrothermal Systems

The geothermal gradient expresses the increase in temperature with depth in the Earth's crust. Down to the depths accessible by drilling with modern technology, the average geothermal gradient is about 2.5-3°C per 100 m. For example, assuming a mean annual temperature of 15°C within the first few meters, which on average corresponds to the mean annual temperature of the external air, the temperature will be about 65-75°C at 2000 m depth or 90-105°C at 3000 m. However, there are vast areas in which the geothermal gradient is far from the average value. For example, in areas in which the deep rock basement has undergone rapid sinking, and the basin is filled with geologically "very young" sediments, the geothermal gradient may be lower than 1°C per 100 m. On the contrary, the geothermal gradient of some active geothermal areas is even higher than ten times the average value.

Geothermal systems can be found in regions with a normal or slightly above normal gradient, but especially in regions around plate margins where the geothermal gradients may be significantly higher than the average value. In the first case, the system will be characterized by low temperature at economic depth (usually not higher than 100°C). In the second case, the temperatures could cover a wide range from low to very high, even up to and above 400°C. In terrestrial geothermal systems the circulation of waters can reach depths of approximately 5 km, lasting from tens of thousands up to about millions of years (Pirajno 1992).

Modeling the geothermal history of fossil hydrothermal systems is important for understanding maturation, migration, and accumulation of petroleum, formation of ore deposits, and diagenetic processes. Many world ore deposits are located at the sites of ancient geothermal systems, which must have been similar to recent ones, concerning size, chemistry and behavior. Past groundwater temperatures and pressures governed ore deposition (White 1981, Henley and Ellis 1983, Hedenquist and Henley 1985). With the help of mathematical modeling it is possible to look at the distant past of a geothermal system.

In order to get insight in evolution and structural features of the field, the starting point for any exploration program of recent hydrothermal systems is to do stocktaking of the geology and hydrogeology. Both are important in subsequent phases of geothermal research, right up to the positioning of exploratory drillings and production boreholes. They also provide the background information for constructing a realistic model of the geothermal system and for building up a chemical model. The geochemical survey itself provides useful data for planning exploration and its costs are relatively low compared to geophysical surveys.

In the last years, great effort has been invested in reservoir stimulation, which can be hydraulic as well as chemical stimulation. Reservoir stimulation is based on the premise that hot rock formations containing fluids may often have such low permeabilities that the fluids are unable to circulate and no geothermal system can develop. This situation could be due to the nature of the rock formation or may be a consequence from partial sealing of existing fractures or pore space. The most effective means of stimulating the reservoir is by hydraulic fracturing, but under certain conditions the permeability of the rock can be increased or its original permeability restored, by injecting acid solutions for example. For this kind of stimulation, reactive transport models can be used for predictive modeling and to fix the dimensions of the stimulation test.

1.2 Hydrogeothermal Energy Use

Archeology proves that primeval man used geothermal water from natural pools and hot springs for cooking and bathing and keeping themselves warm, for more than 10,000 years (Cataldi 1999). Written history depicts that, among others, Roman (Cataldi and Burgassi 1999), Turk (Özgüler and Kasap 1999), Japanese (Sekioka 1999), Russian (Svalova 1999), Icelandic (Fridleifson 1999), French (Gibert and Jaudin 1999), and Maori people in New Zealand (Severne 1999) have used geothermal resources. Cataldi et al. (1999) present an extensive overview on the world's geothermal heritage.

After the Second World War many countries were attracted by geothermal energy, considering it to be economically competitive with other forms of energy. Electricity generation is the most important form of utilization of high-temperature geothermal resources today. Medium to low temperature resources are suitable for many different types of application. Lindal (1973) published a diagram, listed in Table 1.1, which shows the various use of geothermal energy.

Table 1.1. The Lindal classification of low to medium temperature geothermal energy use (Lindal 1973)

T [°C]	Geothermal energy application	Steam	Water
180	power, evaporation of highly concentrated solutions, refrigeration by ammonia absorption, digestion in paper pulp	♦	
170	heavy water via hydrogen sulfide process, drying of diatomaceous earth	♦	
160	drying of fish meal, drying of timber	♦	
150	alumina via Bayer's process	♦	
140	drying farm products	♦	
130	evaporation in sugar refining, extraction of salts	♦	
120	fresh water by distillation, multiple effect evaporation	♦	♦
110	drying and curing of light aggregate cement labs	♦	♦
100	drying of organic material (vegetables etc), washing of wool	♦	♦
90	drying of stock fish, intense de-icing operations		♦
80	space heating, greenhouses by space heating		♦
70	refrigeration (lower temperature limit)		♦
60	animal husbandry, greenhouses by space or hotbed heating		♦
50	mushroom growing, balneological baths		♦
40	soil warming		♦
30	swimming pools, biodegradation, fermentations, warm water		♦
20	hatching of fish, fish farming		♦

Most geothermal resources can be used for space heating applications (e.g. urban district heating, fish farming, and greenhouse heating). Only the hotter systems (>180°C) are used to generate electricity through the production of steam. Commercial extraction of heat from active hydrothermal systems has been growing steadily over the past few decades. Nevertheless, extraction of minerals from fossil hydrothermal systems continues to have a much larger economic significance and provides a major practical impetus for research on hydrothermal systems.

Huttrer (2000) assessed the status of international geothermal power generation from all nations generating or planning to generate electricity and revealed that: (1) geothermally fueled electric power is being generated in 21 nations, (2) the installed capacity has reached 7,974 $MW_{electric}$ which is a 16.7% increase since 1995, (3) the total energy generated is at least 49,261 GWh, and (4) during the last five years, about 1,165 wells more than 100 meters deep were drilled. Huttrer (2000) concluded that greater increases in the total international installed geothermal generation capacity were inhibited by the economic crisis that occurred in Southeast Asia and by the low petroleum prices prevailing within the last decade.

The recent worldwide application of geothermal heat for direct utilization is reviewed for 60 countries by Lund and Freeston (2000), of which 55 reported some form of geothermal direct utilization. An estimate of the installed thermal power at the end of 1999 (1995 in brackets) from the current reports is 16,209 $MW_{thermal}$ [8,660 $MW_{thermal}$] utilizing at least 64,416 kg s^{-1} [37,050 kg s^{-1}] of fluid, and the thermal energy used is 162,009 TJ yr^{-1} [112,441 TJ yr^{-1}]. The distribution of the thermal energy used by category is approximately 37% for space heating, 22% for bathing and swimming pool heating, 14% for geothermal heat pumps, 12% for greenhouse heating, 7% for aquaculture pond and raceway heating, 6% for industrial applications, less than 1% each for agricultural drying, snow melting, air conditioning and other uses. The reported data for number of wells drilled was 1,028 over the five years. Research progress manifests by a total of 841 chemical interpretations of geothermal fields (Freeston 1995, Lund and Freeston 2000).

Schellschmidt et al. (2000) pointed out that by the end of 1999 direct thermal use of geothermal energy in Germany amounted to an installed thermal power of roughly 397 $MW_{thermal}$. Of this sum, approximately 55 $MW_{thermal}$ are generated in 27 major centralized installations. Small, decentralized earth-coupled heat pumps and groundwater heat pumps are estimated to contribute an additional 342 $MW_{thermal}$. By the year 2002 an increase in total installed power of about 120 $MW_{thermal}$ is expected: 82 $MW_{thermal}$ from major central and 40 $MW_{thermal}$ from small, decentralized installations. This would boost direct thermal use in Germany close to an installed thermal power of 517 $MW_{thermal}$. At present no electric power is produced from geothermal resources in Germany, whose annual final energy consumption at present amounts to about 9469 PJ. It is less than the corresponding primary energy because of losses, mainly due to conversion and distribution. This is equivalent to a total consumed power of approximately 300,000 MW per year. Almost 60 % of this energy is required as heat. The total technical potential for the direct use of geothermal energy in Germany is 2125 PJ a^{-1}, with a weighting according to the local variation in the demand for heat; equivalent to a maximum thermal power generation of about 67,380 $MW_{thermal}$. That corresponds to about 22 % of the country's annual final energy consumption, or roughly 37 % of its demand for heat. However, at present only about 6 ‰ of the existing maximum technical potential for direct thermal use of geothermal energy meets the demand for heat. If the vast potential of geothermal energy for direct thermal use was utilized to substitute fossil fuels, roughly 100 million tons less of CO_2 would be released to the atmosphere annually, equivalent to about 10 % of Germany's CO_2 output in 1998.

1.3 Reservoir Exploration and Management

Before drawing up a geothermal exploration program, geophysical surveys are done. The advantage of geophysical data collection from deep geological formations is the fact that they are indirectly obtained from the surface or from depth close to the surface. In the 1920s it was shown for the first time, that for example, electrical resistivity measurements could be made in a well and that the readings were different for different geological layers (Hearst et al. 2000). Due to more and more sophisticated measurements, the discovery and location of reservoirs is done with the help of these subsurface methods. These measurements use electromagnetic fields and waves, acoustic waves, neutron scattering, gamma-ray radiation, nuclear magnetic resonance, infrared spectroscopy, and pressure and temperature sensors, with the aim to characterize in detail the geological structure within the vicinity of a bore. However, there is no single technique adequate to define the structure and properties of a whole reservoir. This is a fact which led to significant efforts and advances in predictive reservoir simulation within the last two decades.

During the 1960s, when our environment was in a healthier state than it is at present and we were less aware of the threat to the earth, geothermal energy was still considered a "clean energy". There is actually no way of producing or transforming energy into a form that can be utilized by humanity without direct or indirect impact on the environment. Similarly, exploitation of geothermal energy also has an impact on the environment, although it must be said that it is one of the least polluting. Models should be used to ensure exploitation of geothermal resources to be as harmless as possible, because any modification to our environment must be evaluated carefully, in deference to the relevant laws and regulations. The danger is that an apparently insignificant modification could trigger a chain of events whose impact is difficult to fully assess beforehand. In most cases the degree to which geothermal exploitation affects the environment is proportional to the scale of such exploitation (Lunis and Breckenridge 1991).

The use of computer modeling in the planning and management of the development of geothermal fields has become standard practice during the last 10 – 15 years. The computer power available in the 1980s limited the size of the computational meshes used and many of them were based on geometrically simple models. In addition to modeling specific geothermal fields, an important application of numerical reservoir simulation has been in the study of generic issues of geothermal reservoir dynamics, and fluid and heat processes. It is also possible to predict the response of recent geothermal systems to various natural and industrial processes. This is important for reservoir management and sustainable exploitation of the resources (O'Sullivan 2001).

Geochemical data are necessary for delineating favorable exploration areas, estimating the recoverable geothermal resources from a given reservoir, and identifying potential pollution, waste disposal, and corrosion problems. The objectives are to study the chemistry and controls on the chemistry of water in geothermal and other subsurface systems to provide basic data needed (Pham et al. 2001).

1.4 Geochemical Models

Economic and sustainable exploitation of geothermal reservoirs requires powerful reactive transport models. The models must cover two aspects: geochemical and hydrodynamic processes. Geochemical processes include, among others, aqueous speciation and redox reactions, interface reactions, precipitation / dissolution of minerals and colloids. Hydrodynamic transport processes mainly include diffusion and migration due to advective forces, leading to dispersion of the chemical species in space and time. Modern development is to consider geochemical and hydrodynamic transport processes interdependently, because geochemistry and hydrogeology are closely entangled.

Geochemist's goal is to describe the chemical states of natural waters, including how dissolved mass is distributed among aqueous species, and to understand how such waters will react with minerals, gases, and fluids of the Earth's crust and hydrosphere. This can be done involving simple chemical systems in which the reactions can be anticipated through experience and evaluated by hand calculation. Facing more complex problems, one must rely increasingly on quantitative models of solution chemistry to find answers.

Geochemists now use quantitative models in order to understand sediment diagenesis and hydrothermal alteration, to explore for ore deposits, to determine which contaminants will migrate from mine tailings and toxic waste sites, to predict scaling in geothermal wells and the outcome of steam-flooded oil reservoirs, to solve kinetic rate equations, to manage injection wells, to evaluate laboratory experiments, and to study acid rain, among many examples. The advantage is that geochemical models allow geoscientists to estimate the results of a hydrothermal experiment or to interpret reservoirs spending less amounts of time and money.

The field of reactive transport within the Earth Sciences has become a highly multidisciplinary area of research. The field encompasses a number of diverse disciplines including geochemistry, geology, physics, chemistry, hydrology, and engineering. A wide variety of geochemical processes including such diverse phenomena as the transport of radiogenic and toxic waste products, diagenesis, hydrothermal ore deposit formation, and metamorphism are the result of reactive transport in the subsurface. Such systems can be viewed as open reactors where chemical change is driven by the interaction between migrating fluids and solid phases. The evolution of these systems involves diverse processes including fluid flow, heat transfer, solute transport, and chemical reactions, each with different characteristic time scales. The ability to quantify reactive transport in natural systems has advanced dramatically over the past decades. Much of this progress is due to the exponential increase in computational power. Taking advantage of this increase, numerous comprehensive reactive transport models have been developed and applied to natural phenomena. These models can be used either qualitatively or quantitatively to provide insight into natural processes. Quantitative models force the investigator to evaluate or falsify ideas by putting real numbers into an often-vague hypothesis. As a consequence, a thinking process is initiated along a path that may result in acceptance, rejection, or modification of the original hypothesis. Used qualitatively, models provide insight into general features of a particular phenomenon, rather than specific details.

One of the major questions facing the use of hydrogeochemical models is whether or not they can be used with confidence explaining history as well as predicting future evolution of natural groundwater or geothermal systems. There is much controversy concerning the validity and uncertainties of non-reactive fluid flow models (Konikow and Bredehoeft 1992). Adding chemical interaction to these flow models only confounds the problem. Although such models may accurately integrate the governing physical and chemical equations, many uncertainties are inbuilt in characterizing the natural system itself. These systems are inherently heterogeneous on a variety of scales rendering it highly challenging to know precisely the many details of the flow system and chemical composition of the fluids and host rocks. Other properties of natural systems such as permeability and mineral surface area, to name just two, may have to be approximated with little precision. Because of these uncertainties, it is important to delineate to what extent a certain numerical reactive transport model is useful in making accurate quantitative predictions. Nevertheless, reactive transport models should allow for predicting the outcome for the particular representation of the porous medium used in the model.

Heat in Earth's crust represents the greatest potential contribution to the world's energy base. Yet this important energy source is at present markedly underutilized. The principal reasons for this situation are the high costs currently associated with geothermal energy production and exploration and the lack of technology to reduce these costs. Many of the significant problems encountered by the geothermal industry reflect complicated chemical interactions between solids, gases and liquids. Adverse chemical effects, such as scale formation, corrosion and noxious gas emission, which can arise from the manipulation of the high temperature natural fluids driving the energy production process, are expensive to control. Mineral precipitation for example, can not only damage plant equipment and wells but also significantly decrease the permeability of the formations containing the geothermal fluids, thereby limiting the longevity of the resource itself. The ability to predict these chemical behaviors and the heat content of a reservoir as well as to design optimal operating strategies would significantly increase the cost effectiveness of geothermal energy production. Predicting potential chemical problems will become even more important as deeper, higher temperature geothermal systems with very high development costs, will be utilized to meet future energy needs (Moeller et al. 2000).

In order to understand the chemical processes in fossil or recent geothermal systems or to predict the chemistry during geothermal energy production, it is necessary to understand the thermodynamics and kinetics of the waters within the hosting rock formation or the production processes. Unfortunately, the chemical behavior of these waters, which are often high temperature brines, is a very complex function of their composition, temperature and pressure. Since these variables can change significantly during lifetime of the geothermal reservoir or resource, the past experience of geothermal energy production therefore may not be a reliable guide for future performance. Laboratory simulations are costly and limited to the experimental conditions selected. Reactive transport simulation is a testing strategy to control unwanted behavior in active operations as well as to forecast the value of geothermal reservoirs as potential production sites and last but not least to evaluate geothermal system history.

Finally, it should be mentioned that the numerical solution of the geochemical model equations often is the only recourse to analyze a hydrothermal system where investigations must be carried out over geologic time spans. Without such models it would be impossible to analyze fossil hydrothermal systems, because performing laboratory experiments would take too much time and recently active geothermal systems are most often within a different stage of development.

2 Concepts, Classification, and Chemistry of Geothermal Systems

Main concepts and a classification of different types of geothermal systems are presented in this chapter. Particular attention is given to chemical, physical, and geometric features of the geothermal systems inferred from active geothermal areas or reconstructed from geological observations.

Additionally, different types of water existing in geothermal reservoirs worldwide are reviewed here. They are discussed and related to the basic processes that dominate their chemistry. The chemistry of geothermal waters discharged from wells provides specific information about the deep fluids in geothermal systems and how they relate to natural discharges from springs at Earth's surface. This knowledge can be used to obtain essential information about reservoir behavior before and during exploitation and to set up conceptual models of reservoirs. Derivation of the hydrologic and chemical structure of geothermal systems forms the basis for reactive transport simulation, here and in general.

2.1 Conceptual Model and Classification

Geothermal systems as hosts of resources or potential reservoirs are also called geothermal fields. They are found throughout the world in regions with normal or above normal geothermal gradients, and especially in regions around tectonic plate margins where the geothermal gradient may be significantly higher than the average value. Geothermal systems are encountered in a range of geological settings, and are increasingly developed as an energy source. They can be described schematically as the distribution of waters circulating laterally and vertically at various temperatures and pressures in the upper crust of the Earth. The geothermal system transfers heat from a heat source to a heat sink, usually the free surface (Hochstein 1990). A geothermal system is made up of three main elements: a heat source, a reservoir and a fluid, which is the carrier that transfers the heat. The reservoir is a volume of permeable rocks from which the circulating fluids extract the heat. The geothermal fluid is water, in the majority of cases meteoric water (rain, lake, river), in the liquid or vapor phase, depending on its temperature and pressure (Fig. 2.1). Geothermal fluids often discharge at the surface. Hydrothermal mineral deposits are formed due to circulation of the warm to hot fluids (about 50 to > 500°C) that leach, transport and subsequently precipitate their mineral load usually to the discharge site of the system (e.g. single conduit, fracture network).

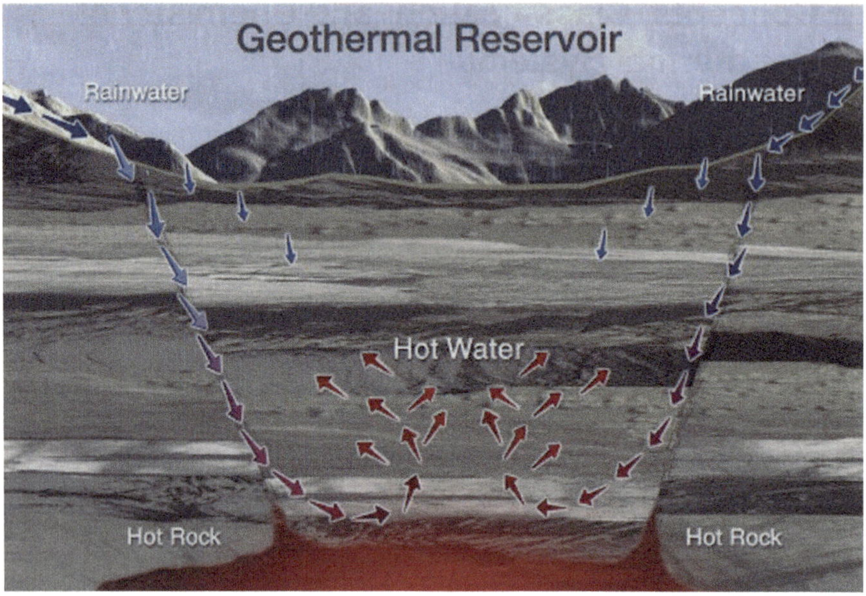

Fig. 2.1. Schematic structure of a geothermal reservoir fed by meteoric (rain) water (after GEO – Geothermal Education Office 2001)

Dynamic geothermal systems arise where input of heat (usually magmatic heat) at depths of a few kilometers, sets deep groundwater in motion. These groundwaters are usually of meteoric origin but in some systems deep fossil marine or other saline waters may be present (connate waters). Systems near the coast may be fed by both meteoric water and seawater. It is possible that the magmatic heat source adds some water and dissolved constituents like HCl, CO_2, SO_2, and HF. But due to dilution and reaction during convective up-flow, this is very difficult to prove (Nicholson 1993).

Geothermal systems occur in nature in a variety of combinations of geological, physical, and chemical characteristics, which are reflected in the geothermal fluids and their potential applications. A fossil system represents the freezing of geological and tectonic settings of a hydrothermal field. Rocks in hydrothermal systems undergo varying degrees of alteration, because the mineral assemblages in the wall rocks are unstable in presence of the moving fluid and tend to re-equilibrate, forming new mineral congregations that are stable under the new conditions. Many ore deposits are localized in vein networks that once hosted hydrothermal fluid circulation, leading to a large number of deposit types and mineralization styles in Earth's crust due to variable geological situations. The most important of these hydrothermal ore deposits involve silver and gold and the sulfides of copper, tin, lead, zinc, and mercury (Ingebritsen and Sanford 1998).

The classification of geothermal systems applied by Nicholson (1993) is based on a series of descriptive terms, primarily targeting geothermal reservoirs. They are referred to as liquid or vapor dominated, low or high temperature, sedimentary or volcanic hosted. Pirajno (1992) subdivided hydrothermal systems focusing on their products in terms of mineral deposits whereas Heiken (1982) concentrated on the geologic settings of the systems. The classification used here, convenient to the following investigation of reactive flow in hydrothermal systems, is a combination of all three schemes (Fig. 2.2).

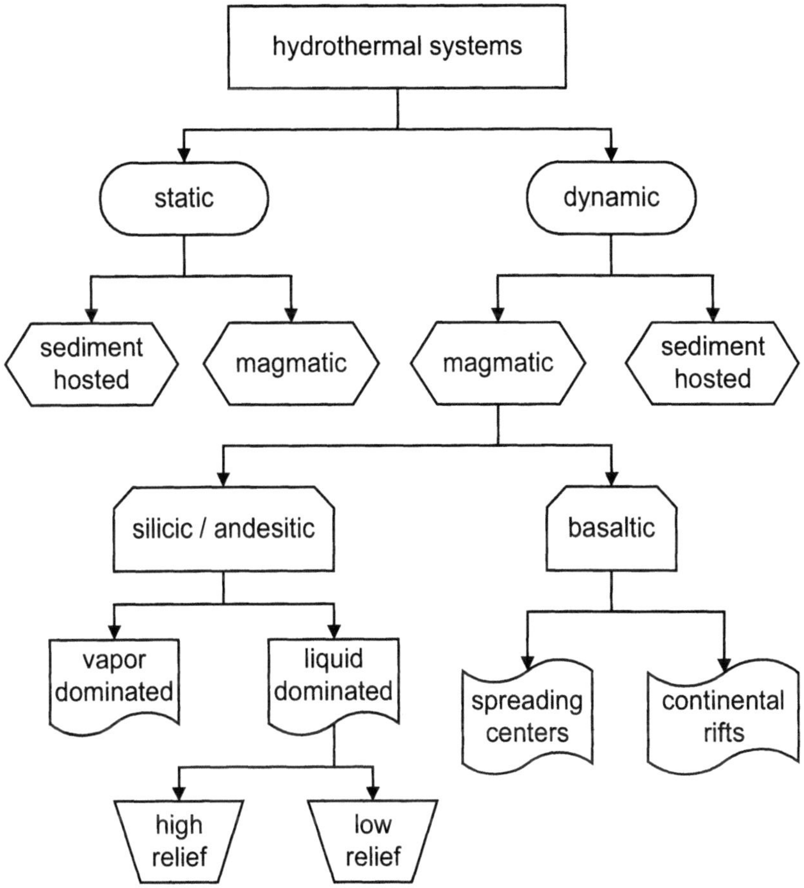

Fig. 2.2. Classification of hydrothermal fields as used within this book, adapted and merged after Nicholson (1993), Pirajno (1992), and Heiken (1982)

The geothermal reservoirs are classified here in the first stage based on their inherent hydraulic conditions. The geothermal reservoirs are primarily divided into dynamic systems, in which fluids are circulating and therefore heat is transferred by convection, and static systems without fluid movement and resulting conductive heat transfer. Both the branch of static as well as the one describing dynamic hydrothermal systems subdivides subsequently into sediment hosted (low temperature) or magmatic (high-temperature) environments. Due to the significance of magmatic dynamic hydrothermal systems for geothermal reservoirs and hydrothermal ore deposits, these are further sorted by their magmatic rock type into silicic / andesitic and basaltic settings. The description of silicic / andesitic and basaltic settings. The description of the basaltic systems distributes into mid-ocean ridge spreading centers and continental rift settings. The silicic / andesitic hydrothermal environments are divided into vapor and liquid dominated systems and the latter additionally into low or high relief settings.

2.2 Static – Conductive Systems

The mean conductive heat flow measured near Earth's surface is approximately 70 mW m^{-2} (e.g. Chapman and Pollack 1975). Correcting for the effects of hydrothermal circulation in the oceanic crust brings the mean global heat flux to 87 mW m^{-2} (Pollack et al. 1993). Integrated over the surface of the globe, this amounts to a heat loss of more than $4x10^{13}$ W. The main source of heat in the crust (shallow heat source) is the continually radioactive decay of long-lived isotopes of uranium (^{238}U, ^{235}U), thorium (^{232}Th), and potassium (^{40}K). However, Earth's main deep heat source is the heat of primordial energy of planetary accretion (Lubimova 1968).

2.2.1 Magmatic Systems

Static, magmatic systems are usually related to shallow or deep-seated granitic plutonism generated by H_2O-rich magmas (> 8 wt.%). These magmas crystallize at depths between a few kilometers to over 10 km but normally do not vent at the surface (Pirajno 1992).

Hydrothermal fluids of such a system are assumed to be generated entirely within the cooling magma body and set up a closed system. This situation results in hydrogen-ion metasomatism (see below) leading to so-called greisen-related deposits. The term greisen refers to an assemblage of quartz and muscovite accompanied by varying amounts of minerals like fluorite, topaz, tourmaline and other F- or B-rich minerals (Burt 1981). Greisen systems are normally associated with Sn, W, Mo, Be, Bi and Li mineralization.

2.2.2 Sediment Hosted Systems

Static systems are characteristically found in strata deposited in deep sedimentary basins. The fluids are derived from the formation waters trapped within the thick sedimentary sequences. These waters attain reservoir temperatures of around 70-150°C at depths of 2-4 km, due to conductive heat flow. Geothermal fields in sediment hosted systems are normally called low-temperature because they are often only suitable for direct energy use and not for power production. The fluids are typically very saline chloride waters or brines, which remain trapped, as the vertical permeability is low within the formations, until released tectonically or by drilling. Examples of such fields are located in North and Eastern Europe, Russia, and Australia.

The following chapters will deal to a great extent with these static systems and additionally with low temperature dynamic systems (see below). Both are of greatest importance for Europe. Especially in Germany, static low temperature systems are the only ones currently under exploitation.

2.3 Dynamic – Convective Systems

Magmatic intrusions, leading to convective heat transfer by geothermal water circulation, are the source of thermal energy to most of Earth's high temperature (> 150°C) hydrothermal systems. However, a few high-temperature systems occur in areas of little or no apparent volcanic activity. These particular systems appear to be caused by deep circulation of meteoric water, leading to convective heat transfer, in areas of above-average conductive heat flow (e.g. Beowawe, Nevada, published by White 1992).

2.3.1 Magmatic - High-Temperature

In geological settings with magmatic or high-temperature systems the geothermal gradient is several times above the crustal average and rock temperatures of several hundred degrees Celsius exist at depths of only a few kilometers. The locations of these geothermal fields is invariably tectonically determined, and they are often found in areas of block faulting, grabens or rifting and in collapsed caldera structures, with reservoir depths of around 1-3 km. Typical settings are around active plate margins (Fig. 2.3) such as subduction zones (e.g. Pacific Rim), spreading ridges (Mid-Atlantic), rift zones (East Africa) and within orogenic belts (Mediterranean, Himalaya).

High-temperature systems are often volcanogenic, with the heat provided by intrusive masses. Geothermal systems also develop on the flanks of young volcanoes. As mentioned, high-temperature fields with a non-volcanogenic or tectonic heat source are less common (Nicholson 1993).

Hydrothermal systems related to volcano-plutonic and volcanic settings start as static-magmatic systems (described above), in the closed system of a plutonic body. The magmatic body rises closer to the surface or even ruptures it forming a stratovolcano. Due to the igneous bodies, providing a powerful heat engine, convection cells form with fluids supplied from meteoric waters. This environment is typical for porphyry or epithermal ore deposits and alteration features known as potassic, propylitic, phyllic, and argillic (all described below). The active time span of such systems may range from 10^5 to 10^6 years (Henley and Ellis 1983).

Fig. 2.3. Hottest known geothermal areas (dark gray) around the world (adapted from GEO – Geothermal Education Office 2001)

Silicic / Andesitic Systems

Following Henley and Ellis (1983) there are four principle settings of silicic / andesitic geothermal systems within a great number of possible scenarios. Hence, silicic or andesitic magmatic terrains can be divided into (1) silicic volcanism, (2) andesitic stratovolcanoes, (3) highland volcanoes, and (4) volcanic islands. The cases (1) and (2) are described as examples within the following section.

All kinds of the mentioned systems are characterized by regions where boiling occurs somewhere within the geothermal field. These boiling events may result in epithermal ore deposits, especially gold but also other precious metal formations (Pirajno 1992).

Vapor-Dominated. Fig. 2.4 displays characteristic features of vapor-dominated systems. Fumaroles, steaming ground and acid sulfate-waters from hot springs are observed at the Earth's surface. The reservoir is composed of steam (with gases) and it is assumed that saline, boiling water feeds the reservoir at depth. In these extensively exploited systems, the undisturbed states are poorly known because deep drillings often do not penetrate the vapor zone. Vapor-dominated reservoirs show a relatively constant temperature with depth of about 236°C, which is the temperature of maximum enthalpy of saturated steam (Haar et al. 1984). The system is convecting due to the steam up-flow, rising from depth and flowing laterally at the top of the reservoir along the base of capping low-permeability rocks.

The steam cools as it flows and eventually condenses and recirculates into the deep reservoir. Less-soluble gases remain more readily concentrated in the steam phase than the more soluble gases. The chemistry of the steam changes with up-flow, lateral flow, and condensation. Oxidation of hydrogen sulfide in the steam and subsequent absorption into the geothermal water will produce acid condensates (acid sulfate waters), whereas condensation of CO_2 results in formation of hydrogen carbonate waters (Fig. 2.4). Vapor-dominated systems are less common than liquid-dominated systems and only three have been well characterized: The Geysers (California, USA), Larderello (Italy), and Kawah Kamojang (Indonesia).

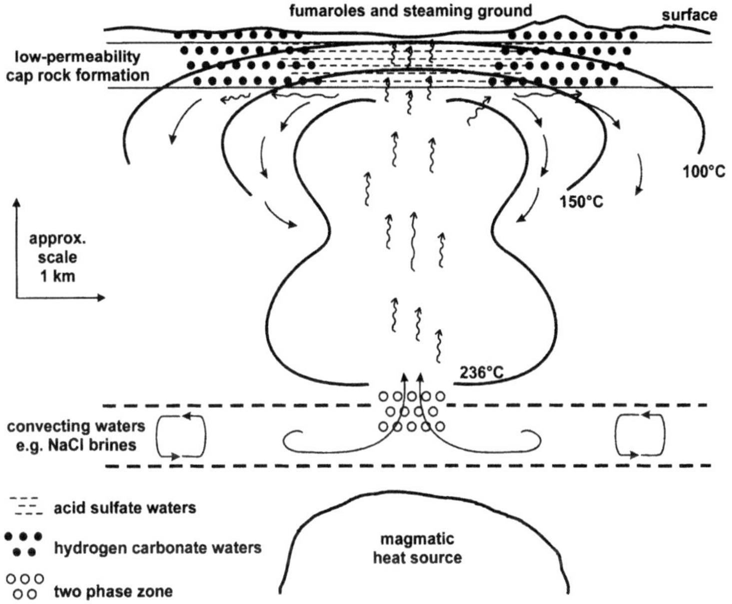

Fig. 2.4. Conceptual model with characteristic features of vapor-dominated geothermal systems (adapted from Nicholson 1993)

Liquid-Dominated. The characteristics of high temperature volcanic hosted and liquid dominated systems are shown in Fig. 2.5 and Fig. 2.6 distinguishing between high relief and low relief terrain, respectively. Liquid dominated geothermal systems in a high relief are typical of andesitic volcanic terrains and in a low relief of silicic volcanic terrains. Many systems display lateral flow structures created by strong hydraulic gradients often caused due to a high relief and a near-surface low-permeability horizon. Cooling by conduction and groundwater mixing are reflected in the chemistry of the discharges. Even in low relief settings ($< \approx 250$ m, e.g. Taupo Volcanic Zone, New Zealand), near surface lateral flows can extend for several kilometers. This is greatly extended in terrain of high relief ($> \approx 1000$ m) where flows are 10-50 km in length.

High relief is common in island arc settings with characteristic andesitic volcanism. The up-flow part of the system is revealed by fumaroles and steam heated aquifers fed by the two-phase zone and supplying the springs from the condensate layers (Fig. 2.5). It is the steep topography that likewise prevents the chloride fluid from reaching the Earth's surface resulting in large lateral flows, often over some 10 km. Over this distance the chloride fluid can be diluted with groundwater or mix with descending sulfate waters from steam condensates. The acid sulfate, chloride, or mixed waters can also emerge down-slope as hot springs, or descend into the system through fractures. Examples of these systems are found in Indonesia, Taiwan, Japan, and the Philippines (Nicholson 1993).

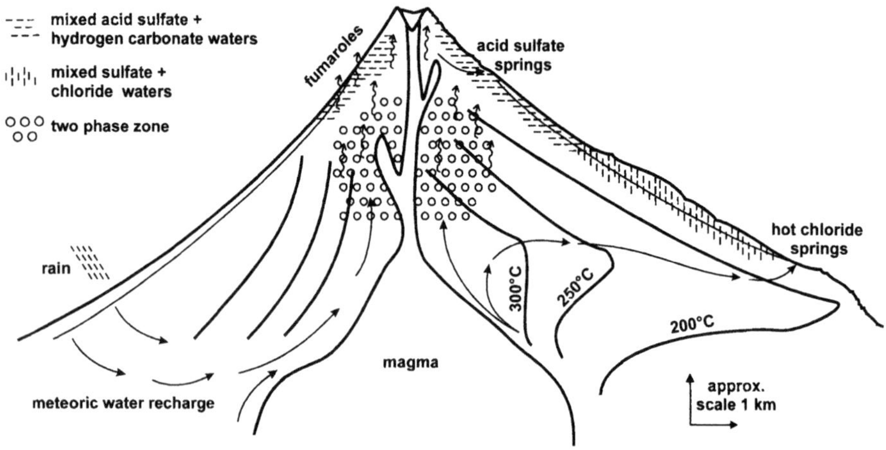

Fig. 2.5. Conceptual model of liquid dominated geothermal systems in a high relief, typical of an andesitic volcanic terrain (modified from Henley and Ellis 1983)

Low relief systems are generally characterized by recharge provided from meteoric groundwater and heat supplied, together with some gases, from deeply buried magmatic systems producing a convective column of near neutral pH chloride water emanating in springs and pools at the surface (Fig. 2.6). The deep geothermal fluid can express at the surface, often close to the up-flow area. Lateral flow is possible but, because of the gentle topography, is not as extensive as in areas of high relief. Two-phase or steam zones are commonly present but are not as thick as in high relief systems. However, these steam zones can increase in depth when fluid removal on exploitation of the systems exceeds natural fluid recharge, as has happened at Wairakei, New Zealand. Oxidation of hydrogen sulfide gas in the steam, together with condensation or mixing of the steam with groundwaters, produces acid sulfate waters. Condensation of carbon dioxide, which is less soluble than hydrogen sulfide, produces hydrogen carbonate rich waters, which are often found on the margins of the field. Because of the low relief over these systems, hot springs of chloride, sulfate, and hydrogen carbonate waters as well as fumaroles, and steaming ground often occur in relatively close proximity to one another. These types of systems are found in New Zealand, USA, East Africa, and Iceland (see below).

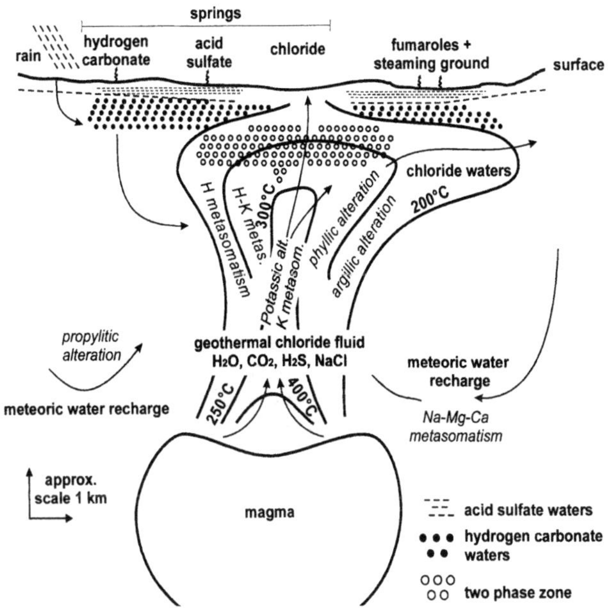

Fig. 2.6. Conceptual model of liquid dominated geothermal systems in a low relief, typical for silicic volcanic terrain (alteration / metasomatism processes in italic letters, adapted from Nicholson 1993 and Giggenbach 1988)

Basaltic Systems

Spreading Centers. Hydrothermal activity at mid-ocean ridges, in sea-floor environments, occurs on a large lateral scale. The recognition of fossil systems of such kind is difficult because oceanic crust is subsequently destroyed at convergent plate boundaries and only fragments may survive. Convective cells result from penetration of seawater to depths between 5 and 10 km. Flow occurs in cracks as well as in the porous media with discharges of the return flow through localized vents or clusters of vents with short life spans of several years only (Pirajno 1992).

Taylor (1983) modeled seawater circulation in oceanic crust based on field mapping for the Samail ophiolites in Oman, translated into the geometry as shown for the general hydrothermal system in Fig. 2.7. The proposed system consists of two circulation schemes. The upper circulation is located above the bird-shaped magma chamber within the region containing sheeted dykes and pillow lavas. The lower part of the system is located beneath the wings of the magma chamber above the ultramafic basement. Both circulation systems act decoupled with a high water-rock ratio in the upper part compared to a low ratio in the lower part.

Geothermal fluid movement within the seawater, above the discharging vents, is important for the development of ores. This hydrothermal system is called a plume. Plumes are diluted hydrothermal fluids rising above the vent producing sulfide particles (black smokers), which settle around the vents. The deposits contain sulfates (anhydrite and barite), talc, calcite, pyrrhotite, sphalerite, chalcopyrite, and galena. These muddy deposits are often rich in organic carbon material and tend to form hydrocarbons.

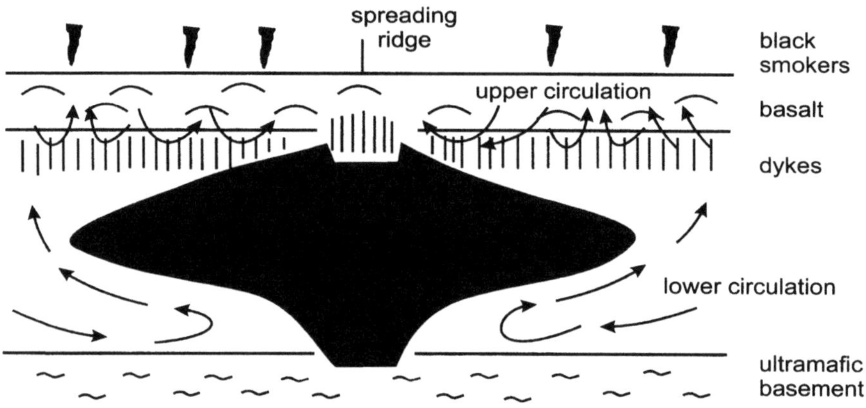

Fig. 2.7. Hydrothermal system within the environment of mid-ocean ridges consisting of two circulation systems, the lower part below the wings of the magma chamber and the upper part above, within a region of sheeted dykes (adapted from Taylor 1983)

Continental Rifts. The geodynamic evolution of rifted basins leads to the activation of hydrothermal solutions followed by their ascent along active faults. The geological settings are manifold and it is beyond the scope of this book to describe all of them. The interested reader is referred to Pirajno (1992). However, studies on recently active rift settings (Red Sea, East African Rift) emphasize that the occurrence of sediment-hosted mineral deposits may be due to hydrothermal systems in continental rift settings.

Hydrothermal system development with accompanying ore deposition is characterized by a reaction continuum from early stages during diagenesis (movement of meteoric water and compaction) to metamorphic processes. Types of deposits thought to be due to ancient rift settings are:

- Sediment-hosted stratiform metals (active modern analogues are deposits formed within the Red Sea brines or within the East African Rift lakes).
- Stratabound carbonate-hosted deposits (like the Mississippi-Valley Type).

During burial and diagenetic compaction considerable amounts of water are released from the sediments. The composition of the waters depends on the composition of the sediments within the basin. With increasing depth the formation waters are enriched with various anions and cations resulting in increasing salinity. Additionally these brines may be heated by a deep seated heat source. Migration of the fluids along the aquifer and up along basin faults result in trapping of the brines below an impermeable cap. Heat flow may drive circulation of the brines within convection cells in the rift setting.

Within the area of the Red Sea rift, linked to geodynamic and magmatic evolution, metalliferrous sediments occur with ores of numerous kinds. Hot brine pools are characteristic for these stratabound and stratiform deposits. The pools are due to active discharges of hydrothermal fluids located at the intersection of fractures and transform faults and are found to be density-stratified. A lower brine layer of high salinity is in contact with the metalliferrous sediments and of slightly higher temperature compared to the upper brine layer (Fig. 2.8). The high salinities of the hydrothermal fluids result because they originate from seawater and additionally due to circulation through evaporitic formations on the shoulders or the floor of the rift basin. The water is heated by the local heat flow and carries the metals leached from the basaltic rocks of the basement. The fluids subsequently discharge at the sea floor where they precipitate metal sulfides, sulfates and silicates.

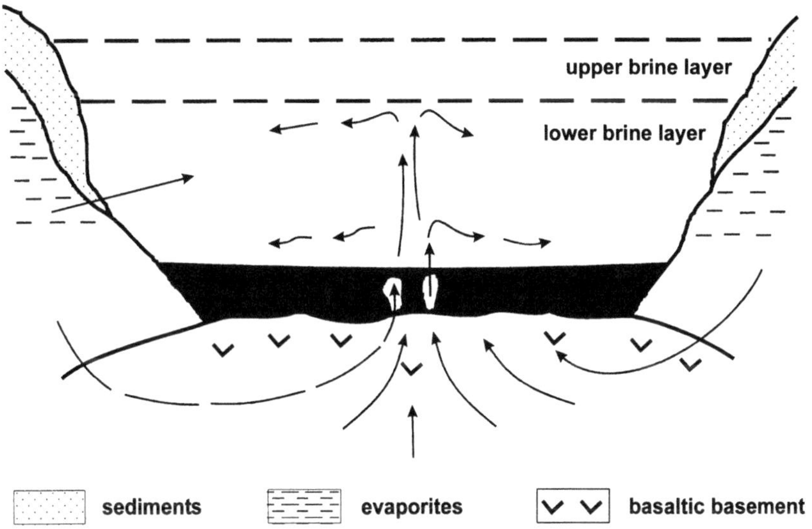

Fig. 2.8. Typical features of a stratified brine pool in the Red Sea rift, resulting from sea water circulating through evaporites and the basaltic basement subsequently emanating from vents and finally leading to precipitation of minerals (after Pottorf and Barnes 1983)

2.3.2 Sediment Hosted - Low-Temperature

Low-temperature systems can occur in a variety of geological sediment hosted settings of both elevated and normal heat flow. Deep fluid circulation through faults or folded permeable strata (Fig. 2.9), tectonic uplift of hotter rocks from depth and the residual heat from intruded plutons can yield low-temperature fields. These are found throughout Europe and Asia, and along some areas of Tertiary volcanism in the Pacific region.

The structure of low-temperature systems cannot be idealized like the geothermal systems discussed previously due to fact that they develop and can be found in a large variety of environments. Low-temperature systems usually discharge dilute waters through warm springs (Fig. 2.9) with temperatures around ≈30-65°C. The geothermal water composition in sediment hosted systems depends on the mineral composition of the host rocks and often on the relative contribution of the inherent formation water, the up-flowing geothermal water, and the recharging meteoric water.

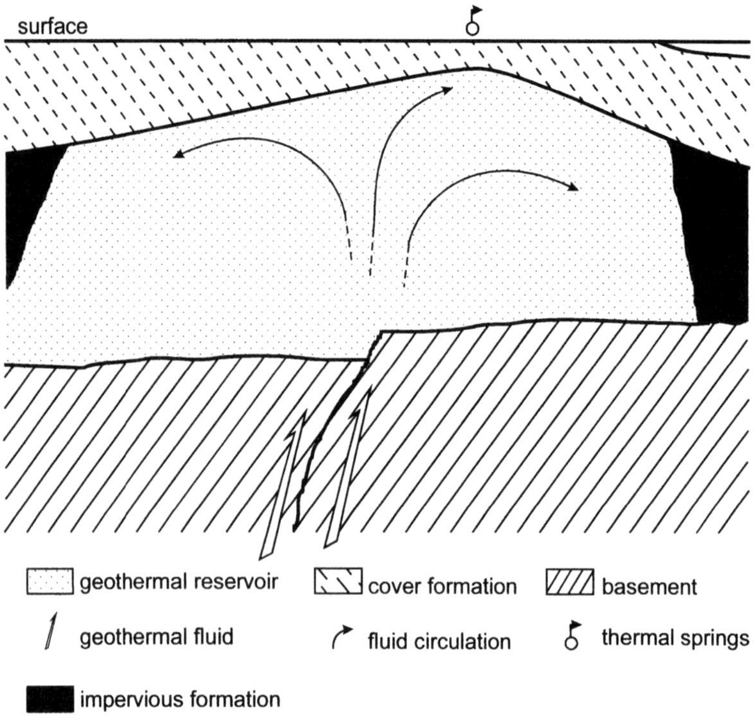

Fig. 2.9. Possible groundwater circulation model of a dynamic system of low temperature

2.4 Geothermal Water Compilation

The composition of a geothermal fluid can be characterized by a number of sources either as a pure or mixed type of water. It may, for example, represent surface (meteoric) water, which has gained depths of several kilometers through fractures and permeable horizons, or it can be water, which was buried along with the host sediments (formation or connate waters). Other sources of water in geothermal systems have been suggested; these include waters evolved during metamorphism (metamorphic waters) and from magmas (juvenile waters), but the importance of these both sources is uncertain (Nicholson 1993).

Most geochemical information regarding temperature, pressure, and chemical conditions within geothermal systems is extracted from analytical data of constituents dissolved in the liquid phase reaching the surface. Due to the importance of the water composition for geothermal power and heat production and the fact that ore deposit exploration focuses on the solid material, the majority of water samples taken from hydrothermal systems are from geothermal reservoirs. Referring to the distribution of geothermal fields in the world (Fig. 2.10) a compilation of existing water compositions is given here. To collect water analysis data an extensive literature review was performed.

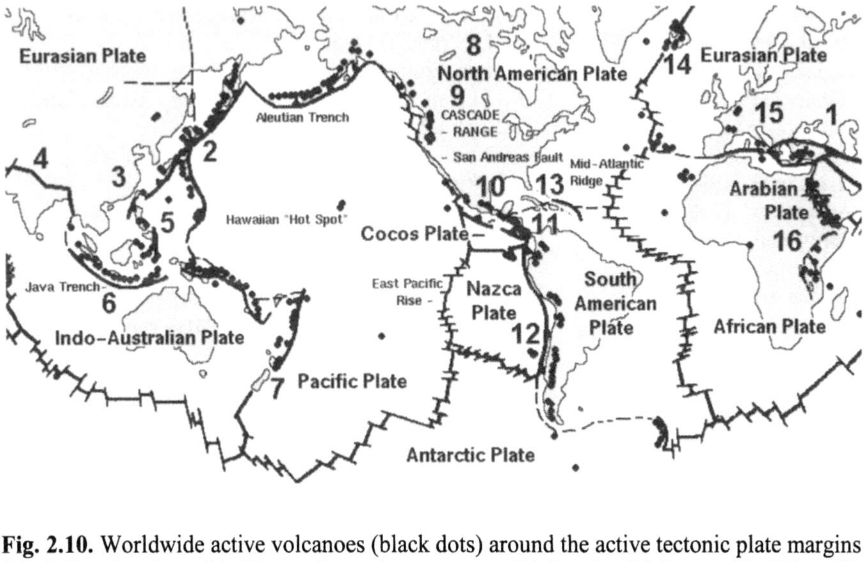

Fig. 2.10. Worldwide active volcanoes (black dots) around the active tectonic plate margins (black lines) and worldwide geothermal energy use; (1) Russia, (2) Japan, (3) China, (4) Himalayan, (5) Philippines, (6) Indonesia, (7) New Zealand, (8) Canada, (9) USA, (10) Mexico, (11) Central America, (12) Andes, (13) Caribbean, (14) Iceland and Atlantic islands, (15) Europe and Mediterranean, and (16) Africa (adapted from Topinka 1997)

About 1500 water analyses from 250 geothermal systems in 33 countries around the world comprise the data set. Data are taken from around 100 published papers with the sources listed in Table 2.1. Firstly, the water analyses were checked for analytical errors. In this step the working database was reduced to only include samples with an electrical balance better than ± 5%. Secondly, the database was checked for completeness concerning Na, K, Ca, Mg, Cl, and SO_4 and further reduced to 807 samples.

Table 2.1. Geothermal waters chosen for the data set are given with authors and title of source sorted by continents. In general the ionic balance is better than ± 5 % deviation

Reference / Continent	Title of source or reservoir
Africa	
Idris (1994)	Dakhla Oasis (Egypt)
Endeshaw (1988)	Aluto-Langano (Ethiopia)
Gianelli and Teklemariam (1993)	Aluto-Langano (Ethiopia)
Beyene (2000)	Wonji, Fantale/Meteka, Dofan (Ethopia)
Tole (1988)	Narosura (Kenya)
Svanbjörnsson et al. (1983)	Olkaria (Kenya)
America	
Ghomshei et al. (1986)	South Meager Creek, British Columbia (Canada)
Lahsen (1988)	Northern, central, and southern zone (Chile)
Marini et al (1998)	San Marcos (Guatemala)
Goff et al. (1992)	Tecuamburro (Guatemala)
Gandino et al. (1985)	Soufrière Caldera, St. Lucia (Lesser Antilles)
Prol-Ledesma et al. (1995)	La Primavera (Mexico)
Lopez and Arriaga (2000)	Los Azufres (Mexico)
Ramirez (1988)	El Valle de Anton, Chitra-Calabre, Tonosi (Panama)
Bath and Williamson (1983)	Cerro Pando (Panama)
Campos (1988)	North, central, southern area (El Salvador)
Nieva et al. (1997)	Chipilapa (El Salvador)
Adams et al. (1989)	Heber (USA)
Sorey et al. (1991)	Long Valley (USA)
White and Peterson (1991)	Long Valley (USA)
Goff and Tully (1994)	Archuleta County, Pagosa Springs (USA)
Kharaka (1986)	Texas, Louisiana, California, Mississippi (USA)
Thomas (1986)	Hawaii (USA)
Goff et al. (1981)	Jemez Springs area (USA)
Sorey and Colvard (1997)	Yellowstone Park (USA)
Fournier (1989)	Yellowstone Park (USA)
Asia	
Grimaud et al. (1985)	Central Tibet (China)
Huang and Goff (1986)	Fuzhou (China)
Zhonghe (2000)	Northern North Basin (China)
Liu et al. (1999)	Nagqu (Tibet)
Zongyu (2000)	Xiaotangshan (Beijing China)
Mahon et al. (2000)	Prospect overview (Indonesia)
Sundhoro et al. (2000)	Sembalun Bumbung (Indonesia)
Saxena and Gupta (1985)	Godavari Valley (India)

Table 2.1. continued

Reference / Continent	Title of source or reservoir
Giggenbach et al. (1983)	Parbati Valley (India)
Moon, Dharam (1988)	Puga Valley, Parbati Valley, West coast (India)
Saxena and Gupta (1987)	Salbardi, Tatapani (India)
Shanker et al. (2000)	Tapoban hot water system, NW Himalaya (India)
Yusa and Ohsawa (2000)	Beppu (Japan)
Noda and Shimada (1993)	Kyushu (Japan)
Abe (1993)	Onikobe (Japan)
Goko (2000)	Ogiri (Kyushu, Japan)
Reyes et al. (1993)	Alto Peak (Leyte Province, Philippines)
Lawless et al. (1983)	Bacon-Manito (Philippines)
Balmes (2000)	Mt. Balut Island, Davao del Sur (Philippines)
Chaturongkawanich et al. (2000)	Changwat Ranong (Thailand)
Praserdvigai (1987)	San Kampaeng, Fang (Thailand)
Hochstein et al. (1987)	San Kamphaeng (Thailand)
Gianelli et al. (1997)	South, Central (Vietnam)
Europe	
Georgieva and Vlaskovski (2000)	Bourgas Basin (Bulgaria)
Fritz (1989)	Bruchsal (Germany)
Lenz et al. (1997)	Allermöhe (Germany)
Bartels and Iffland (2000)	Stralsund, Karlshagen (Germany)
Kühn (1997)	Neustadt-Glewe, Neubrandenburg (Germany)
Merkel (1991)	Waren, Neubrandenburg (Germany)
Adams (1996)	Ascension Island (UK, South Atlantic)
Traganos et al. (1995)	Mygdonia Basin (Greece)
Grassi et al. (1996)	Nea Kessani (Greece)
Szita and Kocsis (2000)	Great Plain (Hungary)
Arnórsson et al. (1983)	Overview (Iceland)
Kristmannsdóttir (1989)	Low temperature (Iceland)
Bortolami et al. (1983)	Acqui Terme, Piemont (Italy)
Marini (2000)	Acqui Terme-Visone (Italy)
De Gennaro et al. (1984)	Island of Ischia (Italy)
Chiodini et al. (1988)	Phlegrean, Naples (Italy)
Dongarra et al (1983)	Pantelleria Island (Italy)
D'Amore et al. (1987)	Sardinia (Italy)
Kralj and Kralj (2000)	Murska Sobota (Slovenia)
Simsek (1985)	Denizli, Sarayköy - Buldan Area (Turkey)
Martinovic and Milivojevic (2000)	Macva (Yugoslavia)
Oceania	
Cox and Browne (1991)	Ngawha Area (New Zealand)
Sheppard and Giggenbach (1980)	Ngawha (New Zealand)
Giggenbach and Glover (1992)	Rotorua (New Zealand)
Sunaryo et al. (1993)	Ulubelu, South Sumatra (Indonesia)
Severne (1998)	Tokaanu-Waihi (New Zealand)
Giggenbach et al. (1994)	Waiotapu (New Zealand)
Wood ct al. (1997)	Wairakei (New Zealand)
Reyes and Giggenbach (1999)	Poihipi sector (Wairakei, New Zealand)
Mahon (1966a)	Natural hydrothermal Systems (New Zealand)
Mroczek et al. (1999)	Wairakei (New Zealand)
ARWB/ARC (1980-1999)	Waiwera (New Zealand)

Data of SiO_2, HCO_3, CO_3, CO_2, and pH and temperature are listed if available but its lack was not an exclusion criterion (constituents of samples are fully listed in the Appendix). Exceptions were made in the cases of several highly acidic waters, for example from acid sulfate springs. If a pH less than 5 occurred an electrical balance worse than –5 % has been accepted due to the fact that high amounts of metals like iron or aluminum have to be expected for the waters. Prerequisite was that the metals were not analyzed within the mentioned samples. Additionally, missing carbonate or hydrogen carbonate values are accepted for these analyses, because they are highly acidic.

Temperature, pH, and concentration range occurring within the selected geothermal waters is given in Table 2.2. Temperature values of 98 % of the samples are available and fall between 6 and 335°C. The pH varies from 2.4 up to 9.6 in 75 % of the waters. The remaining 25 % were cited in the literature without pH. The amounts of Na, K, Ca, Mg, Cl, and SO_4 are available in 100 % of the samples, because samples lacking theses values were omitted. It is obvious that the majority of the geothermal waters do have a pH in the neutral range, because hydrogen carbonate is the dominant carbonate species. Cited silica amounts occur with up to 1436 mg L^{-1} (within 84 % of analyses include SiO_2 data are including).

Table 2.2. Statistical report of constituents in the geothermal waters from the data base

Constituent	Minimum	Maximum	% of Samples
Temperature [°C]	6.0	335	98
pH	2.4	9.6	75
Na [mg L^{-1}]	0.05	95000	100
K [mg L^{-1}]	0.018	3264	100
Ca [mg L^{-1}]	0.124	18100	100
Mg [mg L^{-1}]	0.002	2710	100
Cl [mg L^{-1}]	0.103	1793000	100
SO_4 [mg L^{-1}]	0.25	4109	100
HCO_3 [mg L^{-1}]	0.0	6016	97
CO_3 [mg L^{-1}]	0.0	489	9.6
CO_2 [mg L^{-1}]	0.0	5240	9.5
SiO_2 [mg L^{-1}]	0.0	1436	84

2.5 Chemical Interpretation of Geothermal Waters

Interpretation of the geothermal water quality, of the analyses compiled in the database (previous section), is done here based on individual samples. The first step in water quality assessment is the examination of data accuracy, as mentioned above. Second step is the attempt to estimate their source rocks from the analysis and further investigation concerning potential mineral reactions that may have taken place.

Originally it was thought that magmas are the source of the heat in the geothermal system and also responsible for the constituents of the waters emanating from springs and tapped from wells. In the early 1960s this idea was skipped when

Craig (1963) showed that geothermal fluids are dominantly of meteoric origin. Rock-water interaction is proved to be the major source for many of the solutes in geothermal waters. It is emphasized in several studies (Ellis and Mahon 1964, 1967) that all solutes in geothermal fluids could be derived from reactions between meteoric waters and the host rocks of the geothermal system. Bischoff et al. (1981) and Seyfried and Bischoff (1981) produced experimental results providing a similar explanation for the composition of seawater-influenced geothermal systems as located in Iceland. While there is no doubt that the geothermal fluids are of a predominantly meteoric origin, based largely on stable isotope studies, the data permit at least 5-10 % of the fluid to be from an alternative, possibly a magmatic source (Nicholson 1983).

Recalling the conceptual sketches of dynamic liquid dominated geothermal systems (Fig. 2.5, Fig. 2.6) the evolution of reservoir waters can be summarized as follows: meteoric water circulates to depth and while descending it is heated and may begin to change composition, the ascending water, driven by thermal convection, reacts with the host rocks of the reservoir.

Within the last decades extensive research has been conducted to delineate the mineral alteration processes active in geothermal systems. Since then, the amount of analytical, experimental, and theoretical information pertaining to hydrothermal rock alteration increased significantly. Research started with empirical investigations on the behavior of fluid constituents as functions of a few parameters, stepped on to theoretical models, and is now using complex reactive transport models. However, for routine usage of reactive transport models research is still necessary as outlined later.

2.5.1 Thermal Water Types

It is common practice to classify geothermal waters according to their dominant anion. Although the subdivision into chloride, sulfate, and hydrogen carbonate waters is not a formal genetic scheme, this descriptive classification does permit insight to the likely origins of the waters. The most common type of fluid found at depth in high-temperature geothermal systems is of near-neutral pH, with chloride as the dominant anion. Other waters encountered within the profile of a geothermal field are commonly derived from this deep fluid as a consequence of chemical or physical processes (Nicholson 1993).

Chloride Water

The chloride water type, also termed "alkali-chloride" or "neutral-chloride", discharges from deep geothermal wells and from associated neutral chloride springs (Fig. 2.5, Fig. 2.6). These waters are likely to represent well-equilibrated fluids from the major up-flow zones in geothermal reservoirs. They are of near neutral pH with the dominant anion chloride and concentrations of up to thousands of mg kg^{-1} (Table 2.2). Sulfate and hydrogen carbonate concentrations are variable in chloride waters, but are commonly several orders of magnitude less than that of chloride (Giggenbach 1988, Nicholson 1993).

Sulfate Water

The so-called "acid-sulfate" waters are formed by condensation of geothermal gases into groundwater. Sulfate is the principal anion, formed by the oxidation of condensed hydrogen sulfide. The decrease of the pH in these geothermal waters is due to the following oxidation reaction:

$$H_2S(g) + 2O_2(aq) = 2H^+(aq) + SO_4^{2-}(aq) \tag{2.1}$$

Sulfate waters react rapidly to leach the host rock due to their acidity. They are found on the margins of a field at some distance from the major up-flow zone (Fig. 2.5, Fig. 2.6) or in the primary neutralization zone deep in the reservoir above the magma body (Fig. 2.6).

Hydrogen Carbonate Water

These waters are the product of steam and gas condensation into subsurface groundwater. Such fluids with high CO_2 reactivity can occur in a perched condensation zone overlying the geothermal system and are, like the acid-sulfate waters, common on the margins of the field. Surface features of this type are "soda" springs (Fig. 2.5, Fig. 2.6). The waters are of near neutral pH as reaction with the local rocks neutralizes the initial acidity.

2.5.2 Graphical Interpretation Methods

Giggenbach (1988) showed how the simple Cl-SO_4-HCO_3 ternary plot (ratios of the total anion content in mg kg^{-1}) aids the identification of the above-described waters (Fig. 2.11). The end member chloride, acid sulfate, and hydrogen carbonate waters group towards the corresponding corners of the triangular plot. Whereas the examples of Giggenbach (1988) mainly show the unmixed water types, Nicholson (1993) published additional data of mixed sulfate-chloride waters, including the volcanic gas condensates, mixed chloride-seawaters, and mixed chloride-hydrogen carbonate waters (Fig. 2.12). It is obvious, that in acid spring waters (Fig. 2.11), sulfate, sulfate-chloride, and volcanic condensate waters (Fig. 2.12) the amount of hydrogen carbonate is almost zero. This is due to the fact that the pH value of these waters is too low to allow the hydrogen carbonate anion to exist in the water. Both the examples given by Giggenbach (1988) and Nicholson (1993) depict that mixed waters do occur between two main anion-types only. Besides the mixing types mentioned above and shown in Fig. 2.11 and Fig. 2.12 it can additionally be concluded that there seems to be a lack of mixing between sulfate and hydrogen carbonate waters in geothermal reservoirs or that at least that they are rarely found.

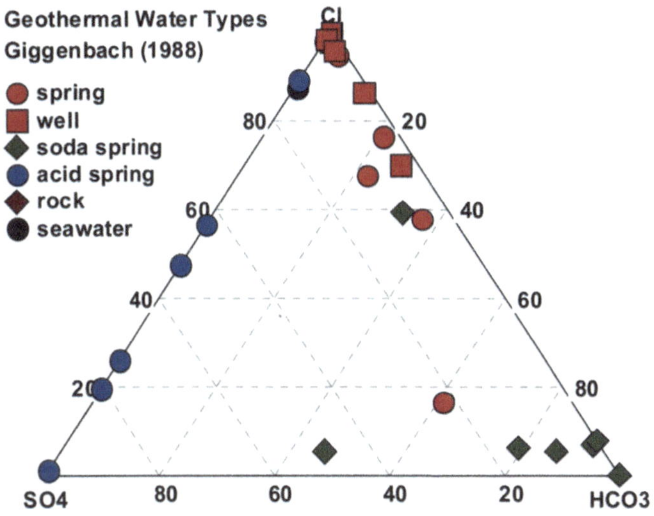

Fig. 2.11. Ternary plot of Cl-SO_4-HCO_3 [mg kg^{-1}] to distinguish the three common end member water types of a geothermal system: acid sulfate, chloride, and hydrogen carbonate (data taken from Giggenbach 1988); seawater is shown for reference

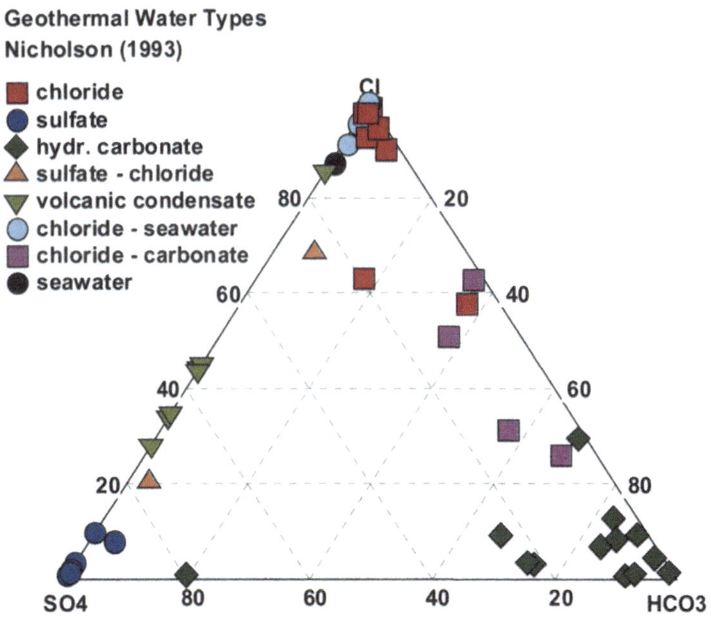

Fig. 2.12. Ternary Cl-SO_4-HCO_3 [mg kg^{-1}] plot with data of Nicholson (1993) who subdivided the three end member waters into: chloride, sulfate, hydrogen carbonate, sulfate-chloride, condensate, chloride-seawater, and chloride-hydrogen carbonate waters; seawater is shown for reference

Another common plot used to compare water quality data on the same diagram is from Piper (1944). It consists of two triangles, one for the cations and one for the anions, and a central, diamond-shaped polygon. Cations are plotted on the (Na+K)-Ca-Mg triangle and anions on the Cl-SO_4-HCO_3 triangle as percentages (concentrations in meq L^{-1}). Both triangles distinguish three dominant water types. Points on the anion and cation diagrams are projected upward to where they intersect on the diamond. From the diamond four main water types can be read, these are chloride waters or brines (towards right corner of the diamond), waters of permanent hardness (towards upper corner), waters of temporary hardness (towards left corner), and alkali – hydrogen carbonate waters (towards lower corner).

The data collections of Giggenbach (1988) and Nicholson (1993) are shown in Piper plots in Fig. 2.13 and Fig. 2.14, respectively. Apart from the fact that the concentration dimensions are different, the triangles of the ternary plots shown above (Fig. 2.11, Fig. 2.12) and the anion triangles of the Piper plots (Fig. 2.13, Fig. 2.14) are identical. They show the division of the geothermal waters in the main types chloride, sulfate, and hydrogen carbonate. On the contrary, within the cation triangles there is only the alkali type water, which dominates especially among the chloride waters (either from deep wells or springs) and these mixed with chloride waters. The sulfate and hydrogen carbonate waters and their mixture (volcanic condensates) do not show any dominant type in their cation distribution. Average seawater plots very close to the geothermal chloride brines in all of the four figures.

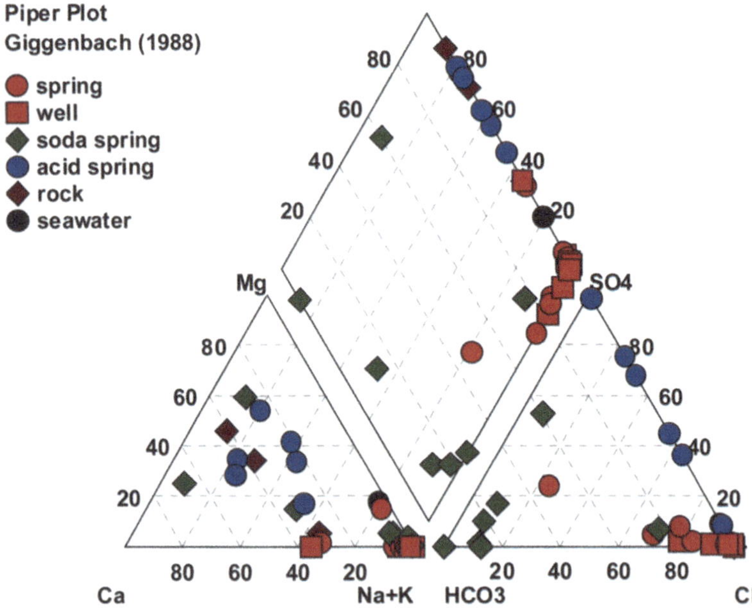

Fig. 2.13. Geothermal waters published by Giggenbach (1988) shown on a Piper plot

Giggenbach (1988) showed data from Taylor (1964) of water compositions produced by isochemical dissolution of crustal rocks, such as basalt, granite, and average crust. Due to the fact, that the only resulting anion in these samples is chloride, they are only visible in the cation triangle of Fig. 2.13. In the anion diagrams of Fig. 2.11 and Fig. 2.13 the printed rock samples are covered behind the chloride waters.

The question now is, is it possible to get distinct information from a geothermal water sample, plotting it in a ternary or Piper diagram as shown above, about the structure of the corresponding geothermal field? From the entire geothermal water database all the samples where taken with distinct information available about their source rocks in the field. The collected samples are subdivided into geothermal fields located in a (1) sedimentary, (2) volcanic, or (3) plutonic environment and mixed lithologies of fields in (4) volcanic – sedimentary and (5) plutonic - sedimentary environment. Geothermal fields in a mixed environment of plutonic and volcanic rocks do not exist.

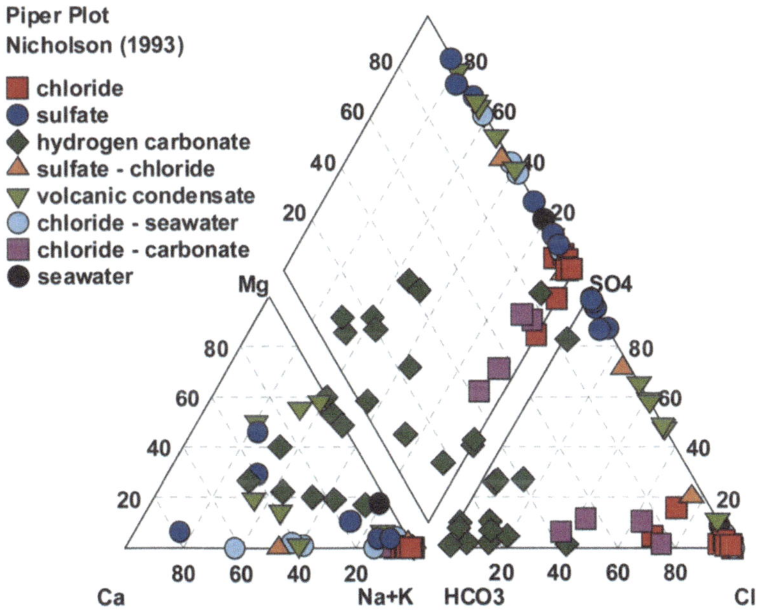

Fig. 2.14. Geothermal waters published by Nicholson (1993) shown on a Piper plot

Fig. 2.15 depicts that there does not exist any tight dependency between the rock types of a geothermal field and the resulting water composition. This corresponds with the findings of Browne (1978) that the parent rock influences hydrothermal alteration at high temperatures mainly through permeability whereas the initial mineralogy of the reservoir rocks seems to have little effect on the resulting secondary mineral assemblage. However, at lower temperatures the nature of the parent rock clearly influences the alteration product. Especially waters of the sedimentary and volcanic type are distributed over the entire Piper plot. However, several tendencies can be observed. The geothermal waters of plutonic origin, for example, group towards the alkali type in the cation triangle and therefore finally in the diamond towards the brines (right corner). Within the anion triangle there is no obvious domination, but it can be seen that geothermal waters of plutonic origin do not show the acid sulfate type. The dominant cation type of geothermal waters from a mixed volcanic - sedimentary environment seems to be of the calcium cation-type and often of the sulfate-anion type. Waters from plutonic – sedimentary systems show a tendency towards the center of the diamond in the Piper plot due to quite even amounts, on the one hand of calcium and alkali ions in the cation triangle and the other hand of chloride and hydrogen carbonate in the anion triangle.

It can be concluded that it is not possible without any further information to determine the source rock tectonic setting of a geothermal field without any doubt, because the geologic settings are manifold. However, water analyses provide valuable data and a starting point for the investigation or interpretation of the structure of geothermal fields.

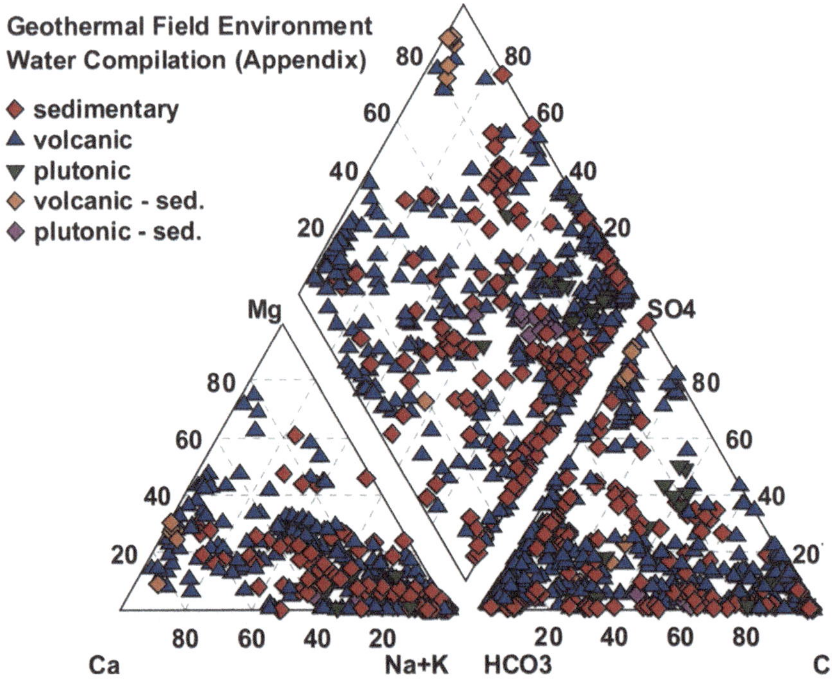

Fig. 2.15. Piper plot of the geothermal water compilation subdivided into waters from geothermal fields in sedimentary, volcanic, plutonic, volcanic – sedimentary, or plutonic - sedimentary environment

2.6 Processes Affecting the Chemical Composition of Hydrothermal Waters

The contemplation of processes affecting the chemical composition of hydrothermal waters is done here distinguishing between reactions playing a dominant role in dynamic magmatic systems of high temperature and static or dynamic sediment hosted systems of low temperature (Fig. 2.2). Both groups are important for understanding the structure of geothermal reservoirs.

2.6.1 Dynamic Magmatic Systems (High-Temperature)

Reactions of thermal waters with rocks are called hydrothermal alteration. Hydrothermal alteration involves mineralogical, chemical and textural changes. Due to the fact that alteration reactions depend on the particular conditions of temperature and pressure, they are prevailed by the temporal and local development of the physical conditions in hydrothermal systems.

Chemical and physical processes influence the composition of both the meteoric water as it descends into the reservoir and the geothermal water as it ascends through the up-flow zone to the surface (Fig. 2.1 and Fig. 2.4 - Fig. 2.6). The main chemical processes are mineral dissolution and precipitation, while the dominant physical process is boiling, although conductive cooling and mixing are also important.

Constituents of thermal waters can be divided into the soluble or non-reactive group (e.g. Cl, B, Br, etc.) and the rock-forming species (e.g. SiO_2, Na, K, Ca, Mg, etc.) providing the basis for water classification (see above) and reservoir interpretation. Solubility of the rock-forming species is controlled by mineral-fluid equilibria. The reactions, which take place, are a function of temperature, pressure, salinity, and host rock composition of the geothermal system (Nicholson 1993). The product of mineral-fluid reactions is an assemblage of secondary alteration minerals discussed in detail by Browne (1978) and Giggenbach (1981, 1984, 1988).

The physical and chemical conditions of recent dynamic hydrothermal alteration systems can be applied to the situation of fossil systems. The overall fluid-rock equilibrium is rarely attained and the fluids are likely to have reached some complex steady-state composition at particular locations. Compared to a composition in thermodynamic equilibrium, representing the most stable mineral assemblage and water composition, a steady-state is characterized by constant values of metastable reaction intermediates within a dynamic equilibrium. Any specific composition reflects the combined effects of initial fluid constituents, kinetics of primary mineral dissolution and secondary mineral deposition at changing temperatures and pressures, vapor loss, dilution, and mixing with fluids of different origin. Two end member processes determine the actual state of chemical equilibrium or non-equilibrium in the hydrothermal system. The reaction sequence starts with (1) isochemical dissolution (in proportions of the original rock) of the rock material in contact with the rising fluid, a more or less hypothetical process and eventually ends with (2) equilibration of the fluid in respect to the thermodynamically stable alteration assemblage resulting from recrystallization of primary rock at a given temperature and pressure, a process likely to come to completion only in stagnant systems of infinite age (Giggenbach 1984, 1988).

Following the concept of dynamic magmatic high temperature systems within a low relief setting (Fig. 2.6) the hydrothermal alteration processes are distinguished here in (1) Na-Mg-Ca-metasomatism or propylitic alteration at the periphery of the system occurring due to descending meteoric waters (metasomatism: the enrichment or depletion of a system in an element or elements by an externally-derived fluid). Subsequently the deep circulating geothermal water may experience (2) H-metasomatism or argillic alteration in the so called primary neutralization zone due to the attack of strong acids like HCl, H_2SO_4, and HF from the degassing magma body. The very slow drop of temperature over deeper, central parts of the major fluid up-flow zone leads to (3) K-metasomatism or potassic alteration characterized by isochemical recrystallization of the rock components. K-metasomatism is often accompanied by (4) silicification or silication. The fluids emerging from the neutralization zone still contain magmatic CO_2 and H_2S, hence (5) H-metasomatism proceeds by CO_2 attack on Ca-aluminum silicates in a secondary neutralization of CO_2. K-metasomatism and H-metasomatism are superimposed in the region of (6) H-K-metasomatism or phyllic alteration (scheme (1)-(6) is also applicable for other dynamic magmatic systems).

Propylitic Alteration (Na-Mg-Ca-Metasomatism)

Propylite is an old term used to describe altered volcanic rocks. Propylitic alteration, especially occurring in zones of descending meteoric waters in the hydrothermal system (or seawater in geothermal fields like Iceland) as stated by Giggenbach (1988), is characterized by the addition of H_2O and CO_2, and locally S, but without appreciable H-metasomatism. The meteoric water entering the hydrothermal system at the periphery (recharge zone) is heated and undergoes Na-Mg-Ca-metasomatism. Potassium minerals, like the K-feldspar microcline ($KAlSi_3O_8$), are preferentially destroyed on contact with descending solutions leading to the formation of assemblages containing albite ($NaAlSi_3O_8$), chlorite ($(Mg,Fe,Al)_6(Al,Si)_4O_{10}(OH)_8$), and epidote ($Ca_2FeAl_2Si_3\,O_{12}\,(OH)$). It can be said that downward flow and increasing temperatures favor assemblages containing minerals more soluble at low temperatures.

Argillic Alteration (H-Metasomatism)

Argillic alteration is characterized by the formation of clay minerals due to intense H-metasomatism, also called acid leaching. As stated above, H-metasomatism may occur within the hydrothermal system in two different cases, either deep in the reservoir within the primary neutralization zone or adjacent to the main up-flow region within the secondary neutralization zone (Fig. 2.6).

H-metasomatism is a widespread and very important type of reaction in hydrothermal alteration processes. It is based on the hydrolysis of H_2O into H^+ and OH^-, with the consumption of H^+ during reaction with silicate minerals and release of metal ions into the water. Hydration, the transfer of molecular water from the solution to a mineral, often accompanies hydrolysis. The alteration of olivine (Mg_2SiO_4) to serpentine ($Mg_3Si_2O_5(OH)_4$) outlines the consumption of H^+, the release of Mg^{2+} and the incorporation of molecular water as an example.

$$2\ Mg_2SiO_4 + H_2O + 2\ H^+ \Leftrightarrow Mg_3Si_2O_5(OH)_4 + Mg^{2+} \qquad (2.2)$$

Argillic alteration may be subdivided on the basis of the composition of the geological formations undergoing H-metasomatism, specifically rocks:

- dominated by feldspars,
- with a prevailing mafic composition, or
- rich in calcium, like carbonates.

Within hydrothermal reservoirs, the circulating geothermal water may absorb strong acids like HCl, H_2SO_4, and HF from the deep magmatic to hydrothermal transition zone. This is a reaction most likely to occur but impossible to observe directly. The absorption of magmatic vapor into the deeply circulating fluid leads to initial rock dissolution. The vapor composition can be estimated of high temperature gases released from basaltic (Giggenbach and Le Guern 1976) or andesitic magmas (Giggenbach 1987). Magma gases contain 20-40 wt.% CO_2, 5-10 % b.w. total sulfur, predominantly in the form of SO_2 at low pressures, but H_2S at higher pressures, and 1-2 % b.w. HCl (Rose et al. 1986). The passage of the geothermal water through the neutralization zone results firstly in an increase of Na, K, Mg, and Ca due to almost isochemical rock dissolution. Secondly, still in the neutralization zone, the amounts of Ca and Mg subsequently decrease by formation of alteration products like amphiboles [e.g. anhydrite ($CaSO_4$), biotite, chlorite ($K(Mg,Fe)_3AlSi_3O_{10}(OH)_2$), tremolite ($Ca_2(Mg,Fe)_4AlSi_7AlO_{22}(OH)_2$), or fluorite ($CaF_2$)].

The fluids emerging from the neutralization zone still contain magmatic CO_2 and H_2S leading to H-metasomatism adjacent to the up-flow region. The CO_2 attack on Ca-aluminum silicates, a more gentle reaction [Eq. (2.3)], leads to the conversion of carbon dioxide to calcite ($CaCO_3$) and “acid” clays

$$\text{Ca-Al-silicate} + CO_2 + H_2O \Leftrightarrow CaCO_3 + 2(\text{H-Al-silicate}) \qquad (2.3)$$

and subsequently to “neutral” aluminum silicates (z = stoichiometric factor):

$$z(\text{H-Al-silicate}) + M^{z+} \Leftrightarrow (\text{M-Al-silicate}) + zH^+ \qquad (2.4)$$

H-Metasomatism of Feldspars. Hemley and Jones (1964) conducted experimental studies of reactions between feldspars and sodium chloride solutions. Investigating the K_2O-Al_2O_3-SiO_2-H_2O system, an example is given here of the conversion of the anhydrous silicate K-feldspar microcline ($KAlSi_3O_8$) to the hydrous K-mica sericite ($KAl_3Si_3O_{10}(OH)_2$) during quartz (SiO_2) formation. Sericite is a varietal name used for fine-grained muscovite, illite or paragonite.

$$3\ KAlSi_3O_8 + 2\ H^+ \Leftrightarrow KAl_3Si_3O_{10}(OH)_2 + 2\ K^+ + 6\ SiO_2 \quad (2.5)$$

The reaction of andesine ($Na_2CaAl_4Si_8O_{24}$) to K-mica due to argillic alteration displays H-metasomatism intersecting with K-metasomatism (see below).

$$\begin{aligned} &3\ Na_2CaAl_4Si_8O_{24} + 8\ H^+ + 4\ K^+ \Leftrightarrow \\ &4\ KAl_3Si_3O_{10}(OH)_2 + 6\ Na^+ + 3\ Ca^{2+} + 12\ SiO_2 \end{aligned} \quad (2.6)$$

Successive reactions following the development of K-mica may be, due to further H-metasomatism, the occurrence of kaolinite ($Al_2Si_2O_5(OH)_4$) or pyrophyllite ($Al_2Si_4O_{10}(OH)_2$). Low temperatures and low cation / H^+ ratios support the development of kaolinite (Meyer and Hemley 1967).

$$2\ KAl_3Si_3O_{10}(OH)_2 + 2\ H^+ + 3\ H_2O \Leftrightarrow 3\ Al_2Si_2O_5(OH)_4 + 2\ K^+ \quad (2.7)$$

$$2\ KAl_3Si_3O_{10}(OH)_2 + 2\ H^+ + 6\ SiO_2 \Leftrightarrow 3\ Al_2Si_4O_{10}(OH)_2 + 2\ K^+ \quad (2.8)$$

The patterns of argillic alteration within the system Na_2O-Al_2O_3-SiO_2-H_2O are transferable to the reactions of K-feldspars. Albite ($NaAlSi_3O_8$) the counterpart to microcline for example converts to paragonite ($NaAl_3Si_3O_{10}(OH)_2$) the sodium counterpart to K-micas and due to further H-metasomatism within a successive reaction also to pyrophillite.

Processes within the system K_2O-Al_2O_3-SiO_2-H_2O-SO_3 are prevailed by the formation of sulfuric acid due to the oxidation of H_2S [Eq. (2.1)]. Hemley et al. (1969) experimentally studied this five component system with special consideration of the stability of K-feldspars, muscovite ($KAl_3Si_3O_{10}(OH)_2$), kaolinite and alunite ($KAl_3(SO_4)_2(OH)_6$). Alunite is a key mineral within this system and is often found in sulfur-rich epithermal gold-silver deposits (Pirajno 1992). The mineral assemblage is determined by the following reactions of K-feldspar to alunite [Eq. (2.9)], K-mica to alunite [Eq. (2.10)] or kaolinite [Eq. (2.11)] and the conversion of muscovite to kaolinite [Eq. (2.12)].

$$3\ KAlSi_3O_8 + 6\ H^+ + 2\ SO_4^{2-} \Leftrightarrow KAl_3(SO_4)_2(OH)_6 + 2\ K^+ + 9\ SiO_2 \quad (2.9)$$

$$KAl_3Si_3O_{10}(OH)_2 + 4\ H^+ + 2\ SO_4^{2-} \Leftrightarrow KAl_3(SO_4)_2(OH)_6 + 3\ SiO_2 \quad (2.10)$$

$$\begin{aligned} &3\ Al_2Si_2O_5(OH)_4 + 2\ K^+ + 6\ H^+ + 4\ SO_4^{2-} \Leftrightarrow \\ &2\ KAl_3(SO_4)_2(OH)_6 + 6\ SiO_2 + 3\ H_2O \end{aligned} \quad (2.11)$$

$$2\ KAl_3Si_3O_{10}(OH)_2 + 2\ H^+ + 3\ H_2O \Leftrightarrow 3\ Al_2Si_2O_5(OH)_4 + 2\ K^+ \quad (2.12)$$

H-Metasomatism of Mafic Rocks. Hydrothermal alteration of mafic rocks is spatially related to the occurrence of basaltic formations, for example at mid ocean ridge spreading centers or continental rift settings. The group of mafic minerals is, in general, comprised of Fe-Mg silicates like biotite ($K(Mg,Fe)_3AlSi_3O_{10}(OH,F)_2$), amphiboles ($(Ca,Na)_2(Mg,Fe,Al)_5Si_8O_{22}(OH)_2$), pyroxenes (orthopyroxene:

$(Mg,Fe)_2Si_2O_6)$, clinopyroxenes $(Ca(Mg,Fe)Si_2O_6 - Na(Al,Fe)Si_2O_6)$, or olivine $((Mg,Fe)_2SiO_4)$.

Hydrothermal activity in sub-sea floor systems is a typical environment for the alteration of mafic rocks. The penetration and descent of seawater through fractures in the oceanic crust leads to heating of the solutions which become more and more reduced due to reactions with the rock forming minerals. A typical reaction is the H-metasomatism of fayalite (Fe_2SiO_4) to magnetite (Fe_3O_4) and pyrite (FeS_2).

$$11\ Fe_2SiO_4 + 2\ SO_4^{2-} + 4\ H^+ \Leftrightarrow 7\ Fe_3O_4 + FeS_2 + 11\ SiO_2 + 2\ H_2O \qquad (2.13)$$

Hydrothermal reactions at spreading centers are described in detail by Edmond and Damm (1983).

H-metasomatism in Ca-Rich Environments. Hydrogen ion metasomatism of carbonates is mainly described for the dissolution of calcite and the production of carbon dioxide resulting in the formation of carbonic acid (H_2CO_3).

$$CaCO_3 + 2\ H^+ \Leftrightarrow Ca^{2+} + CO_2 + H_2O \Leftrightarrow Ca^{2+} + H_2CO_3 \qquad (2.14)$$

The carbonic acid dissociation in turn leads to hydrolysis of silicates, forming clays and liberating silica and metal ions.

Potassic Alteration (K-Metasomatism)

Potassic alteration or K-metasomatism describes exchange reactions in feldspars, specifically Na for K, or K for Na, which in the latter case is called sodium alteration. Both types of exchange are combined under the name of alkali metasomatism. These exchange reactions, with the first one predominant in hydrothermal systems, results in changes of the structural state of the feldspars. This is not to be confused with ion exchange on surfaces where only physical processes occur in contrary to the chemical reactions of alkali metasomatism.

H-metasomatism is subject to rapid conductive cooling or dilution, adjacent to the major up-flow zone and culminating in waters high in hydrogen carbonate, whereas potassium metasomatism occurs in the major up-flow zone (Fig. 2.6). K-metasomatism is the predominant process occurring with the cooling of ascending solutions and is accompanied by silification (see below) and characterized by the formation of K-feldspar, biotite, K-mica or K-rich clays.

Potassic alteration is important in porphyry and epithermal mineralizing systems. The minerals, which are characteristic for K-metasomatism, are K-feldspar and biotite in porphyries, and adularia in epithermal systems. Sulfides like chalcopyrite, pyrite, and molybdenite normally occur due to potassic alteration. In porphyry regions, anhydrite is a common accompanying mineral phase (Pirajno 1992).

The very slow drop of temperature within deeper, central parts of the major fluid up-flow zone leads to recrystallization of rock components over close to stagnant parts of the system and the attainment of full water-rock equilibrium. Waters from deep geothermal wells appear to achieve compositions close to full water-rock equilibrium with respect to the reaction of albite (Na-feldspar, $NaAlSi_3O_8$) and K-feldspar (e.g. microcline).

$$\text{Na-feldspar} + K^+ \Leftrightarrow \text{K-feldspar} + Na^+ \tag{2.15}$$

Phyllic Alteration (H-K Metasomatism)

The regions of K-metasomatism and H-metasomatism are superimposed and combined under the term of phyllic or sericitic alteration. This is one of the most common types of hydrothermal alteration, present not only in Archean volcanogenic but also in recent epithermal systems, responsible for massive sulfide and gold-quartz deposits. A typical mineral assemblage for phyllic alteration zones consists of quartz, sericite, and pyrite accompanied by minerals like K-feldspar, kaolinite, calcite, biotite, rutile (TiO_2), apatite ($Ca_5(PO_4)_3(OH,F,Cl)$), and anhydrite. Phyllic alteration spreads into the potassic type by increasing amounts of K-feldspars and into the argillic type by increasing amounts of clay-minerals (Fig. 2.6, Pirajno 1992).

Giggenbach (1984) stated, accepting fluid and mineral compositions resulting from recrystallization of the parent rock, that a fully equilibrated system corresponds to K-feldspar / K-layer silicate coexistence. For different temperatures, theoretical CO_2 fugacities may be evaluated. For temperatures up to 200°C this is done taking the reaction between laumontite ($CaAl_2Si_4O_{12} \cdot 4H_2O$), microcline, carbon dioxide, calcite ($CaCO_3$), muscovite ($KAl_3Si_3O_{10}(OH)_2$), and silica.

$$\begin{gathered}\text{laumontite} + \text{microcline} + CO_2 \Leftrightarrow \\ \text{calcite} + \text{muscovite} + 4\ SiO_2\end{gathered} \tag{2.16}$$

For temperatures from 200 to 280°C the dominant reaction is between clinozoisite ($Ca_2Al_3Si_3O_{12}(OH)$), microcline, carbon dioxide, calcite, muscovite and silica.

$$\begin{gathered}2\ \text{clinozoisite} + 3\ \text{microcline} + CO_2 \Leftrightarrow \\ \text{calcite} + 3\ \text{muscovite} + 6\ SiO_2\end{gathered} \tag{2.17}$$

At even higher temperatures, wairakite ($CaAl_2Si_4O_{12} \cdot H_2O$)

$$\begin{gathered}\text{wairakite} + \text{microcline} + CO_2 \Leftrightarrow \\ \text{calcite} + \text{muscovite} + 4\ SiO_2\end{gathered} \tag{2.18}$$

or anorthite ($CaAl_2Si_2O_8$) control the reaction.

$$\begin{gathered}\text{anorthite} + \text{microcline} + CO_2 \Leftrightarrow \\ \text{calcite} + \text{muscovite} + 2\ SiO_2\end{gathered} \tag{2.19}$$

Silicification

The best-known and probably most common type of alteration is silicification. During hydrothermal processes silica may either be introduced from the circulating fluids and precipitated due to its prograde solubility or left behind in the form of residual silica as shown for H-metasomatism of feldspars [Eqs. (2.5), (2.6), (2.9) - (2.11)] or mafic rocks [Eq. (2.13)] and the region of H-K-metasomatism [Eqs. (2.16)-(2.19)].

Silication

Silication results in skarn rocks, in which the addition of silica leads to the development of calc-silica minerals. Skarns develop at the contact between plutons with carbonate rocks. Skarns are very important because they host a great variety of ores. Silication is the replacement of carbonate rocks by silicate minerals like the reaction of calcite or dolomite to wollastonite ($CaSiO_3$) or diopside ($(CaMg)Si_2O_6$), respectively.

$$CaCO_3 + SiO_2 \Leftrightarrow CaSiO_3 + CO_2 \tag{2.20}$$

$$CaMg(CO_3)_2 + 2\ SiO_2 \Leftrightarrow (CaMg)Si_2O_6 + 2\ CO_2 \tag{2.21}$$

2.6.2 Static and Dynamic Sediment Hosted Systems (Low Temperature)

The groundwaters in deep sedimentary basins are characterized by long residence times (static system) and the variable mineralogical composition of host rocks and matrix. The North German basin is an example of the occurrence of hydrothermal waters in a static system. Müller and Papendieck (1975) described in detail the genesis of these deep-seated brines. The majority of geothermal waters in the North German basin are low (36 – 150 g L^{-1} total dissolved solids) to high concentrated (280 – 360 g L^{-1}). For the origin and evolution of the brines the following processes have to be taken into account separately or in various combinations:

- Evaporation under surface conditions leading to highly saline brines as by-products. These so called "relict" solutions may remain within the sediments during diagenesis.
- Compaction driven flow of pressurized formation waters and their diagentic alteration.
- Meteoric water intrusion in regions containing evaporites resulting in dissolution of rock salts and the formation of brines.

From intensive evaporation of marine or continental waters salt solutions result at the Earth's surface from which evaporites evolve. Evaporite deposits are important sources of gypsum, halite, and other economically significant minerals. Salts commonly resulting from evaporite evolution of marine origin, composed of the major seawater ions, are given in Table 2.3.

The inherent processes of evaporite evolution from seawater are described by Hardie and Eugster (1970) and Hardie (1996). Spencer et al. (1985a, 1985b) described the relationship between hydrology, salinity, and geochemistry of a continental evaporite system, the Great Salt Lake basin in Utah. Lerman (1979), Berner

(1980), and Boudreau (1997) summarized the role of solute fluxes in and out of pore fluids in aquatic sediments.

Table 2.3. Minerals found in evaporite deposits of marine origin

Mineral	Formula	Mineral	Formula
anhydrite	$CaSO_4$	hexahydrite	$MgSO_4 \cdot 6H_2O$
bischofite	$MgCl_2 \cdot 6H_2O$	kainite	$KMgClSO_4 \cdot 3H_2O$
bloedite	$Na_2Mg(SO_4)_2 \cdot 4H_2O$	kieserite	$MgSO_4 \cdot H_2O$
calcite	$CaCO_3$	mirabilite	$Na_2SO_4 \cdot 10H_2O$
carnallite	$KMgCl_3 \cdot 6H_2O$	polyhalite	$K_2MgCa_2(SO_4)_4$
dolomite	$CaMg(CO_3)_2$	sylvite	KCl
glauberite	$Na_2Ca(SO_4)_2$	tachyhydrite	$Mg_2CaCl_6 \cdot 12H_2O$
gypsum	$CaSO_4 \cdot 2H_2O$	thenardite	Na_2SO_4
halite	$NaCl$	trona	$Na_3H(CO_3)_2 \cdot 2H_2O$

Once evaporites are deposited and start to undergo burial, their physical and hydraulic properties begin to change. Burial compacts the layered evaporites and reduces their porosity. The interstitial brine is forced upward opposite the downward movement of the sediment, into overlying strata. The brines buried within their new host sediments start to react towards chemical equilibrium with the host rocks and cement minerals.

The layered salt bodies tend to bulge upwards through overlying sediments when they are deeply buried, because the contrast between their density (2200 kg m^{-3}) and the density of the surrounding sediments (2500 kg m^{-3}) leads to unstable conditions. The upward migration of salt is accompanied by a reciprocal downward sagging of the surrounding sediments. The salt focuses itself into diapirs or even salt dikes. Salt domes often exist in great numbers in basins that contain deeply buried evaporites (e.g. North German basin). The existing salt structures greatly affect the surrounding groundwater composition, minerals are leached, and conversely the waters influence diagenesis and dissolution of the salt. As shown, for example, by Kühn et al. (2002a) for the geothermal Buntsandstein reservoir of Stralsund (Germany), the formation water is in thermodynamic equilibrium with respect to the matrix minerals calcite and anhydrite. Within the Rhaetian reservoir of Neustadt-Glewe (Germany) barite is in equilibrium with the highly saline brine for reservoir conditions (Kühn 1997).

The possible processes listed above leading to highly saline geothermal brines in deep-seated sedimentary rocks depend chiefly on the regional geology and geologic structure of the system and can not be interpreted like geothermal waters from the elucidated dynamic systems.

2.7 Geothermometer

Geothermometry is the use of a fluid's chemical composition to estimate the temperature at which it equilibrated in the subsurface. It is an important tool in exploring for and exploiting geothermal fields and for understanding the genesis of ore deposits. Generally, geothermometers are applied to characterize deep groundwater flow systems.

Chemical geothermometers are used to reconstruct thermal and chemical conditions within deep structures of hydrothermal systems. They keep characteristics of the temperature conditions prevailing within geothermal reservoirs in element amounts or ratios of constituents. Based on water-rock equilibration reactions they enable estimation of the reservoir temperature or maximum depth of circulation of fluid on condition that (Nicholson 1993):

- element concentrations or species used for calculation are controlled by a temperature-dependent water-rock equilibrium,
- there is an abundance of the mineral in the geothermal system,
- the considered reaction reaches equilibrium in the reservoir,
- there is no re-equilibration during flow of the reservoir fluid to the surface,
- there is no mixing or dilution of the deep fluid (in the case that the geothermometer relies on absolute abundances).

The number of components involved in the chemical reactions occurring in geothermal reservoirs can be used to subdivide the range of geothermometers. The simplest ones are those based on univariant reactions like the silica geothermometer. Mahon (1966b) recognized the relationship between the silica content in the water with respect to a particular mineral phase and the reservoir temperature. In the last decades this geothermometer has been refined resulting in the following equations [Eqs. (2.22)-(2.26)] for varying silica modifications with the SiO_2 concentration in mg kg^{-1} and temperatures below 250°C (Nicholson 1993):

$$T[°C] = \left[\frac{1309}{(5.19 - \log SiO_2)}\right] - 273 \qquad \text{(quartz, no steam loss)} \qquad (2.22)$$

$$T[°C] = \left[\frac{1032}{(4.69 - \log SiO_2)}\right] - 273 \qquad \text{(chalcedony)} \qquad (2.23)$$

$$T[°C] = \left[\frac{1000}{(4.78 - \log SiO_2)}\right] - 273 \qquad (\alpha - \text{cristobalite}) \qquad (2.24)$$

$$T[°C] = \left[\frac{781}{(4.51 - \log SiO_2)}\right] - 273 \qquad (\beta - \text{cristobalite}) \qquad (2.25)$$

$$T[°C] = \left[\frac{731}{(4.52 - \log SiO_2)}\right] - 273 \qquad \text{(amorphous silica)} \qquad (2.26)$$

The major disadvantage of one component geothermometers is their high sensitivity to secondary processes such as dilution (mixing of waters) or increase of concentrations due to the loss of water (boiling).

This handicap may be overcome using concentration ratios of water components as geothermometers. Referring to the reactions elucidated in the previous section the relative amounts of Na, K, Mg, and Ca in geothermal waters may be used as geothermometers if full equilibrium between fluid and minerals is attained. In this case the system is fixed for a given temperature and salinity. Giggenbach (1988) concluded that above temperatures of 200°C, well discharges are close to equilibrium with respect to Na-feldspar as well as K-feldspar. The Na / K ratio responds most slowly to changing water-rock equilibria and, therefore, preserves deep equilibration conditions. But uncertainty arises for acid water compositions due to coincidence in Na / K values with waters in equilibrium at high temperatures. The Na / Mg geothermometer seems to allow for separating non-equilibrated from equilibrated waters, if the temperatures are high enough, due the likely lack of albite equilibrium at low temperatures. The major distinction in the behavior of K / Mg, the third system, from other geothermometers is the fact that solution-mineral equilibrium is attained even for low temperatures and CO_2 rich waters as represented by soda springs. This system appears to adjust most rapidly. Still, the fortuitous coincidence with the composition of some highly acid waters with that expected for equilibrium introduces considerable uncertainties.

Each of the above mentioned geothermometer subsystems, Na / K, Na / Mg, and K / Mg, has its advantages and disadvantages. Giggenbach (1988) pointed out that K-Na and K-Mg are the only two subsystems suitable to provide the basis for a geothermometer and combined them in a ternary Na-K-Mg system to eliminate some of the drawbacks due to separate evaluation of geothermometers. In a thermodynamically stable mineral system evolved by re-crystallization of a dissolved average crustal rock the pair Na-K reaches its equilibrium ($EQ_{K\text{-}Na}$) most slowly as governed by

$$EQ_{K\text{-}Na} = \log (c_K / c_{Na}) = 1.75 - (1390 / T) \tag{2.27}$$

with c_i in mg kg^{-1} and T in K. The equilibrium of the pair K-Mg adjusts much faster and at lower temperatures (<100°C) according to

$$EQ_{K\text{-}Mg} = \log (c^2_K / c_{Mg}) = 14.0 - (4410 / T) \tag{2.28}$$

The correlations among the amounts of Na, K, and Mg can be shown most conveniently in a triangular plot. The construction requires the conversion of absolute values to relative values and in order to render such a plot applicable they are plotted in terms of m_{Na} / 1000, m_K / 100, and $m^{0.5}_{Mg}$ (square root = SQR). From the intersection of the both isotherms [Eqs. (2.27) and (2.28)] a so called "full equilibrium" curve results as shown in Fig. 2.16.

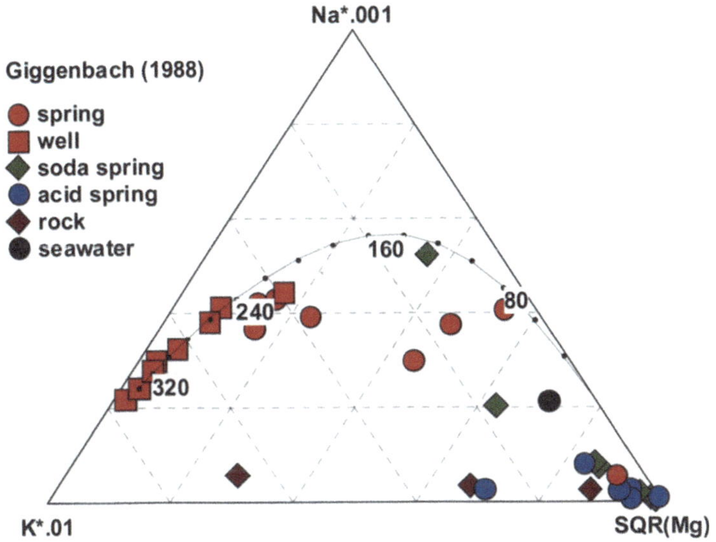

Fig. 2.16. Relative Na-K-Mg contents used as geothermometer after Giggenbach (1988) given with the full equilibration line; the displayed numbers are temperatures at "full equilibrium"

It can be seen that high temperature well discharges (red squares) plot on the full equilibration line, whereas associated spring waters (red dots) are shifted to lower temperatures and away from full equilibrium. This indicates faster re-equilibration of Mg compared to Na, resulting from decreasing temperatures of the rising geothermal fluids. The soda and acid spring waters plot more or less close to the Mg corner. Data points in this area correspond to "immature" waters, generally unsuitable for evaluation of K-Na equilibration temperatures. If the waters are not too acid, K-Mg temperatures may still be valid, reflecting shallower parts of the reservoir due to faster Mg re-equilibration. For comparison, data points referring to dissolution of 1000 g of granite, basalt, and an average crustal rock dissolved in 1 kg of water (brown diamonds) are shown. The major cause of non-attainment of full water rock equilibrium in geothermal waters is the sluggishness of the process supplying the comparatively large equilibrium contents of Na, especially at lower temperatures.

Plotting the geothermal solutions from the compiled database on the Na-K-Mg triangle does not show, as expected, any tight dependencies on the type of source rock of the geothermal system (Fig. 2.17). But nevertheless it is obvious that the number of samples close to the full equilibration line is highest for volcanic settings. A group of sedimentary-hosted waters approaches full equilibrium from the Na-corner. This is the case for seawater-influenced systems or reservoirs with inherent waters evolved through leaching of evaporitic material. However, the majority of waters from sedimentary environments, as well as almost all plutonic fluids, are clearly located away from the equilibrium.

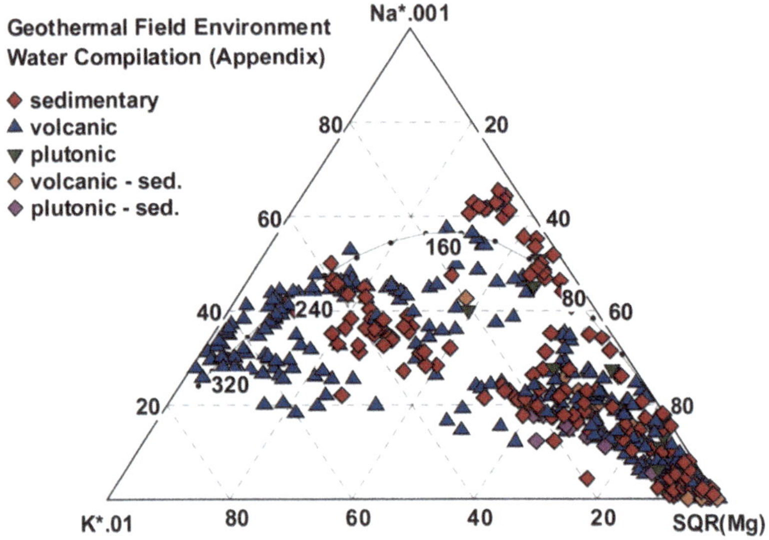

Fig. 2.17. Na-K-Mg geothermometer after Giggenbach (1988) applied to the geothermal water data base separated by source rock environments

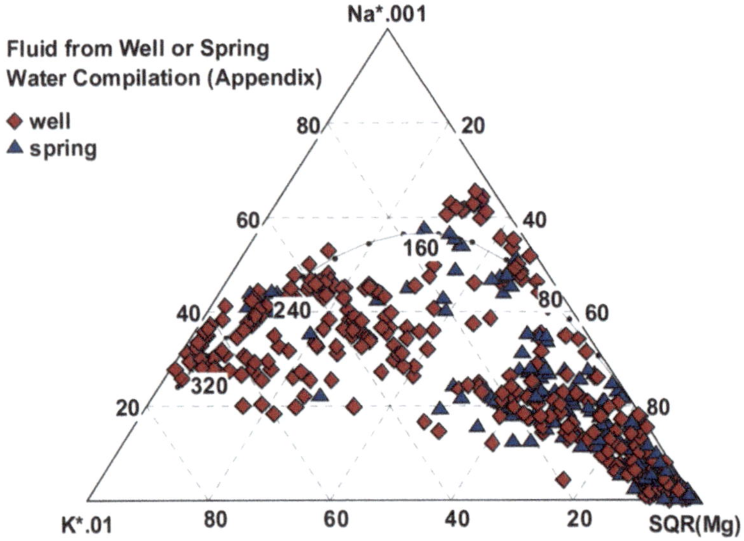

Fig. 2.18. Na-K-Mg geothermometer after Giggenbach (1988) applied to the geothermal water data base separated by samples from wells and springs

Fig. 2.18 shows the Na-K-Mg geothermometer plot for geothermal water samples from the database separated according to sampling sites (wells or springs). It can be seen that most of the samples close to equilibrium are tapped from wells, although a good number of well samples plot significantly off the equilibration line. As expected, high temperature well discharges from deep parts of the geothermal system are the most likely samples to be in thermodynamic equilibrium with the rocks. Waters taken from springs act conversely, with the majority off and only a few close to the equilibration line, indicating acquisition of Mg by the waters in response to decreasing temperatures to be faster than that of Na.

Fournier (1990) stated that there is inherent ambiguity in the use of Na-K-Mg triangular geothermometer diagrams due to uncertainty about which Na / K geothermometer best applies to a certain water composition. This is because a particular geothermometer may work well in one place and not in another. In an expansion of the Giggenbach (1988) model, Fournier (1990) suggests, as shown in Fig. 2.19, to calibrate the Na / K geothermometer for a particular system by additional information (e.g. drilling) and to refer to additional geothermometer of Truesdell (1976) and Fournier and Potter (1979).

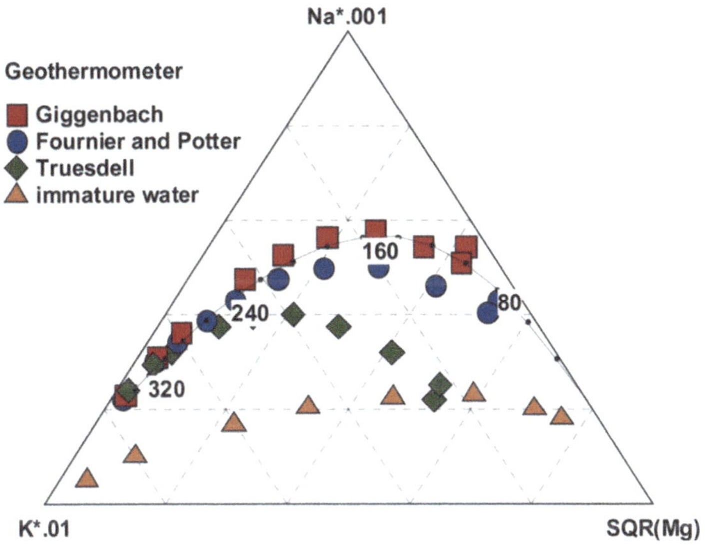

Fig. 2.19. Diagram showing resulting water compositions in water rock equilibrium with respect to the varying Na-K-Mg geothermometers of Giggenbach (1988), Fournier and Potter (1979), and Truesdell (1976); immature waters depict water compositions not suitable for geothermometry

The Na-K-Ca geothermometer, being very popular, show various uncertainties arising from its sensitivity to variations in the CO_2 content of geothermal waters, especially at low temperatures. Rather than forming the basis for a geothermometer it may possibly be used in the formulation of a CO_2 geobarometer.

Calcite is an ubiquitous mineral in geothermal systems (Browne 1978) and geothermal discharges have been shown to be close to saturation with respect to calcite on many occasions (Arnorsson et al. 1983, White 1986, Kühn et al. 2002a). The most important reaction leading to the formation of calcite in geothermal systems is the conversion of Ca-Al-silicates to calcite by CO_2 [Eq. (2.3)]. Therefore the system K-Ca is sensitive to variations of the CO_2 fugacity (f_{CO2}) following the relationship published by Giggenbach (1988).

$$EQ_{K\text{-}Ca} = \log (c^2_K / c_{Ca}) = \log f_{CO2} + 3.0 \qquad (2.29)$$

Before applying Eq. (2.29) to the estimation of the carbon dioxide activity in the deeper parts of a geothermal reservoir the suitability of using the described geobarometer for certain geothermal waters should be checked. The maturity index (MI) of the water, derived from the difference of the K-Mg and K-Na geothermometers, should be greater than 2.0.

$$MI = 0.315\ EQ_{K\text{-}Mg} - EQ_{K\text{-}Na} \qquad (2.30)$$

3 Theory of Chemical Modeling

In order to deal with the complexity of natural systems simplified models are employed to illustrate the principal and regulatory factors controlling a chemical system. Following the aphorism of Albert Einstein: "Everything should be made as simple as possible, but not simpler", models need not to be completely realistic to be useful (Stumm and Morgan 1996), but need to meet a successful balance between realism and practicality. Properly constructed, a model is neither too simplified that it is unrealistic nor too detailed that it cannot be readily evaluated and applied to the problem of interest (Bethke 1996). The results of a model have to be at least partially observable or experimentally verifiable (Zhu and Anderson 2002).

Geochemical modeling theories are presented here in a sequence of increasing complexity from geochemical equilibrium models to kinetic, reaction path, and finally coupled transport and reaction models. The description is far from complete but provides the needs for the set up of reactive transport models of hydrothermal systems as done within subsequent chapters. Extensive reviews of geochemical models in general can be found in the literature (Appelo and Postma 1999, Bethke 1996, Melchior and Bassett 1990, Nordstrom and Ball 1984, Paschke and van der Heijde 1996).

3.1 Geochemical Equilibrium

At the heart of any geochemical model is the thermodynamic equilibrium system, which does not contain any spatial or temporal information. Hence, they are called zero-dimensional models. In closed, open, or isolated systems, chemical reactions tend to reach thermodynamic equilibrium, where all solute concentrations stabilize. The concentrations at equilibrium are governed by thermodynamic principles. At equilibrium, the potential energy, the Gibbs free energy, G, of the chemical system is minimized. The Gibbs free energy is related to enthalpy H, representing thermal energy, temperature, T, in Kelvin, and entropy, S, representing disorder or randomness of a system.

$$G = H - TS \tag{3.1}$$

The partial derivation of the Gibbs free energy with respect to the number of moles of a substance, n_i, corresponds to its chemical potential, μ_i.

$$\frac{\partial G}{\partial n_i} = \mu_i \tag{3.2}$$

Chemical potential is the driving force of chemical reactions and in turn depends on the dimensionless activity, a, of the dissolved solution species via the following equation, with the chemical potential at standard conditions ($\mu°$) and the ideal gas constant, R.

$$\mu_i = \mu_i° + RT \ln a \tag{3.3}$$

Activity calculation and the corresponding determination of mineral solubilities are discussed in the following section.

3.1.1 Activity Calculations and Solubility of Minerals

Precipitation reactions might occur during the operation of a geothermal installation or due to diagenetic processes and are absolutely necessary for development of ore deposits. They can be estimated using saturation indices calculated on the basis of data from chemical analyses. The saturation index, SI, reflects the saturation state of a solution with respect to a mineral phase. It is defined as:

$$SI = \log \frac{IAP}{K_{eq}} \tag{3.4}$$

In Eq. (3.4) the ion activity product, IAP, is the product of the "effective concentrations" of the dissolved species i, the so-called activities, a_i (see below). It is compared to the equilibrium constant, K_{eq}, defined by the solubility product of the mineral, which is the product of the equilibrium activities. If SI is positive, the mineral is supersaturated and may precipitate. A mineral phase is in thermodynamic equilibrium with a solution if the saturation index is equal to zero. However, fluids with saturation indices $-0.2 \leq SI \leq 0.2$ are also called saturated solutions by Langmuir and Melchior (1985). Monnin and Ramboz (1996) specify tighter limits of $-0.05 \leq SI \leq 0.05$. A negative saturation index indicates undersaturation with respect to a mineral. The main task in calculating saturation indices is to determine activities.

A concentration, m_i, in mol kg^{-1} H_2O is related to its activity (a_i) using the dimensionless activity coefficient, γ_i, and the standard state ($m_i°$), 1 mol kg^{-1} H_2O:

$$a_i = \gamma_i \, m_i / m_i° = \gamma_i \, m_i \tag{3.5}$$

The factor $1 / m_i°$ is unity for all species and cancels for practical purposes (Appelo and Postma 1999). In highly diluted solutions (i.e. for vanishing concentrations of m_i), the ionic interactions, reflected by γ_i, become negligibly small ($\gamma_i = 1$). Then the system behaves as an ideal solution. In this case, the activity of a species corresponds to its concentration in the solution. In non-ideal solutions there is electrostatic interaction between ions and therefore $\gamma_i \neq 1$. The Debye-Hückel equation

(Debye and Hückel 1923) provides an approximation for the activity coefficients in dilute solutions by taking into account long-range Coulomb forces:

$$\log \gamma_i = -A\, z_i^2\, \sqrt{I} \tag{3.6}$$

where A is a temperature dependent constant, I is the ionic strength of the solution, and z_i is the specific charge of the ion in question. The ionic strength is given by:

$$I = \frac{1}{2} \sum_i m_i\, z_i^2 \tag{3.7}$$

where m_i is the molality, the concentration in mol of substance per kilogram of water.

The Debye-Hückel equation [Eq. (3.6)] is valid only for very dilute solutions ($I << 0.1$), whereas the extended Debye-Hückel equation [Eq. (3.8)] is suitable for ions in low to moderately concentrated solutions ($I < 0.1$).

$$\log \gamma_i = -A\, z_i^2\, \frac{\sqrt{I}}{1 + B\, a_i^0\, \sqrt{I}} \tag{3.8}$$

Both A and B are temperature dependent constants, listed in Table 3.1, and in addition to Eq. (3.6), a_i^0 is an ion-specific parameter related to the size of the hydrated ion. It can be derived with respect to "mean-salt" activity coefficients (Pytkowicz 1983). In Table 3.2 selected values of a_i^0 are listed after Butler (1998).

Table 3.1. Temperature dependent constants for Debye-Hückel [Eq. (3.6)], extended Debye-Hückel [Eq. (3.10)], and Davies [Eq. (3.9)] equation (Manov et al. 1943)

T [°C]	A	B
0	0.4883	0.3241
10	0.4960	0.3258
20	0.5042	0.3273
30	0.5130	0.3290
40	0.5221	0.3305

Table 3.2. Ion specific parameter a_i^0, related to the size of the ion and its charge, for the extended Debye-Hückel equation [Eq. (3.10)] after Butler (1998)

Charge	a_i^0	Ions
1	3	K^+, NH_4^+, Ag^+, F^-, Cl^-, Br^-, I^-, HS^-, NO_3^-, OH^-
	4	Na^+, HCO_3^-, $H_2PO_4^-$, HSO_3^-
	9	H^+
2	4	Hg_2^{2+}, CrO_4^{2-}, HPO_4^{2-}, SO_4^{2-}
	5	Ba^{2+}, Cd^{2+}, Hg^{2+}, Pb^{2+}, Ra^{2+}, CO_3^{2-}
	6	Ca^{2+}, Cu^{2+}, Fe^{2+}, Mn^{2+}, Zn^{2+}
	8	Be^{2+}, Mg^{2+}
3	4	PO_4^{3-}
	9	Al^{3+}, Fe^{3+}, Cr^{3+}

For brackish waters with ionic strength $I > 0.1$ the Davies equation [Eq. (3.9)] is a better approximation of ion activity coefficients. It is valid up to $I \approx 0.5$ (Stumm and Morgan 1996):

$$\log \gamma_i = -A\, z_i^2 \left(\frac{\sqrt{I}}{1+\sqrt{I}} - 0.3 I \right) \tag{3.9}$$

where A is the same constant from previous equations [Eqs. (3.6) and (3.8)] and listed in Table 3.1. The Debye-Hückel, extended Debye-Hückel, and Davies equation are henceforth referred to as the Debye-Hückel theory.

For calculating species activities in solutions of higher ionic strength, an ion interaction model was developed by Kenneth Pitzer and coworkers in the 1970s (Pitzer 1973, 1975; Pitzer and Mayorga 1973, 1974; Pitzer and Kim 1974). This semi-empirical approach combines the Debye-Hückel equation with additional terms and describes the concentration dependence of the Gibbs energy for non-ideal conditions. The free excess Gibbs energy, G^{ex}, indicating the difference between ideal and real Gibbs energy is written in the form of a virial equation (i.e. a power series expansion):

$$\frac{G^{ex}}{w_w\, R\, T} = f(I) + \sum_i \sum_j m_i\, m_j\, B(I) + \sum_i \sum_j \sum_k m_i\, m_j\, m_k\, \Psi_{ijk} + \ldots, \tag{3.10}$$

with water mass w_W, ionic strength I, and molality m of species i, j, and k. The first term comprises the Debye-Hückel law, which accounts for the dependence on the ionic strength and not on the individual parameters of the solution. It describes the electrostatic far field interactions. The second virial coefficient, B, represents the specific binary, near-field interactions between pairs of components i and j in the solution. The corresponding ternary interactions between components i, j, and k are described by the third virial coefficient, Ψ. For the limiting case of an ideally diluted solution, the second and higher terms vanish, and the equation yields the Debye-Hückel law. The series of virial coefficients can be extended to higher orders but the first three virial coefficients are sufficient to describe solutions of high salinity (Pitzer 1991). The activity coefficient is the derivative of Eq. (3.10) with respect to m_i and is calculated from

$$\begin{aligned} \ln \gamma_M = {} & z_M^2\, F + \sum_a m_a \left(2\, B_{Ma} + Z\, C_{Ma} \right) + \sum_c m_c \left(2\, \Phi_{Mc} + \sum_a m_a\, \Psi_{Mca} \right) \\ & + \sum_{a<a'} \sum m_a\, m_{a'}\, \Psi_{aa'M} + \left| z_M \right| \sum_c \sum_a m_c\, m_a\, C_{ca} \end{aligned} \tag{3.11}$$

for the cations and for anions from

$$\begin{aligned} \ln \gamma_X = {} & z_X^2\, F + \sum_c m_c \left(2\, B_{cX} + Z\, C_{cX} \right) + \sum_a m_a \left(2\, \Phi_{Xa} + \sum_c m_c\, \Psi_{Xac} \right) \\ & + \sum_{c<c'} \sum m_c\, m_{c'}\, \Psi_{cc'X} + \left| z_X \right| \sum_c \sum_a m_c\, m_a\, C_{ca} \end{aligned} \tag{3.12}$$

Here, m is the molality of an ion, with indices M, c and c' for cations and X, a, and a' for anions. The double sums c < c' and a < a' refer to all pairs of different cations and anions. The function F contains an adapted form of the Debye-Hückel equation as well as the derivatives of the second virial coefficient with respect to the ionic strength:

$$F = -A^{\phi}\left[\frac{\sqrt{I}}{1+b\sqrt{I}} + \frac{2}{b}\ln\left(1+b\sqrt{I}\right)\right] + \sum_c\sum_a m_c\, m_a\, B'_{ca} + \sum_{c<c'}\sum m_c\, m_{c'}\, \Phi'_{cc'} + \sum_{a<a'}\sum m_a\, m_{a'}\, \Phi'_{aa'} \tag{3.13}$$

where $A^{\phi} = 1400684\,(\rho_w/(D\,T))^{3/2}$ is the Debye-Hückel parameter, ρ_w water density, and D the static dielectric constant of pure water. The coefficients B [Eq. (3.11) and Eq. (3.12)] and B′ [Eq. (3.13)] are defined as:

$$B_{MX} = \beta^{(0)}_{MX} + \beta^{(1)}_{MX}\, g\left(\alpha_1\sqrt{I}\right) + \beta^{(2)}_{MX}\, g\left(\alpha_2\sqrt{I}\right)$$
$$B'_{MX} = \beta^{(1)}_{MX}\frac{g'\left(\alpha_1\sqrt{I}\right)}{I} + \beta^{(2)}_{MX}\frac{g'\left(\alpha_2\sqrt{I}\right)}{I}. \tag{3.14}$$

For any salt containing a monovalent ion, values for α_1 and α_2 are $\alpha_1 = 2$ and $\alpha_2 = 0$ (Pitzer 1973), while for 2-2 electrolytes and higher valence types the corresponding values are $\alpha_1 = 1.4$ and $\alpha_2 = 12.0$ (Pitzer and Silvester 1976). The functions g and g' are defined as:

$$g(x) = 2\left[1-(1+x)e^{-x}\right]/x^2$$
$$g'(x) = -2\left[1-\left(1+x+x^2/2\right)e^{-x}\right]/x^2 \tag{3.15}$$

where $x=\alpha\sqrt{I}$. Actually, the coefficient C [Eq. (3.11) and Eq. (3.12)] depends on the ionic strength, but there is only one experimental proof for this (Phutela and Pitzer 1986). Therefore, C is defined according to Pitzer (1991) neglecting the influence of ionic strength:

$$C_{MX} = C^{\phi}_{MX}\Big/\left(2\sqrt{|z_M\, z_X|}\right) \tag{3.16}$$

The coefficient Z for C_{Ma} and C_{cX} in Eq. ((3.11) and Eq. ((3.12) is given by:

$$Z = \sum_i m_i\, |z_i|. \tag{3.17}$$

The thermodynamic properties of aqueous solutions containing a single salt (i.e. binary systems) depend only on the interaction parameters $\beta^{(0)}$, $\beta^{(1)}$, $\beta^{(2)}$, and C^{ϕ} which define the variables B and C. The parameters Φ and Ψ [Eq. (3.11) and Eq. (3.12)] as well as Φ′ [Eq. (3.13)] correspond to aqueous mixtures of two salts (i.e. ternary systems). The parameters Φ and Φ′ account for cation-cation and anion-anion interactions, the parameter Ψ for cation-cation-anion and anion-anion-cation interactions.

The parameters Φ_{ij} and Φ'_{ij} are defined by:

$$\Phi_{ij} = \Theta_{ij} + {}^{E}\Theta_{ij}(I) \qquad \text{and} \qquad \Phi'_{ij} = {}^{E}\Theta'_{ij}(I), \tag{3.18}$$

where Θ_{ij} is the only adjustable parameter and is defined for each pair of cations and each pair of anions. The terms ${}^{E}\Theta_{ij}(I)$ and ${}^{E}\Theta'_{ij}(I)$ account for electrostatic mixing effects of asymmetrical (with respect to charge) cation-cation and anion-anion pairs defined by Pitzer (1975). The values of ${}^{E}\Theta_{ij}(I)$ and ${}^{E}\Theta'_{ij}(I)$ depend only on the charge of the ions and the ionic strength of the solution. They are equal to zero if the corresponding cation or anion pairs i and j possess the same charge. The parameters Ψ_{ijk} are introduced for different combinations of two cations and one anion or two anions and one cation. Ψ is derived from data of solutions containing two salts in the same way as Θ_{ij}.

A comprehensive compilation of Pitzer parameters applicable for the system Na-K-Mg-Ca-Ba-Sr-Si-H-Cl-SO_4-OH-HCO_3-CO_3-CO_2-H_2O from low to high temperatures and salinities is given by Kühn et al. (2002b).

3.1.2 Comparison of Ion Activity Calculation Methods

An ideal model for calculating activity coefficients for geochemical or engineering applications would have the following properties: (1) consistency with the laws of thermodynamics; (2) compact mathematical form; (3) high accuracy over wide ranges of temperature, pressure, and concentration; (4) applicability to systems including most of the elements of the periodic table. Unfortunately, none of the currently existing models satisfies all of these requirements.

The Pitzer and the Debye-Hückel models represent different approaches to the complex problem of calculating ion activities. Debye and Hückel described (1) the solvent of an ionic solution as an ideal dielectric fluid without structure, and (2) the solutes as uniform spherical ions with charges located at their centers. The far-field Coulomb forces, which depend only on the ionic charge, cause each ion to be surrounded by a fluctuating group of ions, the ionic sphere. Pitzer extends the Debye-Hückel theory and develops a semi-empirical model of complex solutions which incorporates element-specific ion interactions, based on data of simpler binary and ternary systems containing one salt or two salts, respectively.

Due to its internal consistency and achievable calculation accuracy, the Pitzer equations currently represent the most promising approach for calculating ion activities in brines (Wolery and Jackson 1990).

As a typical example, Fig. 3.1 shows the solubility of gypsum in sodium chloride solutions at 25 °C, calculated according to both the Debye-Hückel theory (here using the Davies equation) and the Pitzer model. Open and full circles represent data of Block and Waters (1968) and Marshall and Slusher (1966), respectively. It is obvious that for concentrations larger than 0.5 mol kg^{-1} of sodium chloride in the solution, only the Pitzer equations yield correct results. Therefore, activity coefficients calculated with the Debye-Hückel equation cannot be applied to chemical reaction modeling of geothermal brines with sufficient accuracy.

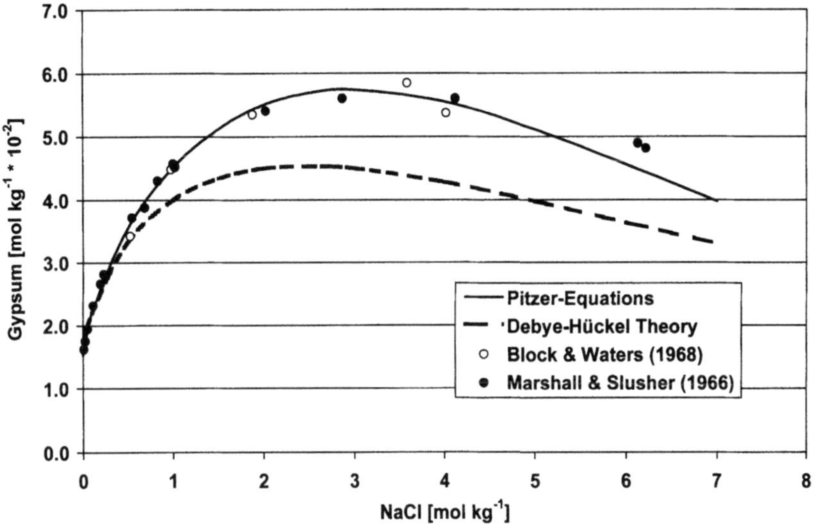

Fig. 3.1. Gypsum solubility at 25 °C as a function of sodium chloride concentration (both in mol kg $^{-1}$) according to the Debye-Hückel model (here using the Davies equation) as well as the Pitzer equations; open and full circles show data of Block and Waters (1968) and Marschall and Slusher (1966), respectively

3.1.3 Batch Models

Speciation-solubility models are based on thermodynamic equilibrium calculations of the ionic species. They are used to determine the ion speciation in solution and the precipitation and dissolution reactions of mineral phases after computation of their state of equilibrium. Most often the speciation calculations are done for closed, static or so called "batch" or beaker-type systems.

Computing a model begins by calculating the initial equilibrium state of the system at the temperature of interest. Depending on the nature of the modeled system the constraints of all chemical constituents within the system are set by the mass of solvent water, the amounts of minerals available, the fugacity of gases, the amount of dissolved components or, directly, the species activities as normally done with H^+, determined by pH.

Such models delineate the concentrations and activities of the dissolved species, the saturation states of the solution with respect to various mineral phases, and the stable species distribution depending on temperature, fluid composition, and mineral assemblage in contact with the water.

Ármannsson et al. (2000) used the program PHREEQC (Parkhurst and Appelo 1999) and its inverse modeling capability to investigate the origin of thermal waters in the Lake Mývatn area of North Iceland. They divided the cold groundwater and geothermal effluent of the area into six distinct groups according to origin and geothermal influence. This division is based on stable isotope ratios, chemical composition and geographical positions. Ármannsson et al. (2000) concluded that the groundwater has apparently two basically separated origins; the local high ground north of Lake Mývatn and the highlands far to the south. No traces of seawater are observed and the concentrations of conservative constituents suggest extensive water-rock interaction. The waters are variably affected by geothermal activity. Simulating the composition of Krafla and Námafjall geothermal water by titrating local groundwater with rock at 205°C and adding volcanic gas yield results in agreement with water analyses data.

3.2 Kinetic Models

For practical purposes, mineral reactions fall into three groups: (1) those in which reaction rates may be so slow relative to the time period of interest that the reaction can be ignored altogether, (2) those in which the rates are fast enough to maintain equilibrium, and (3) the remaining reactions. Only those in the latter group require a kinetic description.

Thermodynamic calculations deal only with the equilibrium state of geochemical systems (see preceding section), but due to the fact that concentrations of reactants and products approach equilibrium depending on time and not instantaneously, kinetic processes may sometimes have to be taken into account in chemical models. Calculations addressing chemical equilibrium are only possible if the geochemical system remains essentially closed long enough for the reactions of interest to approach equilibrium. For a more detailed study of the subject covered in this section, the reader is referred to the comprehensive description of kinetic theory in Earth sciences by Lasaga (1998).

Reaction rates vary tremendously. Some reactions are so fast that they can result in explosions, while others are so slow that geologic time scales are required to measure their progress. General ranges of half life for several common types of aqueous reactions are shown in Table 3.3.

Table 3.3. Approximate ranges of reaction half times for different types of reaction taken from Langmuir and Mahoney (1984)

Type of Reaction	Typical Half Life
Solute - solute	Fraction of a second to minutes
Sorption - desorption	Fraction of a second to days
Gas-solute	Minutes to days
Crystalline solid - solute	Hours to millions of years

Reactions involving more than one phase, for example minerals dissolving into or precipitating from a solution, are called heterogeneous. The kinetics of heterogeneous reactions receive the most attention in reaction modeling, because of the slow rates at which many minerals react and the resulting tendency of fluids, especially at low temperatures, to be in a state of non-equilibrium with the minerals in contact.

Mineral dissolution reactions are often not in equilibrium because during the time it takes to approach thermodynamic equilibrium the reservoir fluid has come into contact with a different assemblage of minerals. When transport or other agents of change are rapid compared to the reaction rate, disequilibrium prevails and kinetic becomes an important factor in the estimation of concentrations. However, even when kinetics are important, equilibrium calculations may still be useful to show where the system is heading in the long run.

Lasaga (1998) makes a useful distinction between "elementary" chemical reactions, which occur as written at the molecular level, and "overall" reactions, which describe a net change that involves several intermediate steps and takes place along several competing, parallel pathways. Carbonate precipitation for example, an important process during diagenesis, can be written as elementary reaction as:

$$Ca^{2+} + CO_3{}^{2-} \Leftrightarrow CaCO_3 \tag{3.19}$$

The rate at which molecules of $CaCO_3$ are produced is directly proportional to the probability that a Ca^{2+} ion will collide with a $CO_3{}^{2-}$ ion in solution. The appropriate rate law in this case therefore is:

$$\frac{dm_{CaCO3}}{dt} = k_{reac} \cdot m_{Ca^{2+}} \cdot m_{CO_3^{2-}} \tag{3.20}$$

in which m_i are the concentrations of each species i, and k_{reac} is a linear rate constant. In fact, for any elementary reaction, the rate is simply proportional to the abundance of reactant species.

In contrast, the rates at which overall reactions occur are commonly non-linear functions of concentration. Surface processes of particles, production of transient, metastable species, or transport of dissolved material toward or away from the reaction interface may influence the rates. As a result, rate laws may be quite complex, reflecting a variety of inhibiting or enhancing processes. Keir (1980), for example, has empirically determined that the rate of calcite dissolution (k_{reac}) in seawater follows the non-linear relationship

$$k_{reac} = B\left(1 - \frac{a_{Ca^{2+}} \cdot a_{CO_3^{2-}}}{K_{eq}}\right)^{4.5} \tag{3.21}$$

in which B is an empirical constant that varies widely depending on the particular system. It is especially a function of the reactive surface. Different approaches have been used to model calcite dissolution in seawater by Sjöberg (1978), Rickard and Sjöberg (1983), and Morse (1983). They can be generalized, as well as Eq. (3.21), by the following empirical rate expression.

$$k_{reac} = B\frac{A}{V}(1 - SI)^{d} \tag{3.22}$$

where A is the reactive surface area, V is the volume of the solution, SI the saturation index [Eq. (3.4)] reflecting the saturation state, and B and d are coefficients depending on the composition of the solution and are obtained by curve fitting observed rates.

The rate constant k_{reac} [Eqs. (3.20) - (3.22)] can be related to the temperature by the phenomenological Arrhenius equation.

$$k_{reac} = A_0 \cdot e^{-E_A/RT} \tag{3.23}$$

Here, A_0 is the pre-exponential factor, E_A an activation energy, R the gas constant, and T the absolute temperature. The values of A_0 and E_A are determined for a given reaction by measuring k at several temperatures.

Soler and Lasaga (1998) presented results of long-term numerical simulations (> 1 Ma) of the formation of bauxite from granitic rocks. Within a one-dimensional reactive transport model they used a kinetic approach to describe the mineral dissolution and precipitation reactions. In agreement with numerous field observations Soler and Lasaga (1998) determined the development of a typical "bauxitic" profile consisting of an upper gibbsite-rich and a lower kaolinite-rich zone. The simulations showed that the infiltrating solutions are closer to saturation with respect to microcline than with respect to albite.

3.3 Reaction Pathways

Reaction path models simulate the successive reaction steps of a system in response to the mass or energy flux. Within these kind of models, some temporal information is included in terms of reaction progress, but no spatial information is contained. Once the initial equilibrium state of the system is known, the model can trace a reaction path.

Chemical reaction path models have long been of use in the interpretation of the chemical evolution of subsurface waters, often to determine the effect of diagenetic reactions in a zone of interest. These models are constructions of hierarchically arranged reaction pathways of batch reactions. The final solution composition or mineral assemblage of the preceding reaction step being the initial solution or mineral assemblage of the proceeding step connects the batch reactions with each other. This is based on the relative abundances of critical components that serially dominate the reactions of the water-rock system (Saripalli et al. 2001). Strictly speaking it is the course followed by the equilibrium system as it responds to changes in composition or temperature. Reaction path models can be divided into:

- Polythermal reaction models
- Titration models
- Systems open to external gas reservoirs
- Flow-through reaction path models

3.3.1 Polythermal Reaction Models

The reaction path in polythermal models is characterized by the application of varying temperatures or a temperature gradient to the system. Polythermal reaction models are commonly applied to closed systems, as in studies of groundwater geothermometry and interpretation of laboratory experiments. But there is no restriction for applying them also in open systems. If a fluid is, for example, sampled at 300°C, but analyzed at room temperature, the thermal reaction path model equilibrates the fluid at 25°C and then carries the closed system to the temperature of the groundwater system or the experiment. One of these cases is the reconstruction of the pH under formation conditions. This is of importance with respect to the investigation of hydrothermal reservoirs.

Reed and Spycher (1984) used chemical analyses and 25°C pH measurements of geothermal waters from Iceland, Broadlands, and Sulphur Bank, hot spring waters from Jemez, Yellowstone and Blackfoot Reservoir to show that most of these waters approach equilibrium with a subsurface mineral assemblage at a temperature close to measured temperatures. The in-situ pH values were recalculated and by means of the mineral saturation indices they determined whether the distinct waters are in equilibrium with their host rocks, the occurrence of probable mineral assemblages, and the most likely temperature of equilibrium.

In a study of seawater-basalt reactions, Reed (1983) showed that in multicomponent equilibrium calculations of tholeitic basalt with seawater at 300°C and subsequent cooling to 25°C the resultant solutions can lead to ore deposition.

3.3.2 Titration Models

The titration model is the simplest open-system model. It is characterized by a reactant, which is gradually added to the solution in equilibrium. A titration can be used to simulate (1) fluid mixing, where the reactant is a second aqueous solution or (2) evaporation, in which the titrating substance, solvent water, is removed from the system rather than added. In most cases the reactant will be (3) a mineral undersaturated with respect to the initial solution. Adding an aliquot of the mineral at each step the mineral will dissolve dependent on the fluid composition. Thereby, the various titration models [(1)-(3)] may cause other minerals to become saturated or even to precipitate or drive minerals that already exist in the system to dissolve. Within the course of the reaction path of the titration model the equilibrium system evolves until the fluid reaches saturation with the reactant or the reactant is exhausted.

Titration models are applied to predict, for example, how rock will react with its pore fluid. In that case, minerals that make up the rock are titrated into the formation water. The solubility of most minerals in water is rather small, so the fluid in such reaction path models is likely to become saturated after a small amount of mineral has reacted.

Lu et al (1992) presented numerical results of skarn formation calculations using the program SOLVEQ (Reed 1982). An impure limestone (90% calcite and

5% each of quartz and kaolinite) was titrated into a solution of a composition constrained by fluid inclusion investigations. The calculations demonstrated that a complete sequence of skarn assemblages forms during progressive equilibration of a very small amount of limestone with the infiltrating fluid.

3.3.3 System Open to External Gas Reservoirs

Many geochemical processes occur in which a fluid remains in contact with a gaseous phase. The calculation of reaction path models open to external gas reservoirs assumes that gas species move to or from the solution in order to maintain a specified fixed-fugacity in the reacting system. The gas, which could be the Earth's atmosphere or a subsurface gas reservoir, buffers the system's chemistry. By dissolving gas species from the buffer or exsolving gas into it, the fluid will, if reaction proceeds slowly enough, maintain equilibrium with the buffer. The process of gas exsolution might be triggered if a fluid boils.

Models of this type may be appropriate for describing weathering at Earth's surface, reactions in soils, geochemical interactions in partially saturated rock formations, boiling in hydrothermal systems, and certain kinds of experimental configurations. The gases most likely to be appropriately treated by this option are O_2 and CO_2, with the possible oxidation reactions of pyrite (FeS_2) to goethite (FeOOH)

$$4\ FeS_2 + 15\ O_2(aq) + 10\ H_2O \Leftrightarrow 4\ FeOOH + 16\ H^+ + 8\ SO_4^{2-} \tag{3.24}$$

or the exsolution of carbon dioxide accompanied by calcite precipitation.

$$Ca^{2+} + 2\ HCO_3^- \Leftrightarrow CaCO_3 + H_2O + CO_2(g) \tag{3.25}$$

Saunders and Schoenly (1995) emphasized, using the program CHILLER (Reed 1982, Spycher and Reed 1992), that a gold enriched fluid has been a principal factor in the genesis of Au-Ag ores. The numerical simulations showed that boiling closely reproduces observed minerals and their relative abundances in bonanza ores.

3.3.4 Flow-Through Reaction Path

Reaction between rock and groundwater moving through it is most appropriately conceptualized by using a model configuration based on the assumption of local equilibrium. Two major types of flow-through reaction path models can be used.

A fluid-centered flow-through system follows the evolution of a particular packet of water as it flows through a medium, which could be a fractured or porous material. Reactants are presumed to line the medium in homogeneous fashion and interact with the fluid packet as it passes by. Alternatively, there may be no reactants, but only a change in temperature or pressure. Within the course, secondary mineral phases may form. As the packet moves on, it physically separates from the masses of secondary phases produced. The result is that the transiently formed products do not have the opportunity to re-dissolve in that particular packet of water.

A solid-centered flow-through system is fundamentally different. This reaction path model focuses on the evolution of solids interacting with a mass of fluid, which is either continuously or discretely recharged by a fresh supply of aqueous solution of fixed composition. This system closely matches the scenario in many flow-through interaction experiments.

Plumlee (1994) applied the reaction path model CHILLER to calculate changes in the fluid chemistry and amounts of minerals precipitated from solution at each of a series of steps along various chemical evolution paths. Around 150 reaction paths were modeled to determine the sensitivity of probable ore deposition from hydrothermal fluid compositions and corresponding temperatures as well as overlying meteoric groundwater compositions and temperatures, ambient pressure conditions, the extent of boiling, and fluid reactions dependent on the mineral assemblages. The modeling results depict that epithermal ore grades and mineral patterns are influenced to a great extent by boiling and fluid mixing in shallow parts of hydrologic systems.

Plumlee et al. (1995a) used reaction path modeling to explain fluorite deposition mechanisms in the Illinois-Kentucky fluorspar district. They applied CHILLER with temperatures, major cation and anion initials and amounts of dissolved gases based on fluid inclusion data. The investigated reaction path mechanisms were simple cooling, reactions of the fluid with limestones, isothermal boiling, mixing, and varying combinations of the enumerated processes. The results indicate that quite acidic waters in the hydrothermal system are additionally rich in fluorine, due to absorption of magmatic gases. The acidity of the geothermal fluids led to extensive dissolution of the host rock limestone.

3.3.5 Reaction Path Models Applied to Hydrothermal Systems

Since a valid reaction model is a prerequisite for reactive transport simulation, the first step in any case is to construct a successful reaction path model for the problem of interest.

Formation of Acidic Fluids in High Temperature Systems

Akaku et al. (2000) set up reaction path models for the high temperature hydrothermal reservoirs of Fushime and Kakkonda (Japan). Using the program CHILLER they were able to explain the observed acidity of the discharged waters without incorporating any acidic volatiles such as HCl or SO_2.

Akaku et al. (2000) assumed a liquid formation fluid fully in equilibrium with the alteration minerals in the rock. The geothermal water of Fushime originates from seawater while the water in Kakkonda is believed to originate from meteoric water. Due to a production-induced pressure decrease in the vicinity of the wells the water begins to boil. Under consideration of heat transfer and in response to the change of physical conditions due to the boiling process the reactions of the fluid versus a new state of equilibrium were investigated. The calculation results emphasize that the acidity of the Fushime water results from sphalerite (ZnS) precipitation during boiling [Eq. (3.26)].

$$Zn^{2+} + 2\,Cl^- + H_2S \Leftrightarrow ZnS + 2\,H^+ + 2\,Cl^- \tag{3.26}$$

The supply of acidity within the Kakkonda field is assumed to be due to the reactions of pyrite (FeS_2) to magnetite (Fe_3O_4) [Eq. (3.27)] and the redox reaction of hydrogen sulfide leading to hydrogen gas [Eq. (3.28)].

$$6\,FeS_2 + 12\,H_2O \Leftrightarrow 2\,Fe_3O_4 + SO_4^{2-} + 11\,H_2S + 2\,H^+ \tag{3.27}$$

$$4\,H_2O + H_2S \Leftrightarrow 4\,H_2 + SO_4^{2-} + 2\,H^+ \tag{3.28}$$

The fluid compositions calculated with the help of the reaction path models are in agreement with the fluids discharged from the wells in Fushime as well as Kakkonda.

Development of Ore Deposits Due to Vein Mineralization

In the Tongonan hydrothermal system (Philippines) the vein minerals sphalerite, galena, chalcopyrite, pyrite, anhydrite, quartz, calcite, epidote-clinozoisite, and chlorite are observed with specks of Au-Ag electrum. Balanque (2000) performed calculations with CHILLER to investigate if either boiling or mixing is the better precipitation mechanism to explain the mineral assemblage within the Tongonan field.

Adiabatic boiling, with temperature changing from 300 to 100°C, lead to an early precipitation sequence of gold with quartz followed by quartz, acanthite, and chalcocite and finally late quartz, sphalerite, galena, acanthite, and bornite.

A steam-heated end member solution resulting from the boiling process has been used for the mixing with a sample of the groundwater. This mixture produced the ore minerals acanthite, bornite, chalcocite, covellite, galena, pyrite, and sphalerite accompanied by the gangue minerals anhydrite, Mg-chlorite, alunite, kaolinite, muscovite, and quartz.

Balanque (2000) concluded that it might be possible, that a combination of the boiling and mixture process may have led to the observed mineral assemblage. The shortcomings of the simulations are the absence of some minerals within the natural environment predicted by the program and the presence of other minerals in the veins not predicted by the program.

Determination of a Reservoir Fluid's Origin

Gianelli and Grassi (2001) set up a reaction path model for Pantelleria Island (Italy) with which they confirmed, that the reservoir fluid is a mix of seawater, meteoric water, and volcanic gas. The code EQ3/6 (Wolery and Daveler 1992) has been applied for the refinement of the conceptual model of the Pantelleria geothermal system.

The system recharge is mainly of marine origin, which intrudes throughout the island. The increasing temperature towards the center of the geothermal field leads to the precipitation of anhydrite, quartz, and clay minerals resulting in decreased amounts of Ca, Mg, and SO_4. The seawater flows through fractures and is heated to 300°C.

In the area of volcanic gas up flow the mixing results in increased quantities of C and S. After reaction with trachyte the pH increases and the fluid reaches saturation conditions with respect to albite, quartz, saponite, K-feldspar and muscovite, in agreement with the natural hydrothermal mineral assemblages.

3.4 Simulation of Transport and Reaction

Reactive mass transport models contain both temporal and spatial information about chemical reactions, a complexity that is desired for real world applications. Basic processes that play significant roles in simultaneous and coupled simulation of transport and reaction are fluid flow, heat transfer, solute transport, and chemical reactions. There are numerous books dealing with these diverse and complex fields (e.g. Appelo and Postma 1999, Fitts 2002, Ingebritsen and Sanford 1998). The essential elements of the theory will be incorporated here, with an emphasis on the links to the other reaction topics within this book.

In this section, the equations, which are required to incorporate a velocity and temperature field, or transport of species into the broad framework of chemical reactions, will be discussed. The discussion will include not only the equations that govern flow, temperature, and transport itself, but also the feedback processes that exist between these equations and the rates of chemical reactions.

In any treatment of fluid dynamics, the fundamental equations always start with the three conservation equations: (1) the conservation of momentum, (2) conservation of mass – equation of continuity, and (3) conservation of energy.

3.4.1 Groundwater Flow

The primary coupling between groundwater flow, solute transport, and heat transfer is through Darcy's law. The average linear groundwater flow velocity calculated by Darcy's law is used to describe fluid flow in porous media. The Darcy velocity is needed later to determine solute transport by advection, mechanical dispersion, and heat transport by convection.

Darcy's Law

In 1856, a French hydraulic engineer named Henry Darcy published an equation for flow through a porous medium that today bears his name. In designing a water treatment system for the city of Dijon, Darcy found that no formulas existed for determining the capacity of a sand filtration system. Consequently, Darcy performed a series of experiments on water flow through columns of sand (Darcy 1856).

The experimental apparatus, shown schematically in Fig. 3.2, allowed him to vary the length (L) and cross-sectional area (A) of a sand-packed column and also the elevations of constant-level water reservoirs connected to the upper (h_1) and lower (h_2) boundaries of the column.

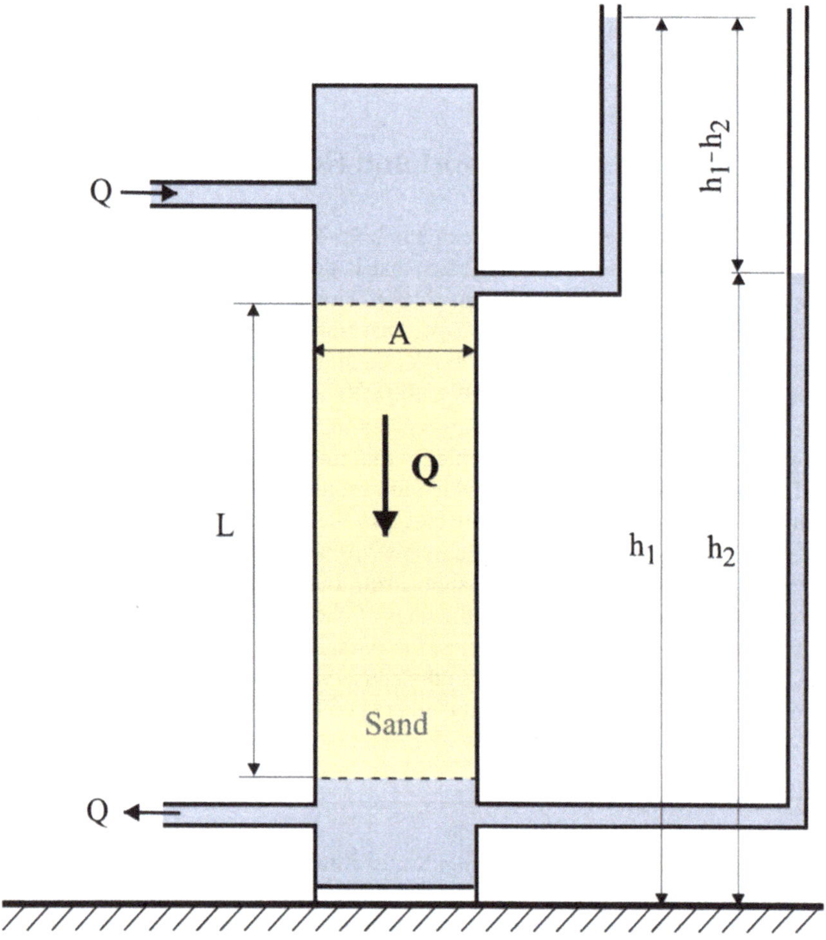

Fig. 3.2. Schematic diagram of the apparatus used in Henry Darcy's sand filter experiments

Under steady flow conditions, the volumetric flow rate through the column (Q) was positively correlated with A and (h_1-h_2) and inversely correlated with L. By introducing a constant of proportionality, K, Darcy's experimental results can be summarized as

$$Q = KA\frac{h_1 - h_2}{L} \tag{3.29}$$

The constant of proportionality (K) is called the hydraulic conductivity [L s^{-1}]. Eq. (3.29) can be rewritten as:

$$\frac{Q}{A} = K\frac{h_1 - h_2}{L} \tag{3.30}$$

Writing Eq. (3.30) in differential form leads to the equation known as Darcy's law:

$$q = \frac{Q}{A} = -K\frac{dh}{dl} \tag{3.31}$$

where q denotes the volumetric flow rate per unit area, called the specific discharge or Darcy velocity. In Eq. (3.31), dh/dl is the dimensionless hydraulic gradient. The negative sign indicates that positive specific discharge (indicating direction of flow) corresponds to a negative hydraulic gradient. Thus, Darcy's Law states that specific discharge in a porous medium is in the direction of decreasing h and directly proportional to the hydraulic gradient. Although q has dimensions of velocity [L s^{-1}], it is not the average groundwater velocity. This would only be the case if the water could flow through all of the unit area A. However, in a subsurface medium only a fraction of the unit area is available for water flow. An aquifer is composed by solid grains and voids in between, thus, the resulting average velocity is higher than the specific discharge. The actual average linear fluid velocity, v, is directly proportional to the specific discharge (q) and inversely proportional to the effective porosity, n_e.

$$v = \frac{q}{n_e} \tag{3.32}$$

The effective porosity is the porosity that is interconnected and available for flow. The average linear velocity represents the mean flow velocity at which a conservative (non-reacting) solute would move through Darcy's experimental column as shown in Fig. 3.2.

Subsequent laboratory-column experiments conducted using a variety of fluids revealed that K expresses a combination of fluid and solid properties. The flow rate is actually proportional to the specific weight of the fluid (ρ g) inversely proportional to the dynamic viscosity of the fluid, μ, and proportional to a property of the solid medium, k, which is called the intrinsic permeability. Thus

$$K = \frac{k\rho g}{\mu} \tag{3.33}$$

where k has dimension of L^2. Hubbert (1940, 1956) as well as Xu and Eckstein (1997) revealed by theoretical considerations and experiments with glass beads or sands of uniform diameter, d_m, that for granular porous media, q, K, and k are proportional to d_m^2. Hazen (1911) proposed the following empirical equation:

$$K = C \cdot d_m^2 \tag{3.34}$$

where K is hydraulic conductivity. If K is measured in cm s^{-1} and the constant C with units of cm^{-1} s^{-1}, than C varies from about 40 to 150 for most sands.

Investigation of Darcy's Law indicates that it fails at sufficiently high volumetric flow rates because above certain thresholds the kinetic energy of the fluid cannot be neglected anymore and significant amounts of energy are lost to turbulence. As a result, Darcy's Law over predicts the flow rate. Such flow rates are rare in the subsurface but can occur in the direct vicinity of a well or in areas of cavernous porosity. The upper limit for application of Darcy's Law is usually estimated on the basis of the dimensionless Reynold's number,

$$Re = \frac{\rho q L}{\mu} \tag{3.35}$$

where ρ is the fluid density, q the Darcy velocity, μ is the dynamic viscosity of the fluid, and L is some characteristic length. In granular porous media, L is commonly related to the grain-size distribution. The transition from Darcian (laminar) to non-Darcian (non-linear) flow appears to take place at $Re \sim 5$ and the transition to turbulent flow occurs at $Re \sim 100$ (Bear 1979, Freeze and Cherry 1979).

The investigation of a possible lower limit for Darcian flow, done for fine-grained materials, led to the suggestion that a threshold hydraulic gradient exists below which flow does not take place (Swartzendruber 1962, Bolt and Groenevelt 1969). Although these findings are still under discussion, the phenomenon is of little importance because the flow rates will be exceedingly small in any case. Schildknecht and Schneider (1987) give a comprehensive description of the state of the art of discussion about the validity of Darcy's Law under low hydraulic gradients in sediments characterized by cohesion.

Driving Forces of Groundwater Flow

In hydrogeologic practice the driving force for groundwater flow is generally expressed in terms of a parameter called hydraulic head or simply head. This is the same quantity indicated by h_1 and h_2 in Henry Darcy's laboratory manometers (Fig. 3.2).

Intuitively, one might tend to think of groundwater as flowing from areas of high pressure to areas of low pressure. But, considering the pressure distribution in a static water column, expressed as:

$$P = \rho \cdot g \cdot z \tag{3.36}$$

where z is the depth below the water surface, this can be seen to not be the case. Hubbert (1940) demonstrated that groundwater actually flows from areas of high energy to areas of low energy.

Two fundamental forms of energy are of interest in this context: kinetic energy and potential energy. Kinetic energy is associated with motion, but in a typical groundwater environment the kinetic energy is negligibly small relative to the total potential energy (assumption for Darcian flow). Potential energy is associated with the work required to move something from one place to another in a conservative forces field. A conservative forces field is a field in which the work done in moving from one point to another does not depend on the path taken. In the groundwater context the most important conservative force fields are gravity and pressure. The gravitational potential energy of a unit volume of liquid water is $\rho_f\,g\,z$, where z is its height above an arbitrary datum and $\rho_f\,g$ is its specific weight. The pressure potential energy per unit volume is simply the pressure P, a force per unit area. For a fluid with variable density (ρ not constant), flow is proportional to the gradient in the quantity $(P + \rho_f\,g\,z)$. For an incompressible fluid (ρ constant), one can divide by the specific weight to obtain the hydraulic head, h:

$$h = \frac{P}{\rho_f\,g} + z \tag{3.37}$$

Both the pressure head $(P / \rho_f\,g)$ and the elevation head (z) have units of length, and the total hydraulic head (h) is equated to the water level observed in a manometer (Fig. 3.2) or well. In this case the driving force for groundwater flow is the head gradient.

Intrinsic Permeability

Permeability is unquestionably the crucial hydrologic parameter. Unfortunately, it is often a parameter very difficult to determine and apply in a meaningful fashion, especially over the enormous space and time scales that apply in many geologic problems. Permeability often appears to be a scale dependent property (Brace 1980, 1984; Clauser 1992).

The measured permeability of common geologic media varies by almost 16 orders of magnitude, from values as low as 10^{-20} m^2 in unaltered crystalline rock, intact shales, and halite, to values as high as 10^{-7} m^2 in well-sorted gravels (Table 3.4).

Permeability versus Porosity. The terms porosity (n) and permeability (k) are often used interchangeably by non-hydrogeologists. Although, this is a mistake, there is a strong positive correlation between the two quantities in many porous and fractured geologic media. For well-sorted, unconsolidated porous media, this correlation is often expressed by the Kozeny-Carman equation (Carman 1956),

$$k = \frac{n^3}{\left[5 s_0^2 (1-n)^2\right]} \tag{3.38}$$

where s_0 is the solid surface exposed to the fluid per unit volume of solid material and the porosity. Solving the Navier-Stokes equations for a system of parallel capillary tubes leads to the Kozeny-Carman relation (more porosity-permeability relations will be presented in the next section).

However, this positive correlation between n and k does not hold for the important classes of clays, clay rich material, and volcanic tuffs. They tend to have very large porosities (~ 0.5) but relatively low permeabilities. Clays as a group tend to be about 10^6 times less permeable than sands despite having higher porosities (n is only ~ 0.35 for a fairly well-sorted sand).

Table 3.4. Permeability ranges for different rock types (Freeze and Cherry 1979)

Rock Type	Permeability [m²]
Unconsolidated rocks	
Gravel	$10^{-7} - 10^{-10}$
Clean sand	$10^{-9} - 10^{-13}$
Silty sand	$10^{10} - 10^{-14}$
Silt, loess	$10^{-12} - 10^{-16}$
Glacial till	$10^{-13} - 10^{-19}$
Unweathered marine clay	$10^{-16} - 10^{-19}$
Consolidated rocks	
Shale	$10^{-16} - 10^{-20}$
Unfractured metamorphic and igneous rocks	$10^{-17} - 10^{-20}$
Sandstone	$10^{-13} - 10^{-17}$
Limestone and dolomite	$10^{-13} - 10^{-16}$
Fractured igneous and metamorphic rocks	$10^{-11} - 10^{-15}$
Permeable basalt	$10^{-9} - 10^{-14}$
Karst limestone	$10^{-9} - 10^{-13}$

(The Darcy is another unit commonly used for permeability: 1 Darcy ≈ 10^{-12} m²).

Scale Dependence. Values of permeability are often determined by one of the following three methods, each of which measures or infers permeability at a different volume-averaged scale. Laboratory tests measure permeability at the drill-core scale, sometimes using methods as simple as that shown in Fig. 3.2 but often using more sophisticated apparatus to make transient (non-steady-state) measurements, impose large head gradients, and/or replicate in-situ pressure and temperature conditions. Regardless of the exact experimental design, the volume of material sampled in laboratory tests is generally very small, almost always << 1 m³.

In-situ or well tests are done by pumping at a steady state rate while monitoring hydraulic head or determining loss of fluid from a well at a fixed head value. Another kind of test is to change the hydraulic head in a well and to monitor its recovery. The head response may be monitored in the perturbed well only (single-well tests) or in the perturbed well and one or more nearby observation wells (multi-well tests). The volume of material investigated by such in-situ tests varies with the size and duration of the perturbation and with the hydraulic properties of the medium, but it generally ranges from < 10 m³ (for a single-well test in a low permeable medium) to perhaps > 10^5 m³ (in a high-permeable medium).

There is no direct way to measure larger-scale permeabilities, but larger-scale "regional" values are often inferred from the results of numerical modeling experiments. In such simulations the unknown values of regional-scale permeability are varied so that the numerical simulation results match known values of hydraulic heads, rates of groundwater flow, solute concentrations, or temperatures. Regional permeability values inferred on this basis are applied to volumes ranging from perhaps 10^2 m^3 to > 10^3 km^3.

Depth Dependence. At any particular site, and for a uniform lithology, permeability is likely to decrease more or less systematically with depth. The depth dependence is due mainly to loss of porosity through increasing confining pressure and effective stress and to temperature and pressure dependent diagenetic and metamorphic processes. In practice it may be hard to distinguish among these various effects. Furthermore, the general decrease in permeability with depth is not necessarily uniform. It may be temporarily reversed by the presence of permeable geologic structures or strata at depth, by anomalous fluid pressures that decrease the effective stress, or by hydraulic fracturing. The general depth dependence of permeability is due to physical and chemical processes common to most geologic settings.

Time Dependence. Because of ongoing deformation, dissolution and precipitation of minerals, and other metamorphic processes, permeability is also a time dependent property. Geologists, who see evidence of episodic fracture creation and healing in fossil hydrothermal systems, have long recognized the transient nature of permeability (e.g. Titley 1990). However, hydrogeologists have rarely incorporated time-dependent permeabilities in quantitative analyses of groundwater flow and transport (see following chap. for application of time dependent permeabilities).

Although time itself is not and cannot be the activating factor, it is nonetheless useful to develop some appreciation of the time scales over which various geologic processes are likely to affect permeability. Some geologic processes (e.g. compaction and/or diagenesis of sediments) cause a gradual evolution of permeability, whereas others (e.g. hydro-fracturing, earthquakes) act very rapidly.

Heterogeneity and Anisotropy. It is obvious from Table 3.4 that permeability might show extreme spatial variability or heterogeneity among different geologic units. Permeability is also generally an anisotropic or direction dependent property. The most important cause of anisotropy is sedimentary or volcanic layering. The anisotropy of permeability motivates introduction of a three dimensional form of Darcy's law written in vector notation and in terms of hydraulic conductivity (compare permeability-hydraulic conductivity relation in Eq. (3.33)).

$$\begin{pmatrix} q_x \\ q_y \\ q_z \end{pmatrix} = -\begin{pmatrix} K_{xx} & K_{xy} & K_{xz} \\ K_{yx} & K_{yy} & K_{yz} \\ K_{zx} & K_{zy} & K_{zz} \end{pmatrix} \begin{pmatrix} \partial h / \partial x \\ \partial h / \partial y \\ \partial h / \partial z \end{pmatrix} \qquad (3.39)$$

Further simplification with $\underline{K}$, indicating that K is a second-order tensor, and the vector operator ∇ leads to

$$q = -\underline{K}\, \nabla h \qquad (3.40)$$

Because $\underline{K}(x,y,z)$ is symmetric and has principal directions, the coordinate axes can be aligned with these directions of $\underline{K}$, so that the off-diagonal terms of Eq. (3.40) become zero. For example, in layered media the coordinate axes can often be aligned with bedding.

The heterogeneous and anisotropic medium is the general case in natural settings. Only in conceptual case studies homogeneous and isotropic media can be assumed.

The Continuum Approach

Darcian groundwater flow and transport is not described at the microscopic level at which individual molecules or even the details of pore-fracture geometries are important. Instead, it is defined for flow and transport phenomena at a macroscopic level, using averaged properties. The domain of interest consists of both solids and void space filled with one or more fluids, and in nearly all cases the distribution of solid-fluid boundaries are not known well enough to use classical fluid-mechanics approaches.

Porosity, for example, is a key property of the medium only definable on a macroscopic scale. At any microscopic point in a domain, porosity will be either close to 0 in the solid material or 1 in a pore space. As one averages over progressively larger volumes, the computed value of porosity will fluctuate over a progressively smaller range. If the medium is sufficiently homogeneous (between V_l and V_u), the volume-averaged value of porosity will eventually become nearly constant (Fig. 3.3). The volume range over which the average porosity remains constant has been termed the representative elementary volume, or REV range (Bear 1979). In a homogeneous material, the REV range could be arbitrarily large. However, all geologic media have some larger-scale heterogeneity, and as one continues to average over larger volumes (above V_u), porosity will eventually depart from its REV-scale average.

It is important to consider if a REV-based continuum approach is justified. The REVs must be large relative to the scale of microscopic heterogeneity (e.g. grain size in a granular porous medium) but small relative to the entire domain of interest. The appropriate REV size will vary dramatically depending on the problem under consideration and the nature of the geologic medium. For example, the minimum REV size needed to represent permeability in a well-sorted coarse sand ($d_m \sim 0.001$ m) would be about 10^{12} times smaller than the minimum REV size needed to represent permeability in a granite where flow is dominated by fractures spaced 10 m apart.

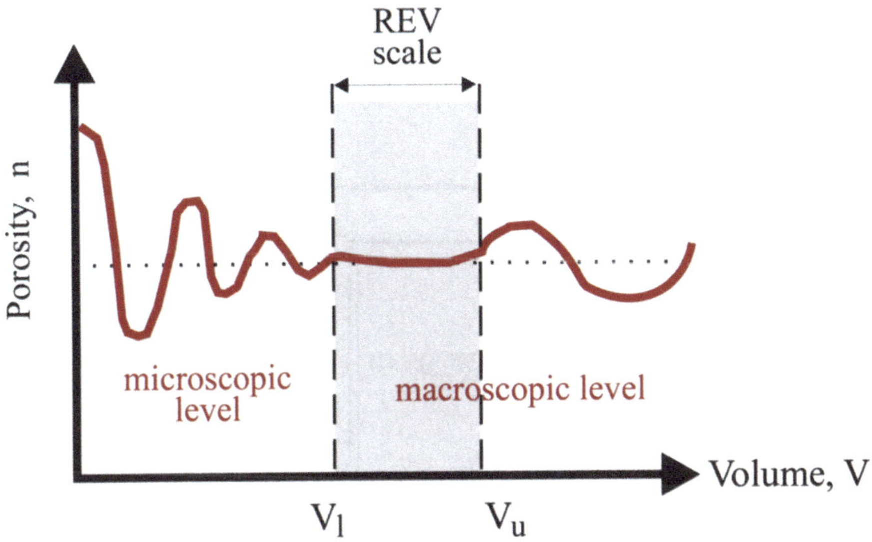

Fig. 3.3. Porosity as a function of averaging volume; over a representative elementary volume (REV, between V_l and V_u), the value is essentially constant

Groundwater Flow Equation

Essential quantities of the groundwater flow equation are the fluid potential, the hydraulic head, the hydraulic conductivity or permeability, and the porosity. The groundwater flow equation is derived on the basis of considering conservation of mass for a REV, a finite control volume fixed in space and centered on a point with spatial coordinates x, y, and z (Fig. 3.4). The change in mass stored within the volume over time must be equal to the difference between the mass flowing into the control volume and the mass flowing out of the control volume. Grouping the flux through opposite faces leads to Eq. (3.41),

$$\begin{aligned} \frac{\Delta\left[n_e \rho_f \Delta x \Delta y \Delta z\right]}{\Delta t} &= \left[\rho_f q_x \Delta y \Delta z - \left(\rho_f q_x \Delta y \Delta z + \rho_f \Delta q_x \Delta y \Delta z\right)\right] \\ &+ \left[\rho_f q_y \Delta x \Delta z - \left(\rho_f q_y \Delta x \Delta z + \rho_f \Delta q_y \Delta x \Delta z\right)\right] \\ &+ \left[\rho_f q_z \Delta x \Delta y - \left(\rho_f q_z \Delta x \Delta y + \rho_f \Delta q_z \Delta x \Delta y\right)\right] \end{aligned} \tag{3.41}$$

where $\Delta[n_e\ \rho_f\ \Delta x\ \Delta y\ \Delta z]$ is the change in fluid mass stored in the volume $\Delta x\ \Delta y\ \Delta z$ over time increment Δt. This term is called the specific storage and indicates that the mass of fluid stored at any point in the system is affected by temporal changes in porosity (n_e) and fluid density (ρ_f). Recalling Darcy's Law [Eq. (3.31)], where q is the volumetric flow rate per unit area ($m^3\ s^{-1}\ m^{-2}$), this quantity must be multiplied by the fluid density ρ_f to get a mass flow rate per unit area ($kg\ s^{-1}\ m^{-2}$) and by the area of the face to arrive at a mass flow rate ($kg\ s^{-1}$).

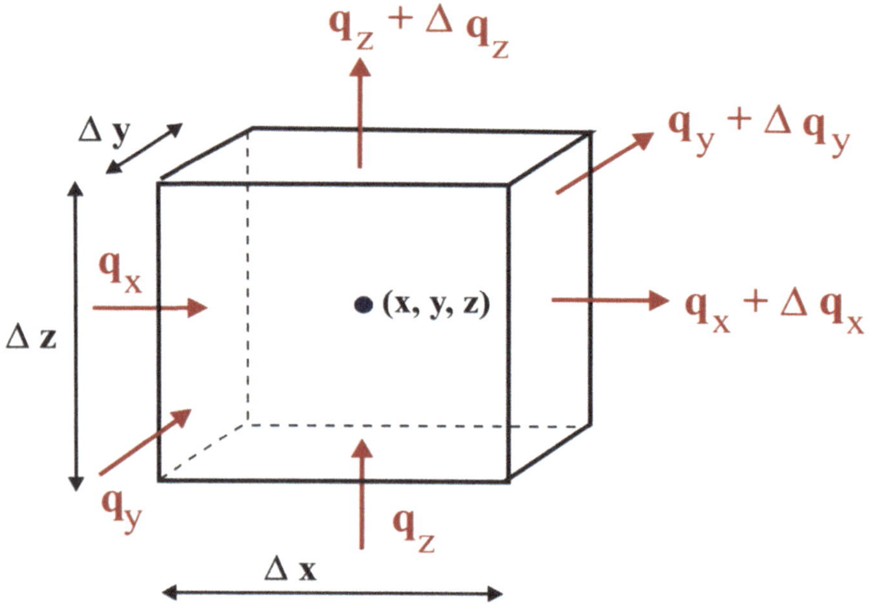

Fig. 3.4. The control volume used to derive the continuity equation [Eq. (3.42)]

Dividing both sides of Eq. (3.41) by $\Delta x\ \Delta y\ \Delta z$ and taking the limits as $\Delta t \rightarrow 0$, $\Delta x \rightarrow 0$, $\Delta y \rightarrow 0$, and $\Delta z \rightarrow 0$ gives Eq. (3.42) which is known as the continuity equation for flow through a porous medium.

$$\frac{\partial(n_e \rho_f)}{\partial t} + \frac{\partial(\rho_f q_x)}{\partial x} + \frac{\partial(\rho_f q_y)}{\partial y} + \frac{\partial(\rho_f q_z)}{\partial z} = 0 \tag{3.42}$$

The continuity equation is a rather general statement of conservation of mass that involves only few assumptions about the nature of the fluid or the geologic medium. From this point onward, various forms of the groundwater flow equation can be derived requiring restrictive assumptions about the fluid or the nature of the flow system. For example, assuming single-phase fully saturated conditions, inserting a general form of Darcy's law, and using the vector operator leads to a general form of groundwater flow equation [Eq. (3.43)]. This equation accounts for the effects of variable fluid properties by calculating fluxes in terms of forces acting on the fluid ($\nabla P + \rho_f\, g\, \nabla z$) by allowing hydraulic conductivity to vary with fluid density and viscosity (for more details about the groundwater flow equation and its derivation compare Freeze and Cherry 1979, Bear 1972).

$$\frac{\partial(n_e \rho_f)}{\partial t} = \nabla\left[\frac{\underline{k}\rho_f}{\mu}(\nabla P + \rho_f g \nabla z)\right] \tag{3.43}$$

3.4.2 Solute Transport

The transport of solutes in groundwater is of particular importance for many geologic processes like diagenesis, formation of ore deposits, or formation and dissolution of mineral phases. As is the case for groundwater flow, solute transport can be described mathematically by combining principles of mass balance with expressions that relate the fluxes of solute to fundamental driving forces. Solutes spread out by molecular diffusion, advection, and hydrodynamic dispersion. Sources and sinks of solutes (e.g. chemical reactions) may also be incorporated into solute-transport equations. If the solute-transport equations are coupled with the groundwater flow equation as well as chemical reactions, they can be used to quantify reactive-transport processes that occur in the subsurface.

Molecular Diffusion

Darcy's Law describes the flux of a fluid resulting from a gradient in fluid potential or hydraulic head [Eq. (3.31)]. An analogue situation exists for solute transport due to diffusive flux as a result of a gradient in the chemical potential, or concentration. This solute flux is directly proportional to the concentration gradient with the coefficient of molecular diffusion as a proportionality factor. This is known as Fick's first law and can be expressed for single-phase water as

$$q_d = -D_W \frac{dC}{dx} \tag{3.44}$$

where q_d is the diffusive flux, D_W is the coefficient of molecular diffusion in free or open water, C is the concentration of the molecule or ion, and x is the distance along the direction of the concentration gradient. The physical process driving molecular diffusion is simply the random motion of ions in solution. Ions in a region of higher concentration will eventually mix with ions in a region of lower concentration (negative sign with always positive coefficient) to create an equal distribution in space. Table 3.5 lists molecular diffusion coefficients and ion radii for different cations and anions. These values show the general trend of decreasing value with both increasing charge and decreasing ionic radius.

Eq. (3.44) expresses Fick's law in terms of D_W for pure and open water. In the subsurface, diffusion (D_m) occurs within a porous medium. The presence of a solid phase restricts the area through which a solute can diffuse, and the tortuosity, τ, of the flow path increases the distance over which the solute must travel to get from one point to another. Tortuosity is dimensionless, always less than one, and is a measure of how tortuous the flow path is. It can be described as the net straight line length of flow divided by the average actual flow path. De Marsily (1986) reported that tortuosity typically ranges from 0.7 (sands) to 0.1 (clays). Bear (1972) stated a range of $0.56 <$ tortuosity < 0.8 for granular media. The effect of tortuosity is that typical diffusion coefficients for geologic media range one to two orders of magnitude lower, between 10^{-11} to 10^{-10} m^2 s^{-1}, compared to open water [Eq. (3.45)].

$$D_m = n_e \cdot \tau \cdot D_w \tag{3.45}$$

Table 3.5. Diffusion coefficients and radii of some ions in water at 25°C (after Li and Gregory 1974)

Cation	D_W [10^{-9} m^2 s^{-1}]	Radius [pm]	Anion	D_W [10^{-9} m^2 s^{-1}]	Radius [pm]
H^+	9.31	21	OH^-	5.27	
Li^+	1.03	78	F^-	1.46	133
Na^+	1.33	98	Cl^-	2.03	181
K^+	1.96	133	Br^-	2.01	196
Rb^+	2.06	149	I^-	2.00	220
Cs^+	2.07	165	HS^-	1.73	
Mg^{2+}	0.70	78	HCO_3^-	1.18	
Ca^{2+}	0.79	106	HSO_4^-	1.33	
Sr^{2+}	0.79	127	NO_2^-	1.91	
Ba^{2+}	0.85	143	NO_3^-	1.90	
Ra^{2+}	0.89	152	$H_2PO_4^-$	0.85	
Cr^{3+}	0.59	64	CO_3^{2-}	0.96	
Mn^{2+}	0.69	91	SO_4^{2-}	1.07	
Fe^{2+}	0.72	82	HPO_4^{2-}	0.73	
Fe^{3+}	0.61	67	CrO_4^{2-}	1.12	

The direct analog of the groundwater flow equation is the diffusion equation, Fick's second law. It is derived from Fick's first law combined with a mass-balance equation [Eq. (3.41)]. When one substitutes Fick's first law into the mass flux, one obtains Fick's second law:

$$D_m \nabla^2 C = n_e \frac{\partial C}{\partial t} \tag{3.46}$$

This diffusion equation can be solved for a system with typical geological distances and coefficients. Results provide that because the typical magnitude of the molecular diffusion coefficient in porous media is relatively small, it can take from thousands to million years for chemicals to migrate significant distances. Thus, diffusion alone cannot account for the transport of chemical mass over the long distances or relatively short time frames required by many geologic processes.

Advection

The second transport mechanism for solutes is advection. Advection is simply the movement of solutes with flowing groundwater, such that solutes move at the same mean velocity as the groundwater and no concentration gradient is required for transport to occur. The one-dimensional advection equation is:

$$\frac{\partial C}{\partial t} = -v_x \frac{\partial C}{\partial x} \tag{3.47}$$

Eq. (3.47) describes the translation of a concentration distribution in x direction at a velocity of v_x. The negative sign indicates that the forward advection of a positive concentration gradient (concentration increases in x direction) leads to a decrease in concentration at that point.

Mechanical Dispersion

The microscopic heterogeneity (compare REV in Fig. 3.3) of porous media creates groundwater velocity fields. These heterogeneities create a variance in the groundwater velocity around the average linear velocity or seepage velocity. These variations create an indirect transport process called mechanical dispersion. Because of mechanical dispersion, a concentration front that originally is sharp will spread out or disperse as it is transported by advection with the groundwater.

Compared to molecular diffusion (see above), there is no proven model of mass flux for mechanical dispersion. However, mechanical dispersion can be mathematically described like molecular diffusion. Both processes result in spreading concentration fronts within the flowing groundwater (Fig. 3.5).

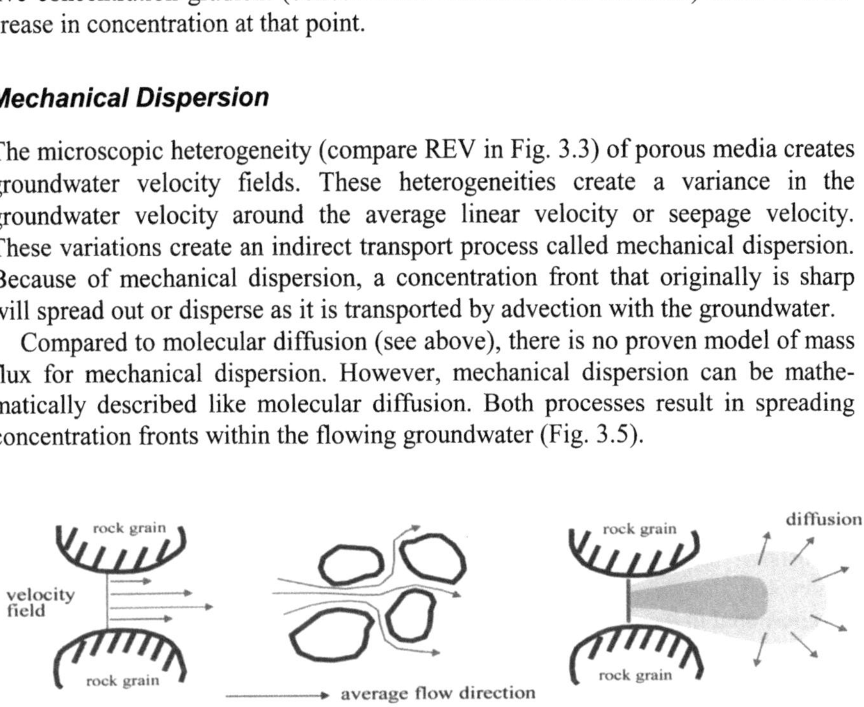

Fig. 3.5. Causes of dispersion on the microscopic scale

This is the reason why mechanical dispersion and molecular diffusion are lumped together in the dispersion coefficient within the transport equation. However, both processes differ from each other. Molecular diffusion depends only on a concentration gradient and is dominant at lower velocities, and can be treated as a scalar constant. Mechanical dispersion dominates at higher velocities. Mechanical dispersion is generally treated mathematically as a second-order tensor (compare permeability, Scheidegger 1961).

At the pore scale, diffusion and mechanical dispersion are interconnected and can only be artificially separated (Bear 1972). The combined effects of mechanical dispersion and molecular diffusion are called hydrodynamic dispersion. In an isotropic medium the dispersivity reduces to just two components: α_L, the dispersivity of the medium parallel to the groundwater flow direction, and α_T, the dispersivity of the medium transverse to the flow direction. For example, under two-dimensional transport conditions in a one-dimensional flow field the dimensional transport conditions in a one-dimensional flow field the dispersion coefficient reduces to

$$D_{xx} = D_L = \alpha_L v_x + \frac{D_W}{n_e} \tag{3.48}$$

and

$$D_{yy} = D_T = \alpha_T v_x + \frac{D_W}{n_e} \tag{3.49}$$

where D_L and D_T are the longitudinal and transverse dispersion coefficients, respectively. For a more thorough description see Bear (1972).

The dispersion coefficient as described above is normally applied in transport simulations as macrodispersivity. Macrodispersivity is not a true physical property of the porous medium, but is used as a fitting parameter to allow for simulation of solute dispersion. The choice of this value depends of the scale of modeled transport as shown by Gelhar et al. (1992) and the resolution of the specific simulation model. For example, heterogeneity effects of the calculated area, which cannot be reproduced in the model, are represented within the macrodispersion coefficient (REV statement). Conversely, in homogenous media the dispersion coefficient is dominated by the microdispersivity, which can be related to physical properties of the porous media. Xu and Eckstein (1997) showed that the porosity and uniformity of grain sizes are the two most important factors affecting the values of microdispersivity. Microdispersivity is directly proportional to the uniformity coefficient and inversely proportional to porosity. Microdispersivity is also directly proportional to the median grain size of homogenous clastic materials.

Microdispersivity is often used to describe transport in laboratory experiments, whereas macrodispersivity is applied to field scale dispersion. Nevertheless, microdispersivity can be used to calculate macrodispersivity as shown by Gelhar and Axness (1983). Microdispersivity sums up for a specific area to asymptotically approach the value of macrodispersion in the limiting case.

Mass-Balance-Equation

The total solute concentration change at any given point is the sum of the fluxes of advection, molecular diffusion, and mechanical dispersion. This total flux can be substituted into a mass-balance equation to derive the solute transport equation. Upon this substitution, a general case of the advective-dispersive equation is derived:

$$\nabla(n_e \rho \underline{D} \nabla C) - \nabla(n_e \rho v C) + Q_s = \frac{\partial(n \rho C)}{\partial t} \tag{3.50}$$

where $\underline{D}$ is the hydrodynamic dispersion tensor, C is the concentration in terms of mass fraction, and Q_s is a source (+) or sink (-) of solute. The solute source or sink could be due to a net loss or gain of a fluid with a finite concentration or mass. The source or sink term may incorporate any number of chemical reactions.

3.4.3 Heat Transport

General models of hydrothermal circulation in Earth's crust must accommodate a large temperature range. Taking varying temperatures into account, the groundwater flow equation must be combined with heat transfer processes. Groundwater flow and heat transfer can be described by a set of coupled equations expressing mass and energy conservation.

Two basic types of heat transfer will be considered here. These are conductive and convective (or advective) heat transfer, whereas radiation and thermal dispersion, the third and fourth basic types of heat transfer, respectively, are neglected. Conductive heat transfer is the motion of heat through a stationary medium by vibration of atoms, whereas convection of heat takes place by advective processes, the transport of heat by a moving medium.

Conductive Heat Flux

Conductive heat transfer is governed by Fouriers law, which says that heat flux is proportional to the temperature gradient, and the higher the gradient the greater the amount of heat flow. Fourier's law of heat conduction is:

$$q_h = -\lambda_e \frac{dT}{dz} \tag{3.51}$$

where q_h is the heat flux, λ_e is the effective thermal conductivity of the medium and T is temperature. Fourier's law is similar in form to Darcy's Law [Eq. (3.31)] as well as Fick's law of diffusion [Eq. (3.44)]. Each of these laws describes a linear relation between a flux and a gradient in potential.

Whereas the hydraulic conductivity of crustal materials varies over approximately 16 orders of magnitude, the thermal conductivity of the upper crust generally varies by less than a factor of 5. Rock thermal conductivity can be up to 10 times greater than water thermal conductivity.

Especially at stagnant conditions, when heat transfer is due to conduction, it has to be taken into account that the thermal conductivity of water is different from that of rock. Therefore, the effective thermal conductivity, λ_e, is defined by:

$$\lambda_e = n\lambda_f + (1-n)\lambda_r \tag{3.52}$$

where n is porosity, λ_f fluid conductivity, and λ_r conductivity of the rock. Although thermal conductivity is treated here, and following, as a scalar, it has to be kept in mind that it is a value which is generally an anisotropic or direction dependent property like permeability or dispersivity (see above).

Convective Heat Flux

Far more important in groundwater flow systems is convection, the transport of heat by moving water. Two different kinds of convection exist: (1) forced convection and (2) free convection. In forced convection, the velocity of convective motion is independent on the temperatures in the fluid, it is part of the natural groundwater flow system and therefore dependent on the hydraulic head (see above). Thus, the groundwater movement forces heat energy transport. In free convection, when only buoyancy effects in the fluid drive water velocities, these are related to temperature change through the coefficient of thermal expansion (see next section). In real and deep groundwater systems, there is a mixture of both types of convection whereas in shallow aquifers, temperature and heat transfer are negligible.

The simple expression of heat transport in forced convection is given by

$$q_h = -\lambda_e \nabla T + n_e \rho_f C_{p(f)} T v \tag{3.53}$$

Remembering that q_h is heat flux (W m^{-2}), we see that $C_{p(w)}T$ is the heat content per unit mass of water, $\rho_w C_{p(f)} T$ is the heat content per unit volume of water, $\rho_w C_{p(f)} T v$ is the flux of heat (W m^{-2}) transported by velocity, v, per unit area. Across the entire rock face this flux is reduced by the effective porosity, n_e.

Conservation of Energy

The starting point for understanding temperature distribution in aquifers is conservation of energy (energy inflow rate – energy outflow rate = change in energy storage with time).

It should be remembered from earlier that the net flux of a vector across a test volume is given by the divergence of the vector, so in the case of heat transport the first two terms of the equation are simple. If no heat is advected, then Fouriers law defines that the energy inflow rate minus energy outflow rate is given by

$$-\nabla q_h = \nabla(\lambda_e \nabla T) \tag{3.54}$$

The net change in stored energy with time is simply

$$\rho C_p \frac{\partial T}{\partial t} \tag{3.55}$$

so that conservation of energy is then

$$\nabla(\lambda_e \nabla T) = \rho C_{p(r)} \frac{\partial T}{\partial t} \tag{3.56}$$

But when heat is advected [Eq. (3.53)], the conduction-convection equation is

$$-\nabla\left(-\lambda_e \nabla T + n_e \rho_f C_{p(f)} T v\right) = \rho C_{p(r)} \frac{\partial T}{\partial t} \tag{3.57}$$

where the term C is used for the specific heat of water. Note that when the velocity is zero or perpendicular to the temperature gradient, this reduces to the simple diffusion equation.

3.4.4 State of the Art of Hydrothermal Reactive Transport Simulation

The simulation of reactive transport requires that the equations described in the previous sections be solved. This can be done either analytically or numerically. Analytical solutions of transport processes of real-world problems do have a limited applicability, because heterogeneity and multidimensionality are reality and additionally, the real world models do have complex constraints which cannot be reflected in analytical solutions. But with the advent of computers, the numerical solution of complex non-linear partial differential equations became possible. However, the application of these techniques to model geothermal systems lagged behind their application in groundwater or oil and gas reservoirs, because especially the coupling of mass and energy transport adds considerable complexity and the economic driving force of the geothermal industry has been smaller.

Groundwater flow, heat transfer, and transport are commonly calculated based on three different numerical solution techniques, finite differences, finite volumes, or finite elements, each of which has particular advantages and disadvantages when applied to the problem of interest. A description of these methods is beyond the scope of this book, but such descriptions have been published in detail elsewhere (e.g. Wang and Anderson 1982, Huyarkon and Pinder 1983, Fitts 2002).

Van der Lee and De Windt (2001) reviewed in general the present state of simulation of geochemistry and hydrogeological systems. Their focus has been on reactive transport processes in groundwater systems neglecting the effect of temperature, which is of great importance for the investigation of geothermal reservoirs. O'Sullivan et al. (2001) presented the state of the art of geothermal reservoir simulation and concluded that the use of computer modeling in planning and management has become standard practice during the last 10-15 years. They published a comprehensive compilation of more than 100 computer modeled geothermal fields worldwide. The restriction is that recent standard models of geothermal fields only considers fluid flow and heat transfer. There are only a few cases investigating geochemical evolution and mineral recovery. Hence, O'Sullivan (2001) predicts that the aim is still a fully coupled 3D mass and energy transport model with detailed chemical interactions between aqueous fluids, gases and mineral assemblages, considering thermodynamic equilibrium as well as kinetic effects. Such kinds of models would provide a more realistic description of the coupled physical and chemical processes inherent in geothermal reservoirs and more generally in hydrothermal systems.

At first it has to be stated that there already exists a variety of programs calculating fluid flow, heat transfer and mass transport or fluid flow, mass transport and chemical reactions, but only a small number of codes are able to simulate all the processes necessary for a comprehensive study of hydrothermal systems, which are at least fluid flow, heat transfer, transport, and chemical reactions. Additionally it is of importance that all the four processes have to be solved as intensively coupled as necessary. Following is list and brief description of the capabilities of available programs. All are specialized in particular fields with resulting inherent advantages and disadvantages. At present there is no program favored over others:

- TOUGH2-Family (Pruess 1991): IFD (integral finite differences) code for simulating flow and heat including 3 phases (water, vapor, organics) developed for geothermal reservoir exploration and exploitation, nuclear waste dumping sites, and hydrology in saturated and unsaturated media. Various modules for reactive transport simulation are built up on the program by several authors:
  - TOUGH / EWASG (Battistelli et al. 1997): The calculated system includes water, salt (NaCl), and gas and additionally calculates dissolution and precipitation and resulting porosity-permeability changes.
  - CHEM-TOUGH (White 1995): TOUGH2 extended for transport of reactive species under consideration of Debye-Hückel activities within the chemical calculations. The program supports a fixed and limited number of reactive components (solutes and minerals) and calculates permeability changes due to chemical reactions.
  - TOUGHREACT (Xu and Pruess 2001): Generalized chemical module based on Debye-Hückel activity coefficient calculations. Porosity-permeability interactions are monitored but do not feed back on the flow.
- 3DHYDROGEOCHEM (Cheng and Yeh 1998): FE (finite elements) program for saturated and unsaturated zone modeling including species transport and chemical reactions (Debye-Hückel, equilibrium and kinetics) as well as microbiological processes. No porosity-permeability coupling or monitoring.
- RST2D (Raffensberger and Garven 1995a, 1995b): Program for coupled simulation of fluid flow, heat transfer, transport, and chemistry (equilibrium and kinetics) with a fixed number of reacting components.
- FRACCHEM (Durst 2002): FE system for modeling coupled flow, heat, transport, and reactions in fractured systems. Chemistry set up under special consideration of and limited to the site Soultz-sous-Forets (France).
- SHEMAT (Bartels et al. 2003, Clauser 2003): FD (finite differences) simulation code for fully coupled investigation of flow, heat, transport, and reactions. The chemistry (equilibrium and kinetic) is applicable for dilute solutions (Debye-Hückel theory) to highly saline brines (Pitzer formalism). The chemical reactions are coupled on the flow via porosity-permeability relationships based on varying methods. The number of reactive components is variable and can be extended by the user.

3.5 Uncertainty, Usefulness, and Limitations of Models

The purpose of mathematical modeling is to develop a computer model that reflects essential features of the phenomenon considered or represents a real system. Mathematical models are abstractions that replace objects, forces, and events by expressions that contain variables, parameters, and constants (Krumbein and Graybill 1965). However, one should remember that models provide only approximate solutions.

As the final product of the modeling process, a computer model includes all simplifications and assumptions made at the previous steps, particularly at the starting step when empirical data are conceptualized. Since there is always uncertainty in empirical data, conceptual models may become a major source of error. Dimensional analysis has proven to be efficient for testing conceptual models and for identifying key physical processes. Errors in conceptualizing a problem are easy to make but may be hard to discover. The modeler should begin work by integrating experimental results and field observations into the study. Having successfully explained the experimental or field data, the modeler can extrapolate to make predictions with greater confidence.

Zhu and Anderson (2002) stated that, on the one hand, a model generally must be "verified", what means that it has to be checked if the computer code solves the set of equations correctly and is free of serious bugs. Commonly, the user does not have to bother with the "verification" of a program, because the developers normally "verify" distributed programs. On the other hand, it has to be "validated", if the assumed conceptual model, which provides the basis for the set of equations incorporated in the code, actually represents the natural processes of interest. Here, the modeler has to ensure that the chosen computer code simulates the problem to be solved in a useful way. On the contrary, Konikow and Bredehoeft (1992) argued convincingly that geochemical models couldn't be proven or validated but only tested and invalidated thus following the school of thought espoused by Popper (1959). However, Walter et al. (1994) stated that with increasing model complexity either complete "verification" or "validation" may be impossible, but above all, a model is valuable to get insight into complex processes. Zhu and Anderson (2002) follow this concurrent opinion and say that geochemical and hydrogeological models are the best way to integrate and understand data, concepts, and processes, which are coupled, but from different disciplines, because reactive transport modeling is an interdisciplinary activity involving, among others, numerical mathematics, geology, hydrology, physics, and chemistry. Anyway, the best way to avoid errors is to always critique your own results.

A geothermal system is never at steady state but undergoes various physical and chemical processes. If a real geothermal process is slow, a quasi-static path can approximate it. The evolution of a real system can then be modeled as a succession of equilibrium states (compare Reaction Path Models).

Calculating a geochemical or reactive transport model provides not only results but also uncertainty about the accuracy of the results. Uncertainty, in fact, is an integral part of modeling that deserves as much attention as any other aspect of a study. The modeler, working with computer models, has to be aware of several nuisance effects (Nitzsche et al. 2000):

- numerical problems of the code itself might occur,
- incomplete understanding of the physical and chemical phenomena,
- lack of precision of the geological and hydrogeological input data,
- missing data for a complete set up of a conceptual model,
- experimental uncertainties in the thermodynamic data.

The importance of a quantitative estimate of uncertainty within the thermodynamic database is emphasized by (Nitzsche et al. 2000). They performed a Monte Carlo analysis to show the impact of uncertainty in equilibrium on the elution and break through of uranium from a sand column, which illustrates remediation problems of uranium mines. The variation of the thermodynamic input data, displaying the effect of uncertainty within a database, and resulting transport predictions illustrated that the complexity of a predictive model will be limited to the growing uncertainty with the number input parameter affected by uncertainty. But unfortunately, the Monte Carlo approach requires a very large number of repetitive calculations what is not feasible for 3D real world problems. This is one of the reasons why uncertainties in modeling results are seldom provided. Hence, the modeler should use his results to provide an impetus to determine more accurate thermodynamic data, derive better chemical and physical models, and improve the understanding and the conceptual models of a site (Bethke 1996).

Reactive transport modeling can be extremely useful in understanding the spatial and temporal distributions of solute concentrations and mineral assemblages in the environment. The main target of reactive flow modeling is the simulation on a real time scale with real spatial coordinates, but generally this goal is only partially achievable. The limitations of reactive transport simulation are embedded in the conceptualized set of equations used to best approximate the real situations. But models provide a tool for critical analysis. They are a means to organize our thinking, test ideas for their reasonableness, and indicate which are the sensitive parameters. They point the way to further investigation and help to design new experiments and to critically test hypotheses. Particularly surprising model outputs often provide new insights otherwise inaccessible.

4 Specific Features of Coupled Fluid Flow and Chemical Reaction

In many geologic processes, water-rock interactions are driven by continues supply of reactants provided by flowing groundwater. When water that carries dissolved chemical species moves through a permeable matrix or a network of fractures a variety of chemical reactions can occur. Reactive transport processes are important when there is a potential for fluid flow coexisting with spatial variations in the thermodynamic states of the system. The resulting patterns of dissolution, precipitation, and rock alteration depend on the reaction kinetics and the rate at which reactants in solution can be delivered to the reaction site by advection and diffusion.

Five fundamental end-member types of reactive transport environments, setting up the basis for the discussion of specific features of coupled flow and reaction within this chapter, are distinguished in the following:

- "Flow across mineralogical boundaries" describes the equilibrium approach of a water body moving into a geological environment with a different assemblage of minerals (time independent).
- "Moving reaction fronts" are characterized by a mineralogical boundary propagating in flow direction because of the time dependent depletion of the originally reacting mineral.
- "Reactions within thermal gradients" are caused by variations in temperature across a region where the mineral assemblage might otherwise be constant.
- "Mixing zone environments" exhibit reactions due to variation in aqueous composition. The mixture of two different waters, both in equilibrium with respect to the same mineral assemblage, may result in a mixing disequilibrium due to the nonlinearity of chemical reactions.
- "Local flow enhancement due to faults" depicts fluid transport through otherwise impermeable layers. This focusing of fluid flow may lead to or accelerate precipitation or dissolution of minerals.

These five distinct systems lead to flow induced spatial reaction patterns. However, it is most likely to encounter overlapping end-member types within natural environments. Specific examples, presented here, are preferential flow path development and diagenesis due to thermal convection. Before proceeding to discuss individual processes, the key for understanding the majority of reactive transport phenomena, changes of permeability due to reactive porosity increase or decrease, is investigated. The porosity-permeability relationship represents the basis for the coupling of flow and transport.

4.1 Flow Induced Reaction Patterns

4.1.1 Flow Across Mineralogical Boundaries

When a fluid enters a geological environment, different in composition from where it originates, the water body moves across a mineralogical boundary, resulting probably in disequilibrium among the fluid and the actual mineral assemblage. The concentration of dissolved species instantaneously starts to move toward a new local equilibrium with the solid phase. The extent of the region of disequilibrium and the spatial distribution of the mineralogical changes depend on the speed of flow, the reaction rates, and the solute diffusivity. The relationships among these quantities provide criteria under which local equilibrium between the fluid and the matrix can be expected.

A typical situation of equilibrium approach at a reaction boundary is displayed by dissolution or reaction of a major constituent of a rock formation. When water undersaturated, for example, with respect to the mineral calcite, enters a permeable, porous limestone bed, via a sharp interface (abrupt change of the mineralogical assemblage), dissolution of calcium carbonate will occur. For the ease of examination of this first end-member reaction type it is assumed that the time over which the dissolution process has continued is not so large that the cumulative effect has caused substantial changes in either the porosity or mass per unit volume of the solid. Thus, any feed back on the flow can be neglected. The significance of the equilibrium or saturation length in the case of dissolving minerals expresses the characteristic distance in the flow direction over which the fluid remains in disequilibrium with respect to a particular constituent. Schulz (1988) published an easy laboratory way to determine carbonate dissolution rates by measuring this saturation length in columns filled with sandy aquifer material. The spatial concentration distribution (measured by various sampling points over the column length) moves towards equilibrium between the adjusted average linear flow velocity v [m s^{-1}] and the rate of dissolution k_{reac} [s^{-1}] to be determined. Under steady state conditions the saturation length x_S [m] is the flow path length from the column entry to the spot where the calcium and carbonate concentrations reached 63% of the concentration prevailing at thermodynamic equilibrium. The reaction rate can be calculated by:

$$k_{reac} = \frac{v}{x_s} \tag{4.1}$$

In Fig. 4.1 experimental data (Schulz 1988) are shown compared to results of a numerical simulation using SHEMAT (Bartels et al. 2003). A pure water solution under constant partial pressure of 0.034 CO_2 enters the experimental column (0 m) with a flow rate of 1.7×10^{-5} [m s^{-1}]. The saturation length determined by Schulz (1988) is 14 cm with a resulting dissolution rate of 1.2×10^{-4} [s^{-1}]. Calculated and experimental results of the dissolution reaction coincide very well. Additional simulations with reaction rates 2.4×10^{-4} [s^{-1}] (doubled) and 0.6×10^{-4} [s^{-1}] (half) display the sensitivity of the saturation length with respect to variations of the reaction rate (Fig. 4.1). It can be deduced that experimentally determined reaction rates can obviously be distinguished in a numerical model with a significance of at least ± 50%.

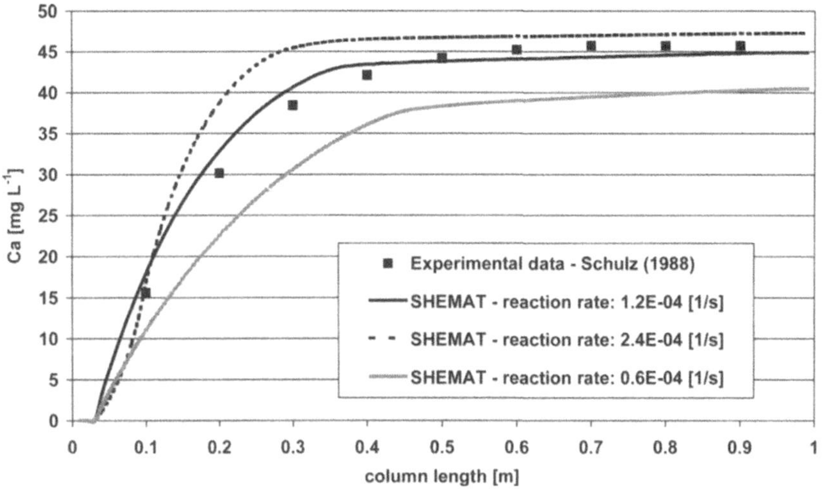

Fig. 4.1. Reaction front developing in a calcite dissolution column experiment representing an equilibrium approach boundary; the solution, initially under constant partial pressure of CO_2 enters the column and dissolves calcite from the sand bed; comparison of experimental data (Schulz 1988) and simulation results yielded with SHEMAT

4.1.2 Moving Reaction Fronts

Reaction fronts may move through space where variations in mineralogy exist and the original mineral can be locally depleted, dependent on time. This is caused, when water moves from one area, where it is in equilibrium, to another area, where it comes into contact with a second mineral assemblage (Fig. 4.2). The difference between "moving reaction fronts" and "flow across mineralogical boundaries" (see above, Fig. 4.1) is an infinite amount of mineral in the latter reaction environment with respect to the investigated time period (immobile front quasi independent on time).

When, as considered in the preceding section and described by Lichtner (1985) and others, fluid enters a permeable matrix across a mineralogical boundary, reaction fronts may originate at that border and propagate away from it. In a reaction the product replaces the original mineral near the interface. When the original mineral is depleted, the reaction cannot continue and the fluid concentration remains almost at its entering composition until it reaches a position farther downstream, where reactants are present again and reaction occurs. As a result the zone of reaction gradually moves downstream, leaving the assemblage of solid products behind. The solute resulting from the reaction is carried away and the patterns remain. They separate an unaltered region ahead from that behind where the reaction has proceeded to a different equilibrium between the altered host rock and the fluid.

Reaction fronts can lead, for example, to the specific process of reaction infiltration instability, discussed in one of the following sections under special consideration of anhydrite cemented sandstones.

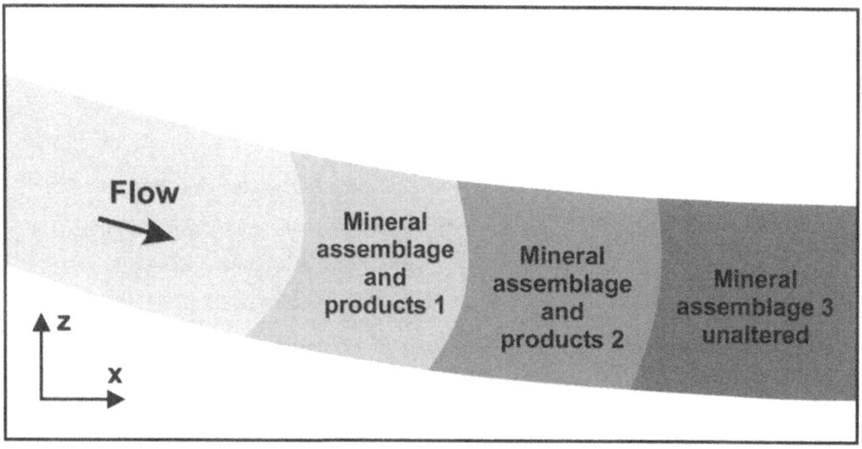

Fig. 4.2. Moving reaction front characterized by changing mineral assemblages

4.1.3 Reactions Within Thermal Gradients

A third type of environment is caused by variation in temperature across a region where mineral and aqueous compositions might otherwise be constant, in contrast to "flow across mineralogical boundaries" and "moving reaction fronts". The changes in temperature result in changes of the chemical equilibrium as the water, carrying the solutes, moves through a thermal gradient. In this case, the thermal gradient is the rate at which temperature increases or decreases within flow direction. This may lead to dissolution or precipitation of mineral phases towards a new state of equilibrium.

The reaction fronts described in the preceding sections arise at mineralogical boundaries or interfaces and propagate in flow direction, where the fluid is close to equilibrium with the new assemblage beyond the equilibrium length at the ambient pressure and temperature. However, the equilibrium concentration changes along the flow path if varying temperatures occur. Hence, the composition of the fluid may remain the same over time at each point, but it varies along the flow path. Solute is continually added to or dissolved from the mineral assemblage. As a consequence, the mineral composition is gradually altered along the flow path. This is a process, which occurs throughout those parts of the system where the fluid flow crosses isotherms.

Reaction environments within thermal gradients may be the reason for diagenetic mineral precipitation reactions as presented in the chapter about thermal convection. In Fig. 4.3, as an example, a closed system model cooled at the top and heated from below is shown in which two convection cells established. The heat transfer due to convection is depicted by the isotherms (lines), the flow direction (arrow heads) and magnitude of flow (arrow length) by the displayed arrows. If the chemical composition of the moving fluid always remains close to equilibrium with respect to a specific mineral phase over the entire system, the temperature distribution, resulting in a distribution of chemical equilibrium states, will lead to changes in the abundance of that mineral. In some areas the mineral amount will increase and in some it will decrease.

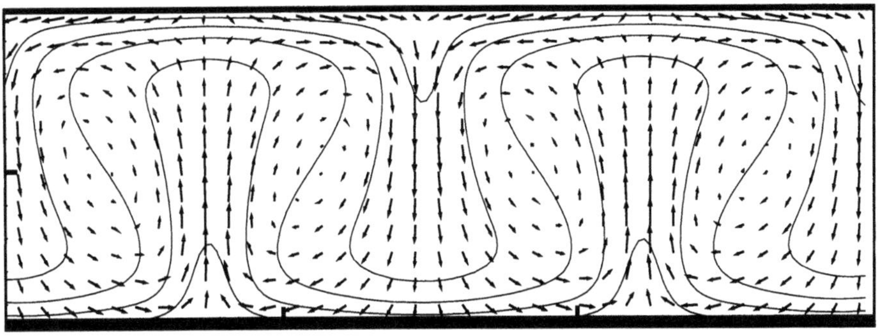

Fig. 4.3. Thermal convection cell displaying temperature isotherms (lines), flow direction (arrows heads), and magnitude of flow (arrow length); the system is heated from below (thick, black solid line) and cooled at the top (medium, black solid line)

4.1.4 Mixing Zone Environments

The fourth type of reaction environment is caused by a variation in aqueous composition whereas the previously described types where characterized by changes either in mineral composition ("flow across mineralogical boundaries" and "moving reaction fronts") or temperature ("reactions within thermal gradients"). Two different waters could both be in equilibrium with respect to the same mineral composition in a region with constant temperature, but upon their mixing the new, mixed water will be out of equilibrium. This type of mixing disequilibrium is caused by the nonlinear nature of the law of mass action governing chemical reactions.

Sanford and Konikow (1989a, 1989b) discuss a common example of reactive transport in mixing zones. They simulated the geochemical reactions and variable-density solute transport associated with a dynamic transition zone between seawater and freshwater in a coastal aquifer. Freshwater-seawater mixing zones have been identified as potential sites for dolomitization (Hanshaw et al. 1971) and dissolution of carbonate rocks (Back et al. 1979). The objective of the numerical study of Sanford and Konikow (1989a, 1989b) was to estimate the quantity of calcite dissolution under typical hydrodynamic and geochemical conditions to assess changes of the rock properties porosity and permeability, because resulting permeability changes in turn affect the flow system. The results illustrate the use of a fully coupled reaction-transport model for analyzing diagenetic processes. They indicate that porosity development within the mixing zone would not be evenly distributed. Porosity develops on the freshwater side of the transition zone. Porosity also develops faster at the base or toe of the mixing zone and at its top, in the discharge area at the coastline. Including a permeability-porosity feedback allowed the flow system to respond to dissolution over time. As porosity is enhanced on the freshwater side of the mixing zone, the resulting permeability enhancement causes the transition zone to migrate landward over time.

4.1.5 Local Flow Enhancement due to Faults

The characteristic feature of the fifth reaction environment is caused by particular hydrodynamic conditions. Fault and fracture zones provide pathways for fluid transport through otherwise impermeable layers. In already permeable zones, planes or surfaces of high permeability attract focusing of fluid flow. This local flow enhancement in turn may accelerate the process of deposition or dissolution that becomes part of the geological record. Deposition leads to veins of mineralization and dissolution to enlargement of the interstices and possibly increased flow rates.

When a fracture traverses a relatively impermeable layer between two more permeable layers, it opens a fluid pathway that allows for mixing of interstitial fluids from one region to the other and provides the opportunity for chemical reactions between the entering fluid and its new environment. This type of reaction system is investigated later as one opportunity of mineral deposition within a deep seated aquifer located in Northern Germany (Chap. 5).

4.2 Porosity and Permeability (Reduction) Models

Permeability is unquestionably the crucial hydrologic parameter. Unfortunately, it is often a very difficult parameter to determine and apply in a meaningful fashion. This is especially due to its enormous variation over space and time in natural systems.

In many geologic problems permeability must be regarded as a time-dependent parameter (see above), being increased or decreased over time by mineral dissolution and precipitation, by changes in effective stress that result in consolidation or hydraulic fracturing, and by thermoelastic effects. Permeability changes due to mineral reactions will be discussed in detail in the following. Permeability reduction due to consolidation effects where studied, for example, by Kühn et al. (2002c) for a geothermal reservoir under exploitation and by Gambolatti et al. (1996) for utilization of subsurface fluids in general.

Almost all diagenetic reactions as well as many metamorphic reactions are coupled to fluid flow. Geochemical reactions that lead to the dissolution or precipitation of a solid mineral phase result in changes of the pore space structure of the porous medium. Usually there is a strong positive correlation between both porosity and permeability. An increase of porosity will therefore lead to an increase in permeability and vice versa. Permeability in turn affects the flow system through Darcy's law. Because of this feedback between porosity, permeability, and flow the relation between porosity and permeability is of major importance for the understanding of diagenetic processes.

Early studies relating porosity-permeability changes under consideration of reactive transport processes are rare, but this field of research is developing during the last decade. Zarrouk and O'Sullivan (2001) gave a review of the effect of chemical reactions, mainly arising from geothermal applications, on the porosity of a porous medium and resulting permeability changes. They concluded that every simulation code for reactive transport should be adaptable concerning the applied porosity-permeability relationships, thus, it is possible to use any relation. The simulator Processing SHEMAT / SHEMAT is an example, where varying relationships can be applied. The selection is based on Zarrouk and O'Sullivan (2001) and includes the Eqs. (4.3)-(4.8) listed in the following. All of them depend on the field of study and their particular application case, but a general procedure capable of describing a wide variety of natural systems is not available until now. The simulator codes CHEM-TOUGH and TOUGH/EWASG are further examples of reactive transport models applicable for geothermal systems in which permeability changes resulting from chemical reactions are considered. Whereas in CHEM-TOUGH permeability is assumed to vary with porosity to the power of three (Phillips 1991, Eq. (4.3) with $D_f = 3$), in TOUGH/EWASG the permeability change is described far more complex. TOUGH/EWASG provides the opportunity to calculate permeability changes based on either a straight capillary tube model, or a model consisting of alternating segments of capillary tubes with larger and smaller radii, or for parallel-plate fracture segments of different aperture in series (Pruess et al. 1999). The first model simplifies to a relationship in which permeability varies with porosity to the power of two (Eq. (4.3) with $D_f = 2$). The models of "tubes in series" and "fractures in series" depend on additional parameters beside porosity and permeability and are therefore not discussed here.

A relationship between permeability and porosity found by Pape et al. (1999) assumes that the shape of the internal rock surface follows a self-similar rule. Thus the theory of fractals can be applied. The fractal relationship between permeability k and porosity ϕ is based on the Kozeny-Carman equation and is expressed by Pape et al. (1999) as a general three-term power series in porosity where the exponents $D_{f,i}$ (i=1, 2, 3) depend on the fractal dimension of the internal surface of the pore space:

$$k = A\,\phi^{D_{f,1}} + B\,\phi^{D_{f,2}} + C\,\phi^{D_{f,3}}\,. \tag{4.2}$$

The coefficients A, B, and C need to be calibrated for each type of sedimentary basin or pore space modification, i.e. porosity change due to chemical reactions.

Eq. (4.2) reflects the fact that in different intervals of porosity different processes dominate the changes in porosity and permeability. This can be approximated by Eq. (4.3) defining different exponents for different porosity intervals. In Eq. (4.3), k_0 and ϕ_0 denote the initial values, which represent the same information as the coefficients in Eq. (4.2):

$$k = k_0 \left(\phi/\phi_0\right)^{D_f} \tag{4.3}$$

In the porosity-permeability relation of Eq. (4.3), permeability k is a function of the porosity change only, because the initials k_0 and ϕ_0 as well as the fractal exponent D_f are defined as constant values. For the applicability of any porosity-permeability relation this is of major importance in reactive transport modeling. That applies also for the further k-ϕ-relations [Eqs. (4.4) - (4.8)], which are in some kind specific cases of Eq. (4.3) and are summarized in Zarrouk and O'Sullivan (2001).

Weir and White (1996) published an equation [Eq. (4.4)] for the calculation of permeability changes due to deposition on spheres in dense, rhombohedral packing. Below the critical porosity ϕ_c permeability vanishes:

$$k = k_0 \left\{ 1 - \left[1 - \left(\frac{\phi - \phi_c}{\phi_0 - \phi_c} \right)^{1.58} \right]^{0.46} \right\} \tag{4.4}$$

Eq. (4.4) was used by Arihara and Arihara (1999) for the modeling of silica scaling in injection wells.

The Blake-Kozeny equation [Eq. (4.5)] for flow in packed columns and applied permeability changes due to matrix acidizing in hydrocarbon wells situated in limestone and sandstone was used by McCume et al. (1979). The same equation was applied by Olivella et al. (1996) for reactive transport calculations in unsaturated salt rocks.

$$k = k_0 \left(\phi/\phi_0\right)^3 \left(\frac{1 - \phi_0}{1 - \phi} \right)^2 \tag{4.5}$$

Lichtner (1996) adapted the Blake-Kozeny equation [Eq. (4.5)] for the dependence of permeability on porosity in a mixture of potassium-feldspar, gibbsite, kaolinite and muscovite [Eq. (4.6)].

$$k = k_0 \left(\phi/\phi_0\right)^3 \left(\frac{1.001 - \phi_0^2}{1.001 - \phi^2} \right) \tag{4.6}$$

Itoi et al. (1987) applied the Kozeny-Stein equation for calculating the effect of silica precipitation in the vicinity of injection wells [Eq. (4.7)].

$$k = k_0 \left(\phi/\phi_0\right)^3 \left(\frac{1-\phi_0}{1-\phi}\right)^2 \left\{\sqrt{\frac{\phi_0-\phi}{3\left(1-\phi_0\right)}+\frac{1}{4}}+\frac{\phi_0-\phi}{3\left(1-\phi_0\right)}+\frac{1}{2}\right\} \tag{4.7}$$

Eq. (4.8) developed by Schechter and Gidley (1969) has been used for modeling permeability changes of limestone due to surface reactions induced by dilute hydrochloric acid. This procedure of matrix acidizing was performed in hydrocarbon wells.

$$k = k_0 \left(\phi/\phi_0\right)^2 e^{2(\phi-\phi_0)} \tag{4.8}$$

In Fig. 4.4 permeability is shown as a function of porosity dependent on the porosity-permeability relationships given in Eqs. (4.3)-(4.8). Initial porosity ϕ_0 and permeability k_0 used within the comparison are 0.15 and $1.0\text{x}10^{-13}$ m², respectively. The scanned porosity range from minimum 0.01 to maximum 0.3 is most common to aquifer properties found.

Obvious is that the results, applying k-ϕ relations following McCume et al. (1979), Lichtner (1996), and Itoi et al. (1987), are almost identical with an exponent D_f of 3 applied in Eq. (4.3), referring to a fractal dimension of 2 of smooth shaped grains. Especially in the range below ϕ_0, with decreasing porosity and permeability, they coincide very well. Above ϕ_0, the relation published by Schechter and Gidley (1969) plots close to the curve using a fractal exponent D_f of 3, whereas below this value it curves with a significantly smaller gradient. This is because the equation of Schechter and Gidley (1969) is set up especially for increasing porosities.

Eq. (4.4) of Weir and White (1996) is valid only for precipitation (here below ϕ_0 of 0.15) and above a critical porosity below which any fluid movement breaks down (here ϕ_c is 0.05). The application of a fractal exponent of 5.0 in Eq. (4.3) leads to similar results.

The application of Eq. (4.3) with fractal exponents D_f of 3.0, 5.0, and 12.0 delineates the possible range of k-ϕ relations. The exponent D_f 3.0 most often used, also in the Eqs. (4.5)-(4.7), represent clean sandstones with smooth shaped grains. An exponent of 5.0, more precise 4.85, has been determined for anhydrite precipitation found in rock samples of deep geothermal aquifers from Northern Germany. The mineral deposit developed in this case in geological time scales. On the contrary, an exponent of 12.0 has been determined in core flooding laboratory experiments, representing the technical time scale, where anhydrite relocated (dissolved and subsequently precipitated) within a temperature front (Bartels et al. 2002).

It can be concluded that the most general and simple k-ϕ relation [Eq. (4.3)], presented above, is suited to describe the other specific relations determined in laboratory experiments or deduced theoretically. The fractal exponent can be chosen to represent distinct minerals and systems referring to geological as well as technical time scales.

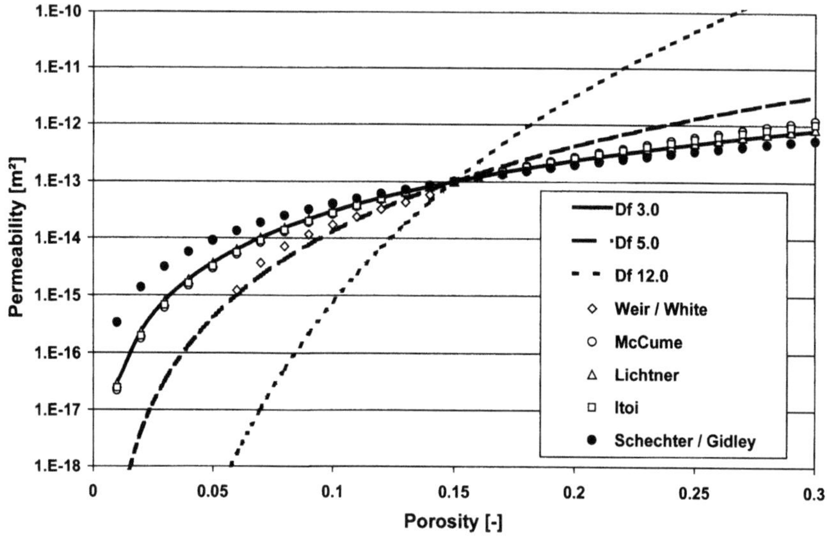

Fig. 4.4. Permeability as a function of porosity change dependent on the k-ϕ relationships given in Eqs. (4.3)-(4.8); initial porosity and permeability and critical porosity [Eq. (4.4)] are 0.15, 1.0×10^{-13} m² and 0.05, respectively

4.3 Reactive Infiltration Instability

Alteration through weathering, diagenesis, metasomatism, or metamorphosis results in specific reaction zones. Within these zones fluid flow may lead to propagating reaction fronts as mentioned above. Such fronts are often fingered even though related features do not previously appear in the unaltered rock. Ortoleva et al. (1987) explained these phenomena by geochemical self-organization. Self-organization denotes that systems may organize themselves into structures not dictated by initial mineralogical or geologic conditions.

Permeability and flow are closely related by a non-linear feedback on a large range of spatial scales through the dissolution and precipitation of minerals in sedimentary basins or metamorphic systems. At regional scale the development of karst in carbonate rocks is the result of massive dissolution by subsurface fluid flow (e.g. White 1988). At the scale of a platform reef the very heterogeneous texture of the carbonate matrix is due to both the coral formation and to coupled reaction and flow within the reef (Schroeder and Purser 1986, Rougerie et al. 1991). At the centimeter scale, experiments on anhydrite dissolution for the study of well stimulation illustrate the positive feedback between flow and reaction leading to the development of preferential flow path (Baermann et al. 2001b).

Prerequisites for preferential flow path formation are described in the literature. Ormond and Ortoleva (2000) showed that interaction between mineral reaction and mass transport in rocks can lead to reaction front instability. The development of channel-like voids occurs if the characteristic Peclet and Damköhler numbers, describing the reactive flow system, fall in a specific range.

4.3.1 Peclet and Damköhler Number

Flow coupled with transport of solutes generated by mineral dissolution can lead to self-organized enhancement of the heterogeneity in the rock (Ortoleva 1994). If the initial rock texture is perfectly uniform, the resulting reaction front will remain planar while advancing down stream with time. However, if the rock texture is initially even slightly non-uniform, the resulting enhancement of the permeability will be uneven and a fingering reaction front may form (Wei and Ortoleva 1990).

Fluid flows preferentially in regions with higher permeability. If the fluid is undersaturated with regard to a specific mineral phase, mineral dissolution occurs. Locally restricted, dissolution is faster in regions of higher permeability. The increase in porosity leads to an increase in permeability. The permeability increase in turn causes higher flow rate of undersaturated solution. Due to that positive feedback loop, preferential flow paths arise.

Compensation of the concentration gradient via dispersion or diffusion counteracts the formation, because the undersaturation, which is the driving force of the process, decreases (Ormond and Ortoleva 2000).

Whether such preferential flow paths may develop, can be determined by evaluating the dimensionless Peclet (Pe) and Damköhler (Da) number.

$$\mathrm{Pe} = \frac{\mathrm{v} \cdot \mathrm{l}}{\mathrm{D}} = \frac{\mathrm{v} \cdot \mathrm{l}}{\alpha_L \cdot \mathrm{v} + \mathrm{D}_m} \tag{4.9}$$

In Eq. (4.9) v is the mean flow velocity, l the characteristic length (in this case the thickness of the system or cross section drained), and D the dispersion coefficient composed of the dispersivity α_L and the diffusion coefficient D_m. Additionally in Eq. (4.10) k_{reac} is the kinetic coefficient of the mineral reaction (compare Chap. 3).

$$\mathrm{Da} = \frac{k_{reac} \cdot \mathrm{l}}{\mathrm{v}} \tag{4.10}$$

The relation between dispersion and advection determines the preferential flow path length. If dispersion evens out the concentration within the flow path, the growth of the channel stops. The Peclet number [Eq. (4.9)] expresses the relation between advection and dispersion. With Peclet numbers above 10 preferential flow paths develop and if Peclet numbers are infinite preferential flow paths grow infinitely long.

The Damköhler number [Eq. (4.10)] describes the relation between the time a chemical reaction requires to reach equilibrium and the time the fluid needs to flow through the characteristic length. Where $\mathrm{Da} < 1$ the reaction rate cannot keep up with the advection term and local disequilibrium results. Planar reaction fronts result from $Da < 10^{-2}$ (Steefel and Lasaga 1990). When $Da >> 1$ the rate of chemical reaction is much larger than the rate at which the solute is transported by flow and the solution remains close to equilibrium. If the kinetics of the chemical dissolution and precipitation reactions are neglected, which means thermodynamic equilibrium is assumed, the Damköhler number becomes infinite.

4.3.2 Example of Preferential Flow Path Development

Kühn and Stöfen (2001) and Kühn (2003) demonstrated in numerical experiments the development of preferential flow paths. The simulations are based on laboratory experiments to determine mineral dissolution rates in anhydrite ($CaSO_4$) cemented sandstone samples (Baermann et al. 2000b). Both the numerical and laboratory experiment should serve here as an example of reaction front instability to introduce the following systematical parameter analysis.

Laboratory Experiments

An old, shut-in oil-field borehole (Allermöhe 1, near Hamburg, Germany) was deepened to install a geothermal space heating system (Baermann et al. 2000a). However, the pore space with an original porosity of up to 20 % is filled to a large extent by anhydrite. The amount of water (3 $m^3\ h^{-1}$), which can be produced from the aquifer, is insufficient for an economical use of the resource.

The experiments performed by Baermann et al. (2000b) used original anhydrite-cemented sandstone samples from the well, 8 cm long and 6.5 cm in diameter. The experiments were performed to assess the feasibility of gentle stimulation (no use of chemicals except water, no hydraulic stimulation) of the Allermöhe aquifer. They first observed only a slight permeability increase, until a preferential flow path suddenly breaks through, resulting in a three orders of magnitude increase of permeability. Mineralogical analyses showed that this permeability increase was caused by the dissolution of anhydrite and the associated increase in porosity. Fig. 4.5 (Baermann et al. 2000b, experiment P6) shows the variation of permeability, calcium concentration at the outflow, and the amount of dissolved anhydrite as a function of total water volume flooded through the core. After flooding about 2,600 mL of pure water sudden permeability increase from 0.05 mD to 50 mD occurred. Simultaneously, a decrease of the calcium concentration at the outflow is observed, from values representing almost thermodynamic equilibrium. The calcium decrease depicts that after the breakthrough either the saturation length exceeds the column length due to an increased flow velocity or the main water volume flows through parts of the core where anhydrite is no longer available.

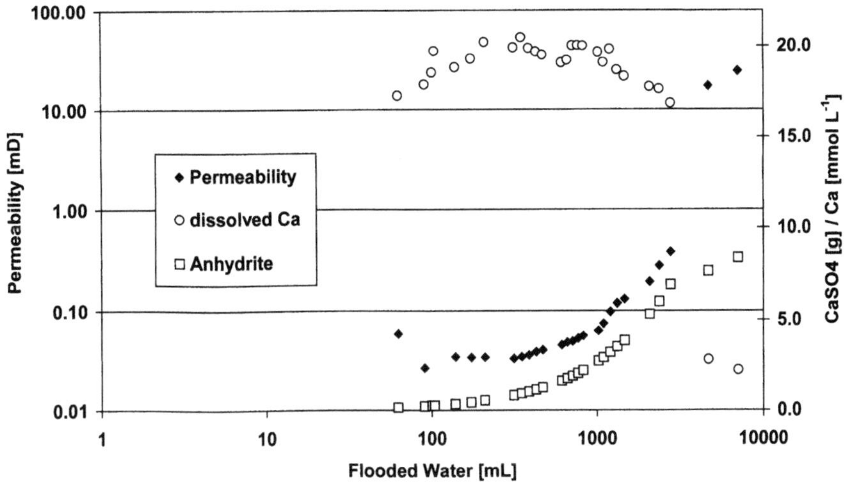

Fig. 4.5. Laboratory core flooding experiment (P6, Baermann et al. 2000b) showing breakthrough of a preferential flow path due to dissolution of $CaSO_4$ (squares); after flooding about 2,600 mL of pure water, permeability (diamonds) increases and the outflow concentration of Ca (circles) decreases

Set-Up of the Numerical Model

The problem considered here is a geometrically simple one in two dimensions (Kühn 2003) based on the laboratory investigations of Baermann et al. (2000b). Objective was to simulate isothermal flow and reaction in a medium in which some percentage of the rock is reactive (e.g. carbonate, here anhydrite cement) while the remainder is treated as inert (e.g. quartz sandstone at low temperature).

The relationship between permeability and porosity used is based on findings of Pape et al. (1999) (see above). The fractal exponent describing the k-ϕ relation in this specific core sample (P6) is 3.5, deduced from experimentally determined values of porosity and permeability [Eq. (4.3)]. An initial heterogeneity is defined in the central inflow region of the core. The remaining areas have uniform porosity and permeability.

The Peclet number of the numerical model is 81 due to the actual dispersion length. The laboratory experiment showed equilibrium concentration of calcium in the outflow of the core. Thus, kinetics of the mineral reaction is neglected, leading to an infinite Damköhler number. The prerequisites for preferential flow path formation according to Da and Pe are met.

The set-up of the numerical model is shown in Fig. 4.6. The fluid, undersaturated with respect to anhydrite, enters from the left with a net pressure drop across the modeled area. An initial heterogeneity with a porosity of 13.7 % and a permeability of 54.0 mD was defined in the central inflow region. The remaining parts of the core were assumed to be homogeneous for the numerical experiment with a porosity of 1.9 % and a permeability of 0.05 mD. Within the first 5 days of the simulation (Fig. 4.6), two preferential flow paths start to develop from the down stream corners of the initial heterogeneity

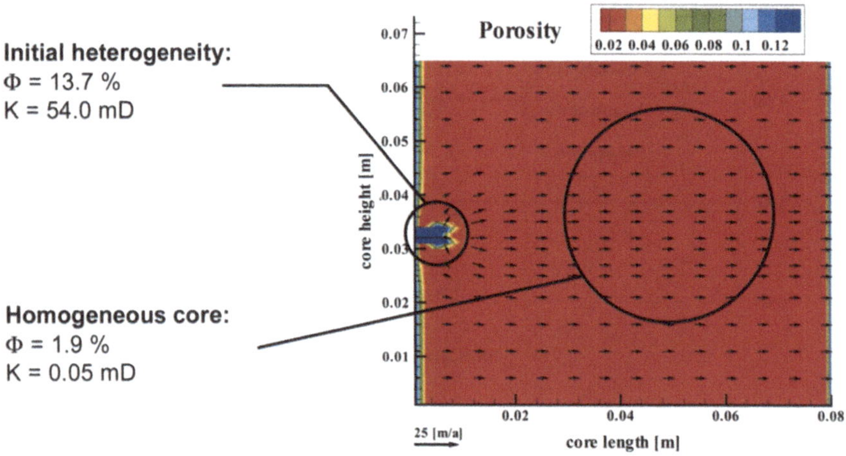

Fig. 4.6. Set-up of the numerical model; undersaturated flow enters from the left with a net pressure drop across the modeled area; an initial heterogeneity (blue contour) was defined in the central inflow region and the remaining parts of the core were assumed to be homogeneous for the numerical experiment (red contour)

Comparison of Laboratory Experiment and Simulation

As an illustration of the growth of a preferential flow path Fig. 4.7 shows the porosity distribution within the core after 20 days (top) and 45 days (bottom). At the beginning of the simulation, two preferential flow paths start growing from the two downstream corners of the initial heterogeneity. After 20 days one finger stops growing (Fig. 4.7 top), while the other one grows further and also towards the center of the core. Finally, after 45 days, this finger swallows the first one, thus forming a single preferential flow path (Fig. 4.7, bottom; Kühn and Stöfen 2001).

As the fingers grow the rate of water flooded through the core increases due to the continuously growing average core permeability (Fig. 4.8). While the numerical simulation qualitatively reproduces both the sudden permeability increase and the reduced increase rate of permeability after the breakthrough, there is an offset between the absolute permeability at the end of the simulation and the end of the laboratory experiment. The simplified assumption of a core with a mainly homogeneous permeability distribution instead of integration of an exact heterogeneity of the sample (not available) seems to be to rough. It is likely that the low permeability regions adjacent to the preferential flow path of the laboratory core participate more in the flow regime than it is the case in the numerical experiment.

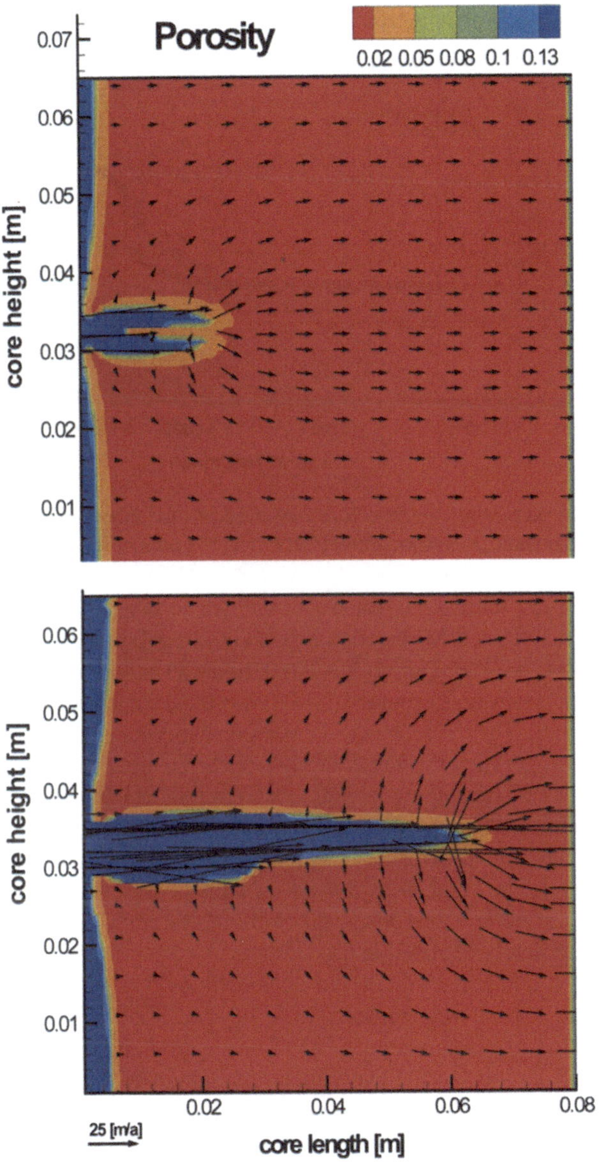

Fig. 4.7. Porosity distribution in the core after 20 days (top) and 45 days (bottom); arrows display the Darcy velocity in [m a^{-1}], scaled according to the reference arrow shown

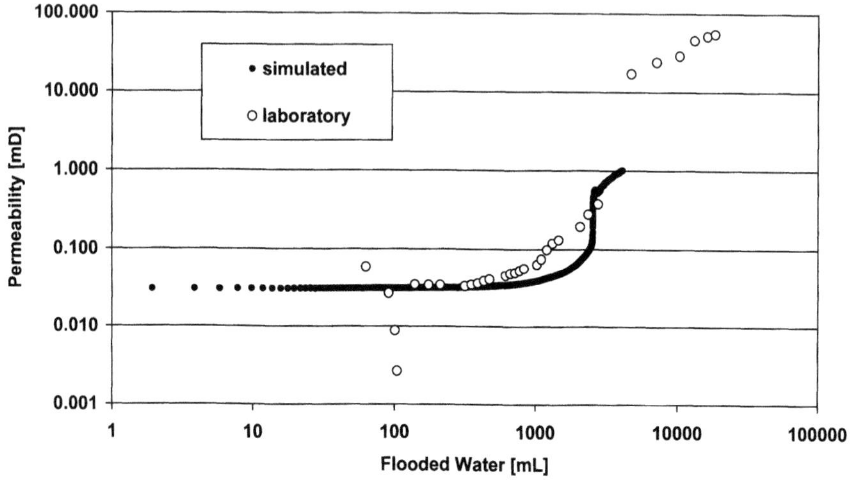

Fig. 4.8. Average permeability of the cores measured in the laboratory (open circles) and determined from the numerical simulation (full circles) as a function of total water volume

The permeability increase is caused by anhydrite dissolution in the core. The total amounts, determined for the simulation and the laboratory experiment, agree almost perfectly (Fig. 4.9). During the simulation, the calcium concentration at the core outlet is constant at 21 mmol L^{-1} (840 mg L^{-1}), corresponding to thermodynamic equilibrium. The value in the laboratory experiment is nearly identical at about 20 mmol L^{-1}. After the breakthrough occurred both the laboratory experiment and the simulation show a rapid decrease in calcium concentration, approaching a constant value of about 2.5 mmol L^{-1}. This is because water, which is still in equilibrium with anhydrite, is diluted by water flowing through the preferential flow path. This water does not get in contact with anhydrite anymore because anhydrite is completely dissolved in the preferential flow path. Therefore, this water is completely depleted in Ca^{2+} ions.

The presented numerical simulation demonstrates that even with the approximations invoked here, a very reasonable fit of core flooding experiment and numerical simulation of reaction front instabilities can be obtained. However, taking into account the duration of the experiment, the length of the core, and especially the exchanged pore volumes until breakthrough occurred it has to be recognized that a gentle stimulation of the well at Allermöhe seems to be impossible in a practical or economical time span. Even the treatment with acids is futile because anhydrite is insensitive to acid additions, conversely to calcite for example.

The comparison between the laboratory and numerically core flooding experiment evaluates the opportunity to use fully coupled reactive transport models to adequately describe the phenomenon of reactive infiltration instability. For a better understanding of the constraints prevailing preferential flow path development a systematical parameter analysis is presented in the following section.

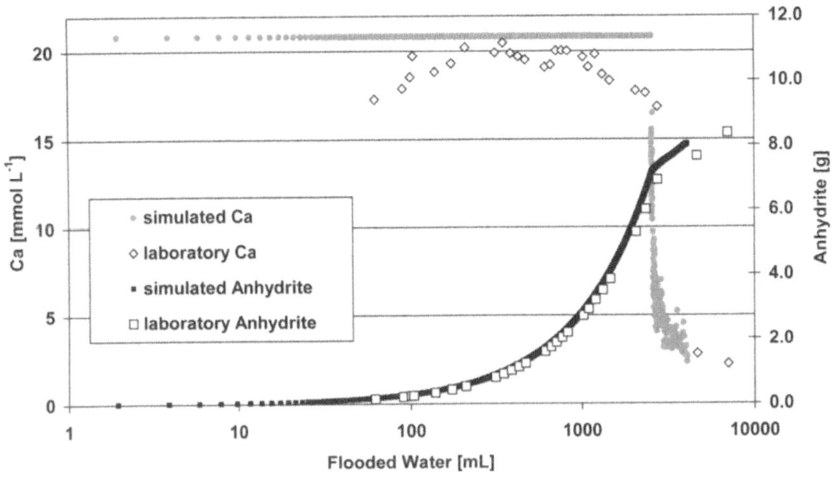

Fig. 4.9. Calcium concentration (open / full diamonds: experiment / simulation) in the solution flowing out of the core and total amount of anhydrite dissolved during the simulation (open/full squares: experiment/simulation) versus the water volume flooded through the core

4.3.3 Parameter Analysis of Reaction Front Instabilities

In the previous section the effect of coupling chemical reaction and fluid flow in space and time has been investigated using a 2D numerical simulation. Resulting evolution of rock dissolution patterns has been shown. This example focused on the non-linear, positive feedback of chemical reactions on flow, which arises through the permeability of the medium. The principal geochemical need for flow focusing in porous media is a fluid out of chemical equilibrium. In a system where dissolution of a mineral phase occurs, porosity and permeability will increase. The aftereffect within this region is an increased flow rate. A larger flux causes increased dissolution, which enlarges again porosity and permeability, which in turn attracts greater flow-through in a runaway process. This instability leads to channeling of flow and the reaction front develops in "fingers" rather than propagating as a planar front. The purpose here is to investigate the interaction among physics, chemistry, and reaction infiltration instability. It is important to note that preferential flow path can form just induced by a small heterogeneity, but in the absence of externally imposed or preexisting periodicities. They arise spontaneously from the nonlinear feedback mechanism and depend so far on the Peclet and Damköhler criteria (Ortoleva et al. 1987).

Two fundamentally different regimes can exist: (1) those characterized by transport-controlled reaction where the reaction rate constant is much faster than any of the transport processes involved (advection, diffusion, Da >> 1) so that the length scale over which a moving fluid comes to equilibrium is small. The existence and amplitude of channels in this reaction regime depend primarily on the ratio of flow velocity to the dispersion coefficient (Pe number controlled). And (2) those characterized by kinetic rate-controlled reaction where equilibrium between the fluid and the reacting mineral occurs over some distance (Da < 1, Hoefner and Fogler 1988). In this case permeability change is more diffuse. With decreasing Da the efficiency of individual channel propagation decreases.

Steefel and Lasaga (1990) investigated in their numerical simulations the propagation of a single finger dependent on the fixed ratio of advection and dispersion (Pe = 100) and dependent on the ratio between the reaction rate and advection both at a given and fixed length scale (10 m). They found that in cases of Da numbers less than 0.01 any perturbation of a planar reaction front decays away. The permeability change occurs over such a large distance that a channel cannot propagate efficiently. However, in the transport-controlled regime, when the Da number is large enough, the critical parameter is the Pe number, which ultimately determines whether a channel propagates at all. Steefel and Lasaga (1990) concluded that in cases of small advection to dispersion ratios channeling may develop only on scales recognizable by regional mapping [compare Eq. (4.9)].

Ormond and Ortoleva (2000) performed simulations at the experimental scale with large Pe and Da numbers, above 100 and 1, respectively. They actually varied the width of the domain as well as the measure and number of the initial heterogeneity. They found that size and growth rate of fingers are independent of the width of the initial heterogeneity. If the initial heterogeneity is wider than around $6.0\text{x}10^{-3}$ m two fingers develop from its downstream corners. But, those two fingers do not grow too close to each other. In that case one finger stops growing and the remaining finger becomes wider (compare example of Allermöhe core flooding). Their second result is that the final finger width is independent on the measure or number of initial heterogeneities but ≈ 4 times the thickness of the reaction front (≡ saturation length). Investigations of a homogeneous matrix perturbed by an initial white noise at the domain's inlet revealed that several fingers start to grow. The elongation of every finger depends in the following on the competition between them to capture the fluid flow. Finally, width and number of developing preferential flow paths are determined by the flow and reaction characteristics of the system (Pe and Da). The progress of any finger itself is directly related to the magnitude of flow. However, the length of an elongated channel depends on the balance between advection and dispersion: If dispersion equilibrates concentration inside the channel before the fluid reaches the channel tip, growth stops.

Based on the systematic studies of Steefel and Lasage (1990) and Ormond and Ortoleva (2000) the interaction of the Pe and Da numbers on the development of preferential flow path is investigated here.

Model Components for the Parameter Analysis

Concerning rock properties, the problem formulated here for the parameter analysis is essentially the same as the one considered above (Allermöhe core P6). Thus, preferential flow path development within the Allermöhe cores and in general is understood in more detail.

The simulations conducted comprise a range of dimensionless numbers. The Peclet number varies between 1 and 500 and the Damköhler number from 0.003 to infinite (thermodynamic equilibrium). The systems investigated measured 2.0 respectively 3.0 cm in height. The model length is 8.0 cm for all studies and they are discretized in cells measuring 1 by 1 mm. The systems investigated are listed in Table 4.1.

Table 4.1. Systems used for the parameter analysis, characterized by the total number of cells, Pe and Da numbers, and the applied heterogeneity; cell dimensions are 1 x 1 mm in all cases

Number of cells	Pe number	Da number	Heterogeneity
20 x 80	1, 2, 5, 10, 20, 30, 40, 50, 60, 70, 80, 90, 100, 200	0.003, 0.03, 0.06, 0.09, 0.12, 0.15, 0.3, 0.6, 0.9, 1.2, 3.0, ∞	0.002 m x 0.004 m
30 x 80	3, 5, 10, 30, 50, 100, 300, 500	0.15, 0.3, 1.5, 3.0, ∞	white noise

The investigated systems are used to conduct simulations for different purposes. Systematic studies are done to get insight in the following topics:

- Development of one channel induced by a single heterogeneity measuring 2 x 4 mm at the center of the inflow within a homogeneous core (20 x 80).
- Onset of reaction front instabilities, development either of a planar front or preferential flow paths, dependent on the Peclet and Damköhler numbers induced by a heterogeneous distribution of the core inlet or within the entire region of the core (30 x 80).

Development of One Channel

The first study done with varying Damköhler (Da) and Peclet (Pe) numbers is to show the development of a single channel from a single rectangular heterogeneity. The heterogeneity is placed at the core inlet (no anhydrite, 2 x 4 mm) within a homogeneous system of 20 x 80 cells measuring 1 x 1 mm each. Da numbers investigated and displayed in Fig. 4.10 and Fig. 4.11 are 0.03, 0.3, 3.0, and infinite in combination with Pe numbers 1, 5, 10, 20, 30, 50, 70, and 100.

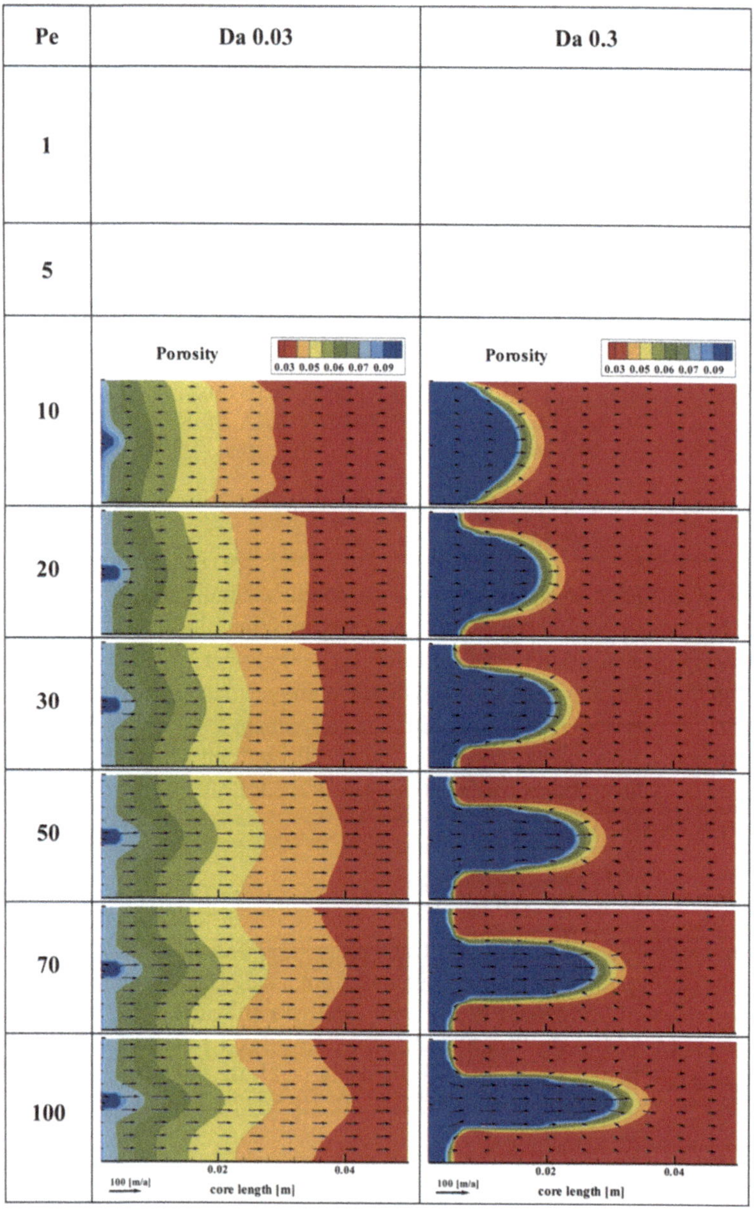

Fig. 4.10. Development of preferential flow paths due to mineral dissolution dependent on Pe and Da from a 2 mm x 4 mm heterogeneity in a column homogeneously filled by anhydrite; contours display porosity between 0.02 and 0.1 and the arrows flow direction and magnitude of flow

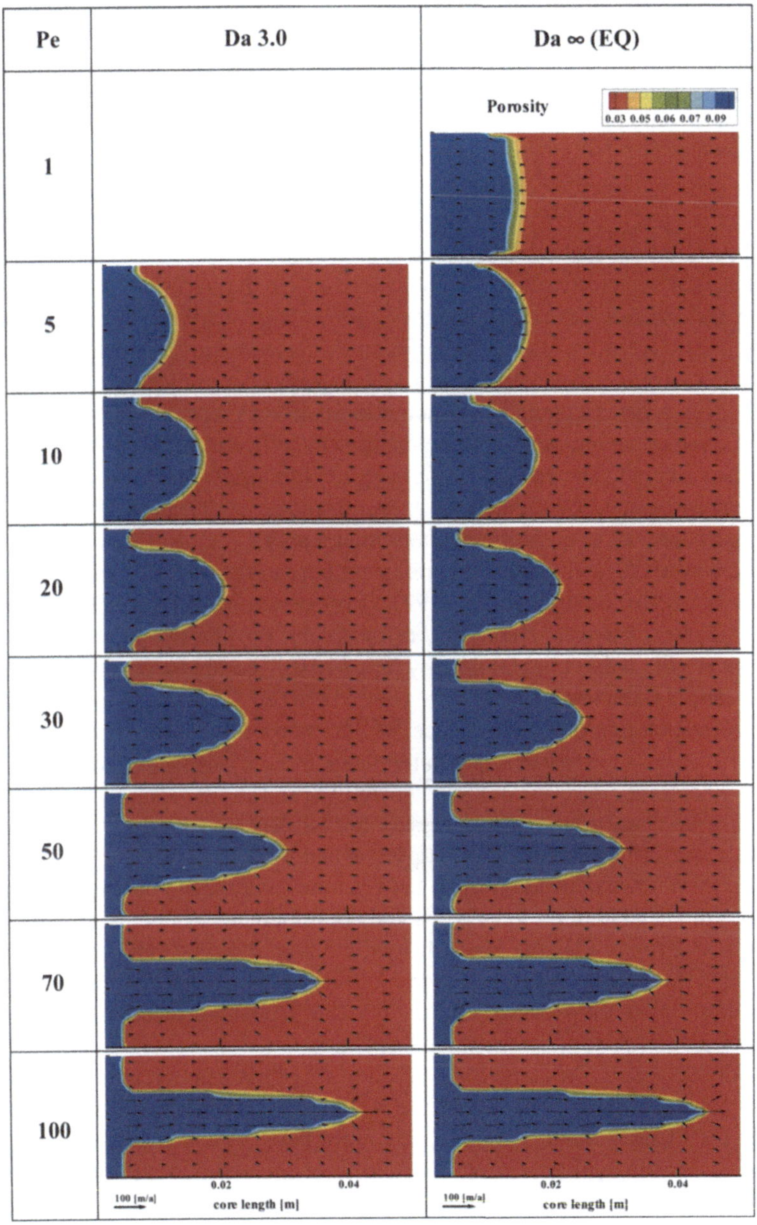

Fig. 4.11. Development of preferential flow paths due to mineral dissolution dependent on Pe and Da from a 2 mm x 4 mm heterogeneity in a column homogeneously filled by anhydrite; contours display porosity between 0.02 and 0.1 and the arrows flow direction and magnitude of flow

Fig. 4.10 and Fig. 4.11 show the development of the channel due to anhydrite dissolution. Dissolution patterns are depicted by the contour colors representing the porosity (red = initial, blue = final). The initial mineral amount within the simulated core is 1500 mol m^{-3}. The simulated time span with totally 8 days is within the order of magnitude of the example of the previous section. All numerical experiments lasted for the same time to enable direct comparison between the simulations. On the contrary to the example above the infiltrated water is brine with a sodium chloride content of 1.7 mol L^{-1}. This is to speed up dissolution and with it the growth of the preferential flow path, because the solubility of anhydrite is at maximum for that salinity (Kühn et al. 2002b).

This first systematic study is characterized by the variation of the dispersion length and the reaction rate (compare Chap. 3) to get varying Pe and Da numbers, respectively. Recalling Eq. (4.9) it becomes obvious that the dispersion length is the only variable quantity for such a system with a fixed hydraulic gradient and a fixed characteristic length. This assumption is true as long as the investigated system is advection-controlled and the diffusion coefficient is small compared to the dispersion length. The characteristic length is the inflow area of the core, the area that can be drained by the system or the thickness of the zone wherein the fluid and matrix are out of equilibrium (saturation length).

The applied dispersion lengths, referring here to the microdispersivity in correspondence with Xu and Eckstein (1997), vary between 0.02 and $2x10^{-4}$ m resulting in Pe numbers between 1 and 100, respectively. For the Da number the reaction rate is here the only variable quantity [Eq. (4.10)]. To obtain Da numbers between $3x10^{-3}$ and 3.0 the reaction rate was varied between $1x10^{-8}$ and $1x10^{-5}$ mol m^{-2} s^{-1}. The corresponding internal surface of the porous medium is assumed to be $0.35x10^{6}$ m^2 m^{-3} ($\approx$ 149 m^2 kg^{-1}) determined from Allermöhe sandstone samples. The concentration of Ca^{2+} and SO_4^{2-} in equilibrium is 63.1 mmol L^{-1} at 25°C (surface and ion concentration are necessary to calculate k_{reac}, Eq. (4.10), in dimension of [s^{-1}]). An infinite Da number is given when the reaction rate is infinite, representing thermodynamic equilibrium between the solution and the mineral phase.

The initial conditions force the active development of one single finger. It is obvious from Fig. 4.10 and Fig. 4.11 that the higher the Pe number the longer but thinner is the developing finger. If the Pe number is smaller than 10 the development of a channel is hindered and the reaction front spreads across the entire domain width and thereby remains almost planar.

With decreasing Da number the reaction front spreads over an increasing area. In the case of Da numbers of 0.03 the width of the reaction front spreads via half of the core length, what is significantly more than the characteristic length of the modeled domain, and as a result, reaction front fingering does not occur. With Da numbers of 3.0 the reaction front does not spread out more than for the case of infinite Da numbers. Hence, Da numbers of 3.0 and above represent already the transport-controlled reaction regime.

In Fig. 4.12 the preferential flow path length is shown dependent on the Da number after the total simulated time of 8 days for each calculation. Displayed are the flow path lengths of simulations with varying Da and Pe numbers. It can be seen that the development of fingers start with Da numbers above 0.03 and that their length increases with increasing Pe numbers.

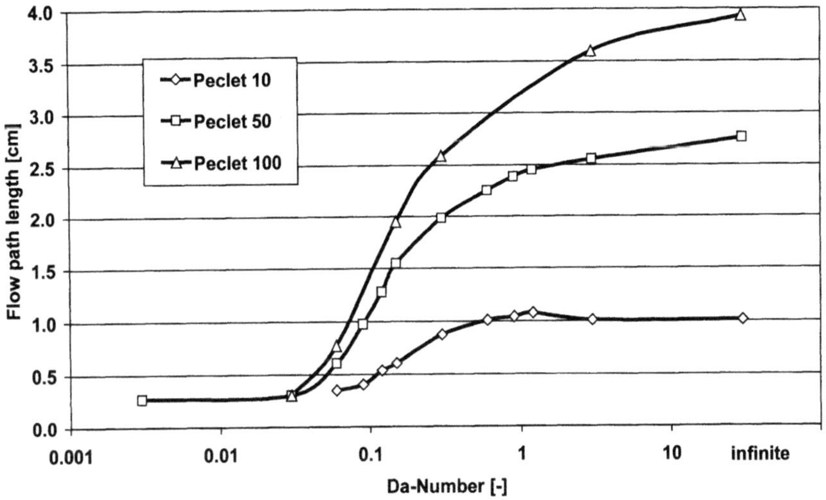

Fig. 4.12. Preferential flow path length versus Damköhler number; at Damköhler numbers below 0.07 channels do not propagate

Steefel and Lasaga (1990) observed in their numerical simulations with Pe 100, that the channel length in the transport-controlled regime is independent of the reaction rate. Within the calculation results shown here this seems only to be true for Pe 10. But in the cases of Pe 50 and 100 the channel lengths are not constant for high Da numbers (transport-controlled regime). Additionally, they stated that a Da number increasing from 0.01 to 0.1 results in significantly decreasing channel lengths (≈ 30%) what cannot be observed here.

With respect to the characteristic, dimensionless Pe and Da numbers it is anticipated that observed channel development should be the same in numerical experiments where both numbers are applied with identical values. However, the results according to the flow path length presented here and published by Steefel and Lasaga (1990) deviate significantly as mentioned above. Explication for this finding is the fact that the Pe and Da numbers [Eq. (4.9)and Eq. (4.10)] are solely defining the range within which reaction front instability occurs at all. In contrast, the preferential flow path growth rate depends on the actual dissolvable mineral amount, the applied porosity-permeability relation [compare Eqs. (4.3) - (4.8)], and the degree of undersaturation of the up-stream water with respect to the reactive mineral phase. Whereas the degree of disequilibrium of the up-stream water is at its maximum in both investigations and Steefel and Lasaga (1990) used a comparable porosity-permeability relation, a smaller amount of reactive cement mineral (5 % compared to 7.3 %) has been applied. This conceptual difference is the reason for the discrepancies between the described numerical case studies.

Onset of Reaction Front Instabilities

In the following the system with 30 x 80 cells is studied in more detail. Simulations are done with a homogeneous matrix perturbed by an initial white noise on the permeability and porosity distribution at the core inlet (identical for all simulations performed). The calculations are performed for Da numbers between 0.03 and infinite and Pe numbers between 1 and 500. The resulting fingers (preferential flow paths) for the maximum and minimum Damköhler and Peclect numbers, respectively, are shown in Fig. 4.13.

As already seen above, a Pe number of 1 is too small for finger growth even if the Da number is infinite. That applies also for Da numbers of 0.03 (Fig. 4.13). The reaction front spreads over a distance larger than the characteristic length resulting in a planar reaction front. Even with Pe 500 the reaction front remains planar. A Pe number of 500 and an infinite Da number results in the evolution of two long and thin fingers.

Fig. 4.14 and Fig. 4.15 exhibit that a Pe number greater than 10 is essential for preferential flow path development. With Pe numbers smaller or equal to 10 the resulting reaction front is planar. As observed earlier the value of the Da number determines the width of the reaction front, which is identical with the saturation length or with the distance the fluid remains in disequilibrium concerning the mineral phase anhydrite. Additionally, it can be seen, that with increasing Da number the efficiency of the propagation of the fingers increases.

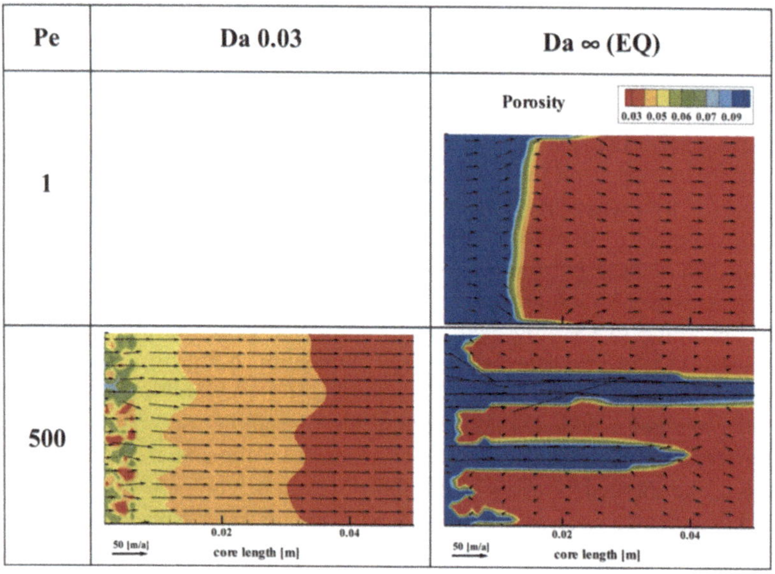

Fig. 4.13. Preferential flow path developments in the system measuring 3.0 cm x 8.0 cm with a heterogeneous mineral distribution at the core inlet; displayed are the simulations with the maximum and minimum Pe and Da numbers

Generally, it can be observed that the finger length increases with increasing Pe number. Simultaneously, the finger width is decreasing. The higher the Pe number the thinner are the developing preferential flow paths.

The first and necessary requirement for preferential flow path development is a Peclet number above ten. If this is the case a reactive flow system is eligible for self-organized reaction front instabilities resulting in channel development. The second but self-sufficient requirement, the Damköhler number, defines the scale on which fingering reactions are possible. The characteristic length of a system, identical with the maximum drainable area, must be larger than the width of the reaction front or the saturation length of the particular reaction. In other words, if the reaction rate is small, the characteristic length must be sufficiently large (e.g. in this case the system must be of adequate height) to get a Da number in the system under consideration that is at least above 0.03 (Fig. 4.12).

Within the numerical experiments with Pe numbers smaller or equal to ten and Da numbers of 1.5 and 3.0 (Fig. 4.15) it can be observed that fingering seems to occur at the upper and lower boundary of the model, although preferential flow paths should not develop (Pe $\leq$ 10, see above). To further investigate this phenomenon, additional numerical experiments were conducted. The upper and lower model boundaries were assumed as impervious so far, according to a laboratory core flooding experiment. Fig. 4.18 displays the results of two numerical simulations where the overall time has been split into four periods of two days each. Conversely to the previous calculations one simulation was performed with "enabled" dispersive and diffusive transport via the model boundaries. Such an experiment corresponds to a natural system with a permeable layer surrounded by material of low permeability but not impervious. Both simulations were conducted with a Pe number of ten and an infinite Da number. In comparison to the previously performed simulations with totally impervious boundaries there is no fingering at the model edges observable in this case (Fig. 4.18). Hence, it can be inferred that impervious boundaries, leading to decreased dispersivities and resulting in increased Pe numbers [Eq. (4.9)], might cause fingering processes directly at the border of the modeled system. This indicates especially for laboratory core flooding experiments that channel development at the margins has to be taken into account.

Finally, simulations of preferential flow path development were done in a fully heterogeneous system according to the mineral, porosity, and permeability distribution (Fig. 4.16 and Fig. 4.17). The same number of preferential flow paths develops at identical locality. The only difference to the previous observations and conclusions is the fact that the increased heterogeneity of the cores leads to a decreased efficiency of the channel propagation. The fingering process is slower, another argument against the feasibility of gentle stimulation of the Allermöhe well (see core flooding experiment above)

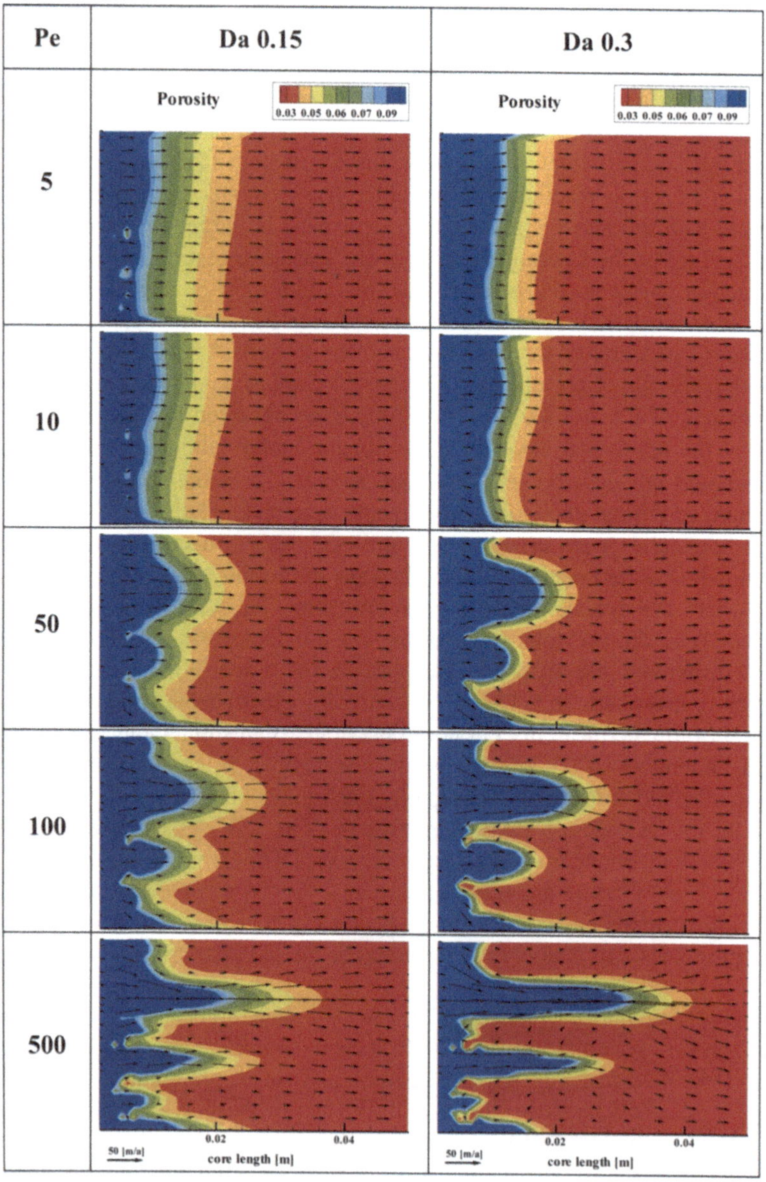

Fig. 4.14. Development of preferential flow paths due to mineral dissolution dependent on Pe and Da from a heterogeneous mineral distribution (white noise) at the inlet; the column is homogeneously filled by anhydrite; contours display porosity between 0.02 and 0.1 and arrows flow direction and magnitude of flow

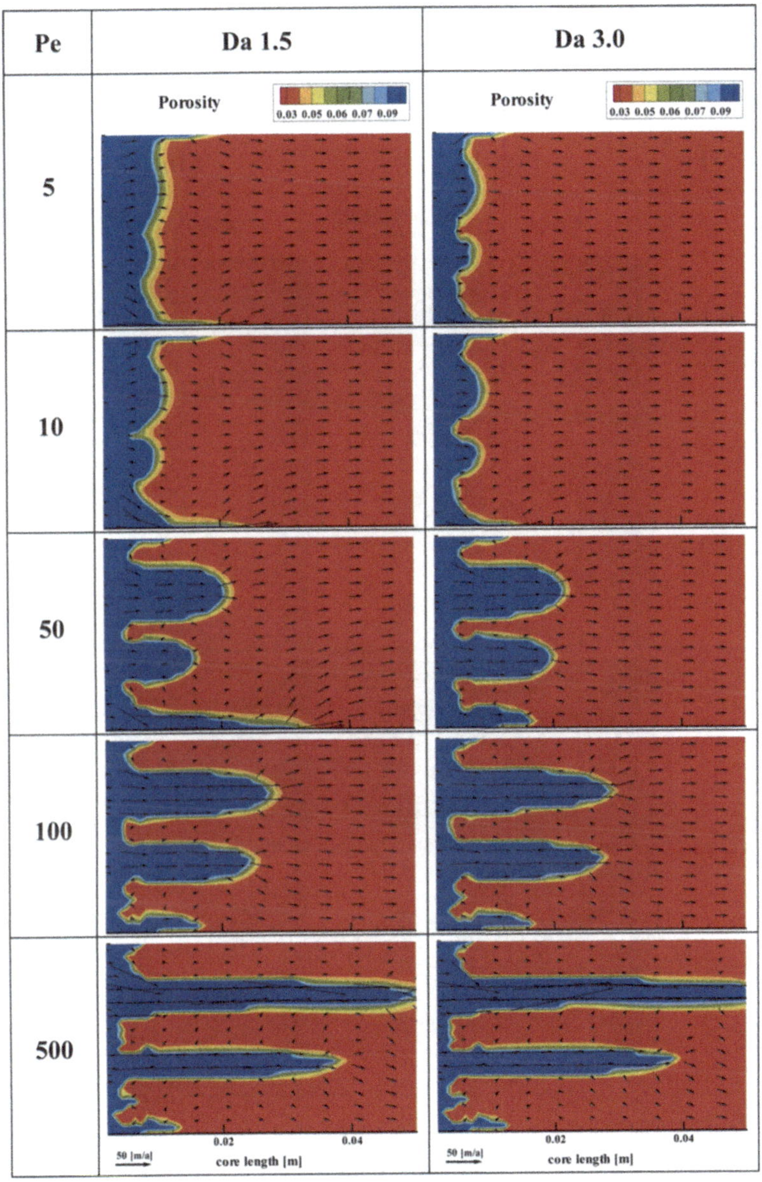

Fig. 4.15. Development of preferential flow paths due to mineral dissolution dependent on Pe and Da from a heterogeneous mineral distribution (white noise) at the inlet; the column is homogeneously filled by anhydrite; contours display porosity between 0.02 and 0.1 and arrows flow direction and magnitude of flow

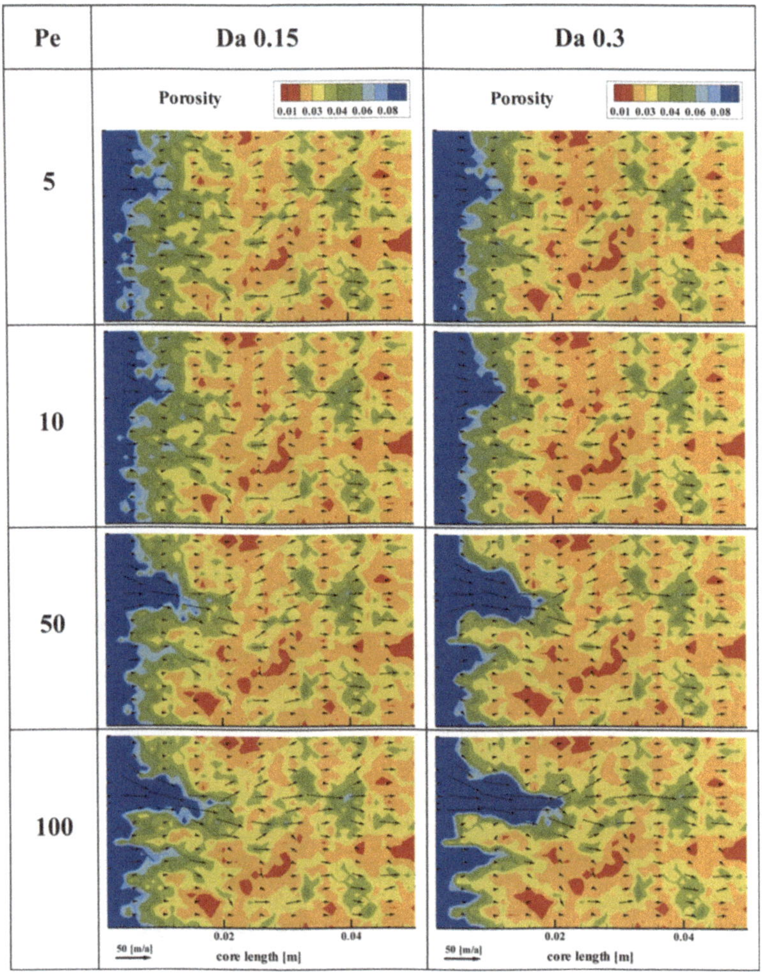

Fig. 4.16. Development of preferential flow paths due to mineral dissolution dependent on Pe and Da from a heterogeneous mineral distribution

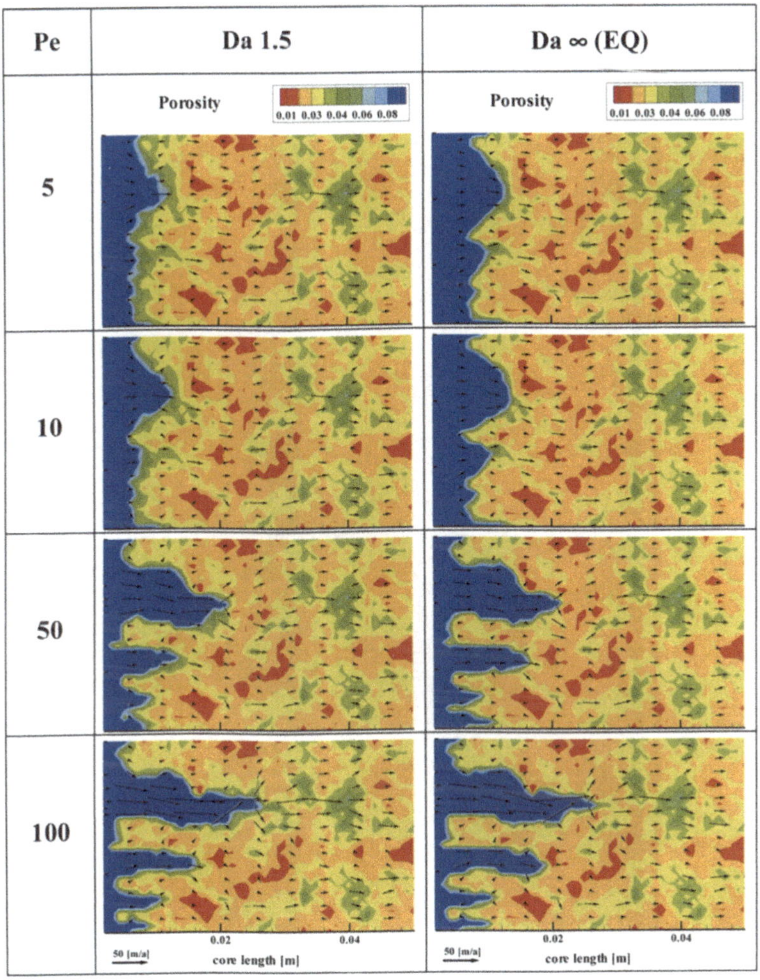

Fig. 4.17. Development of preferential flow paths due to mineral dissolution dependent on Pe and Da from a heterogeneous mineral distribution

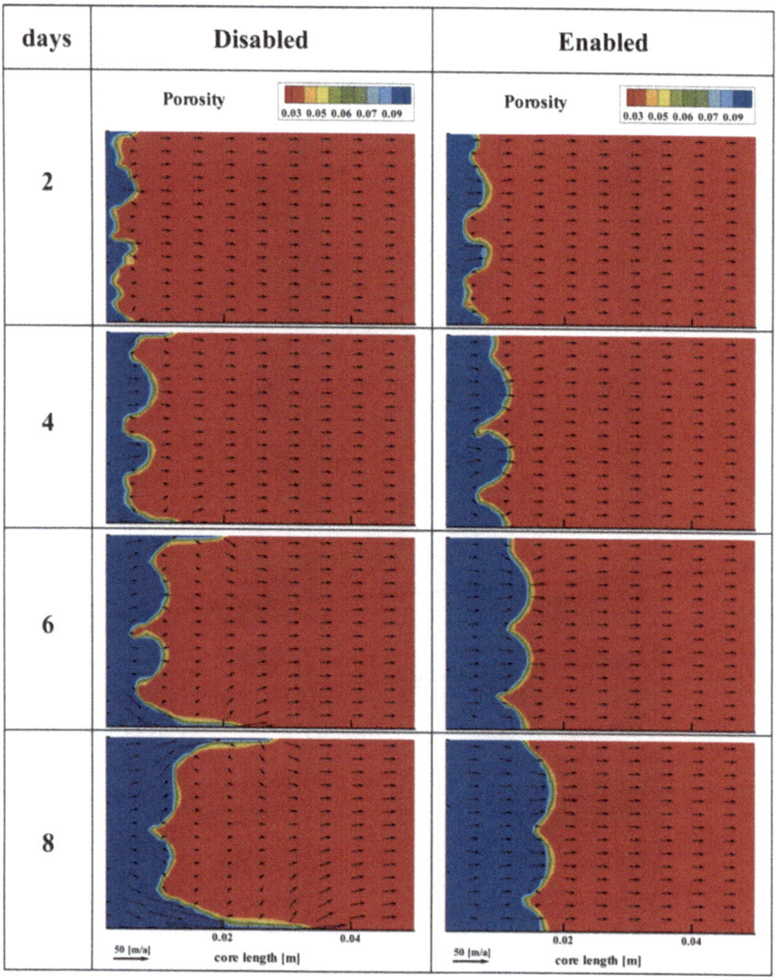

Fig. 4.18. Development of preferential flow path in the system Pe 10 and Da ∞ from a heterogeneous mineral distribution at the core inlet, split into four periods of 2 days each; results are shown with "enabled" and "disabled" dispersion via the boundaries

4.4 Thermal Convection

Several driving mechanisms for large-scale fluid flow in sedimentary basins have been proposed, including: (1) topography- or gravity driven flow (Garven and Freeze 1984); (2) compaction-driven flow during basin subsidence (Cathles and Smith 1983, Bethke 1985); (3) seismic pumping and tectonically driven flow (Sibson et al. 1975, Oliver 1986); and (4) buoyancy-driven flow, including thermal driven or free convection (Cathles 1981, Bjorlykke et al. 1988).

Muffler (1985) considered free thermal convection to be a potential mechanism for mass and heat transport in sedimentary basins. Diagenetic processes in sedimentary basins involve reactions between pore water and mineral phases during which unstable minerals are dissolved and more stable phases are precipitated. These reactions are controlled by thermodynamic stability and reaction kinetics. Spatial variations in temperature and pressure are assumed to be responsible for much of the cementation and dissolution observed in rocks at depth. Muffler (1985) presumed that fluid transport by thermal convection might operate under such gradients to enhance diagenetic cementation processes. Theoretical investigations of heat transport in sandstone layers indicate that common geometries related to geologic structure can give rise to internal convection cells. Most geothermal development to date for example has been carried out in hydrothermal convection systems. Near-surface temperatures are increased in hydrothermal systems by the movement of hot water through pores and fractures (Muffler 1985).

Wood and Hewett (1982) showed that eddy currents of large scale (km) spontaneously arise and persist in porous, fluid-saturated geologic formations, when these systems are subjected to normal geothermal temperature gradients (25-30°C km^{-1}). They assume that the velocity of these fluid currents is on the order of 1 m per year. Such a mass flux may produce significant porosity changes when it prolongs over a period of several million years. Hence, post-depositional reservoir cementation might be due to slowly circulating fluids.

Bjorlykke et al. (1988) studied the onset of free convection in a layered system and concluded that even small (< 1 m) low-permeability layers could effectively split a system and inhibit free convection systems. On the contrary to the propositions of Muffler (1985) and Wood and Hewett (1982) it is currently accepted that free convection is unlikely to occur in most sedimentary basins except under certain unique conditions, such as very high basal heat fluxes or unusually thick and permeable formations (Bjorlykke et al. 1988, Raffensberger and Garven 1995), or in the vicinity of igneous intrusions (Norton and Knight 1977) or salt domes (Hanor 1987, Evans and Nunn 1989). However, in the following section (compare Chap. 5) it will be shown that free convection has to be distinguished into "vertical" and "horizontal" currents. Although "vertical" free convection is most unlikely to occur, free "horizontal" convection has to be taken into account under particular geologic circumstances. Nevertheless, the principles of "vertical" convection systems will be described here to clarify how diagenetic reactions (e.g. cementation) can be explained by reactive transport due to buoyancy-driven flow.

4.4.1 Rayleigh Number

The dimensionless Rayleigh number Ra [Eq. (4.11)] indicates the tendency towards free convection, that is, flow driven purely by density differences. Classic Rayleigh convection theory was developed in the context of an infinite, permeable, horizontal layer bounded at top and bottom by isothermal, impermeable formations. The Rayleigh number is based on the ratio of "buoyant" forces that drive convective fluid flow to the viscous forces inhibiting fluid movement. The value of Ra is given by:

$$\mathrm{Ra} = \frac{\alpha_W \cdot \rho_W^2 \cdot c_W \cdot g \cdot k \cdot L \cdot (T_L - T_U)}{\mu_W \cdot \lambda_m} \tag{4.11}$$

where α_W is thermal expansivity of water (°C^{-1}), ρ_W density of water (kg m^{-3}), c_W isochoric heat capacity (J kg^{-1} K^{-1}), g gravitational acceleration (m s^{-2}), k intrinsic permeability (m^2), L characteristic length (m, formation thickness), T temperature (K, L: lower boundary, U: upper boundary), μ_W dynamic viscosity (kg m^{-1} s^{-1}), and λ_m thermal conductivity of the medium (J s^{-1} m^{-1} K^{-1}, rock formation). Lapwood (1948) showed that the fluid in an infinite horizontal layer would begin to convect at a critical Ra value of $4\pi^2$. There is no critical Ra number for any non-isothermal sloping layer, and all such layers should have fluids circulating at some finite velocity (Ingebritsen and Sanford 1998).

Sorey (1978) noted that if (T_L-T_U) is large enough, the values of relevant fluid properties (α_W, ρ_W, c_W, μ_W) as evaluated at T_L and T_U will be significantly different and the corresponding Ra values at T_L and T_U will be as large as 60 and low as 2, respectively. In these cases a mean Ra number has to be taken into account.

Raffensberger and Vlassopoulos (1999) performed a series of numerical simulations to investigate the influence of varying thicknesses and temperature gradients on the Ra number and resulting flow rates due to occurring free convection. They determined that the correlation between the flow rate q_{free} (m yr^{-1}) and the mean Ra number is best described by a second-order polynomial relationship:

$$q_{free} = 1.2\left(\frac{\sqrt{\mathrm{Ra} - \mathrm{Ra}_c}}{L^{0.375}}\right)^2 + 0.17\left(\frac{\sqrt{\mathrm{Ra} - \mathrm{Ra}_c}}{L^{0.375}}\right) \tag{4.12}$$

where Ra_c is the critical mean Rayleigh number for the onset of free convection ($4\pi^2$=39.478). This relationship allows for predicting the maximum flow rate within an aquifer as a function of layer thickness, intrinsic permeability, viscosity, and thermal properties of the fluid and the rock formation.

4.4.2 Relevance to Diagenesis

Spatial variations in temperature and pressure are assumed to be responsible for causing much of the cementation and dissolution that occurs in rocks at depth. Fluid transport by thermal convection may operate under such gradients to enhance for example cementation of quartz sandstones. Significant temperature gradients within the circulating cells will cause differences in the equilibrium state of the fluid with respect to different minerals. Silica equilibrium, for example, is prograde with respect to temperature below 350°C (Fournier and Potter 1982). Under chemical equilibrium, warmer regions will be undersaturated and dissolve silica, whereas cooler regions will be supersaturated and precipitate silica. Minerals with a retrograde solubility, like anhydrite, will exhibit the reverse pattern.

A convection cell alike to the one of Elder (1967) has been used to exemplarily investigate the reactive transport processes of the minerals quartz and anhydrite according to their diagenetic reaction potential. The numerical simulations were performed with SHEMAT. Fig. 4.19 displays the steady-state temperature distribution within the 2D vertical model measuring 0.3 m in width and 0.1 m in height. The system is cooled from the top with a temperature of 20°C and heated from the bottom with 50°C. The flow velocities (arrows) exhibit the establishment of two convection cells.

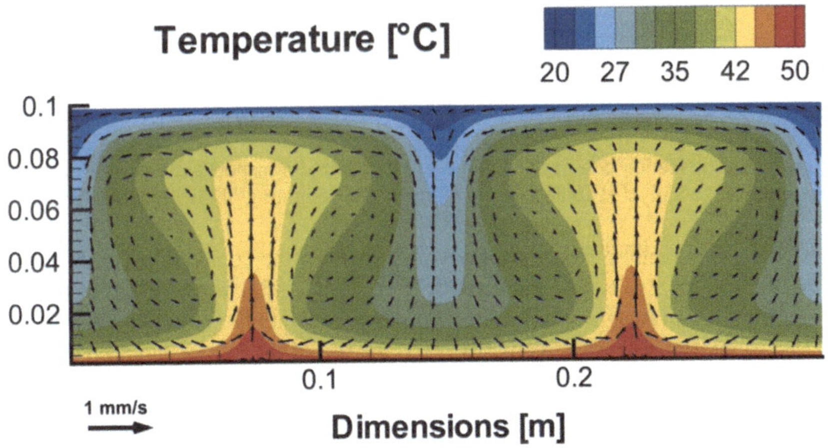

Fig. 4.19. Temperature distribution within the convection cell alike to the experiment of Elder (1967); contour colors display the temperature distribution resulting from cooling with 20°C at the top and heating with 50°C at the bottom of the model; arrows exhibit flow direction and magnitude of flow according to the reference vector shown

The numerical experiment (Fig. 4.19) has been used as a basis to determine how quartz and anhydrite as prograde and retrograde dissolving minerals, respectively, will react dependent on the influence of a convection cell assumed to reflect a hydrothermal reservoir.

Fig. 4.20 shows the quartz distribution after a certain time span. It can be seen that the initially and evenly distributed amount of quartz (green contour) has been relocated. Because quartz is more soluble in hot water than in cold water, the mineral is dissolved (blue contour) at the bottom of the convection cell and silica is transported through the up-flow zones. Ascending within the up-flow zone the water cools and becomes finally supersaturated in silica and quartz precipitates at the upper boundary of the model (red contour). The process described here shows how the transport of silica within a hydrothermal reservoir may lead to sealing (quartz precipitation) at the top of the permeable layer.

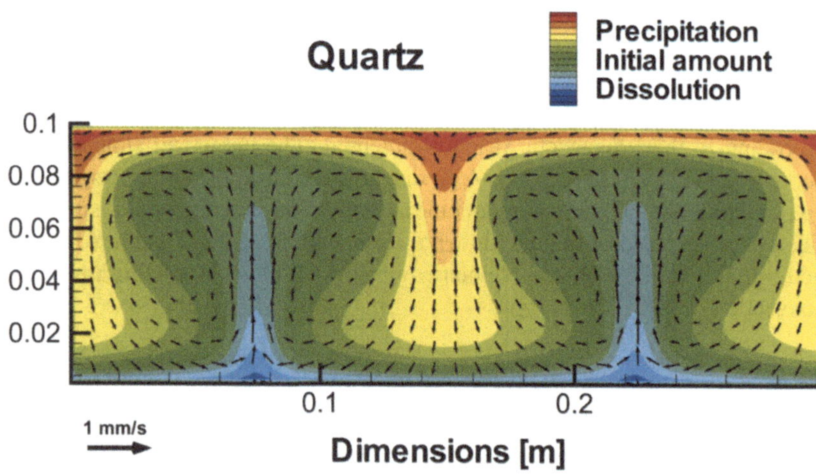

Fig. 4.20. Convection cell in combination with reactive transport processes of quartz; mineral has been evenly distributed (green color) at the beginning of the numerical experiment

The corresponding reaction of anhydrite within the convection cell is shown in Fig. 4.21. The results are displayed in a sequence after three different time periods. Due to its retrograde solubility anhydrite exhibits the opposite behavior compared to quartz. Anhydrite is dissolved at the top of the model, calcium and sulfate are transported with the water, and subsequently anhydrite precipitates at the bottom of the model area.

With proceeding simulation the anhydrite distribution changes significantly. It can be observed that parts of the model are totally freed from anhydrite (blue) and other regions are highly enriched with anhydrite (red). Whereas reactive transport of silica in a hydrothermal system may lead to quartz cementation at the top of the formation, anhydrite cementation occurs at the bottom of the layer. The simulations were conducted with feed back of reaction on flow. As a result of precipitation and dissolution the accompanying porosity changes led to permeability changes and therefore to a significantly changed flow field. The primarily established symmetric convection system (Fig. 4.19), spread out over the entire model, developed towards several smaller structures within the areas of higher permeability due to dissolved anhydrite (blue).

Fig. 4.21 displays specific patterns of mineral distribution created by the flow field of the investigated idealized hydrothermal system. It can be concluded that free thermal convection may lead to relocation of minerals and in turn due to the feed back of reaction on the flow field the structure of the convection cell changes. The direct consequence of the changing flow field is a significant variation of the temperature distribution within the model area (Fig. 4.22).

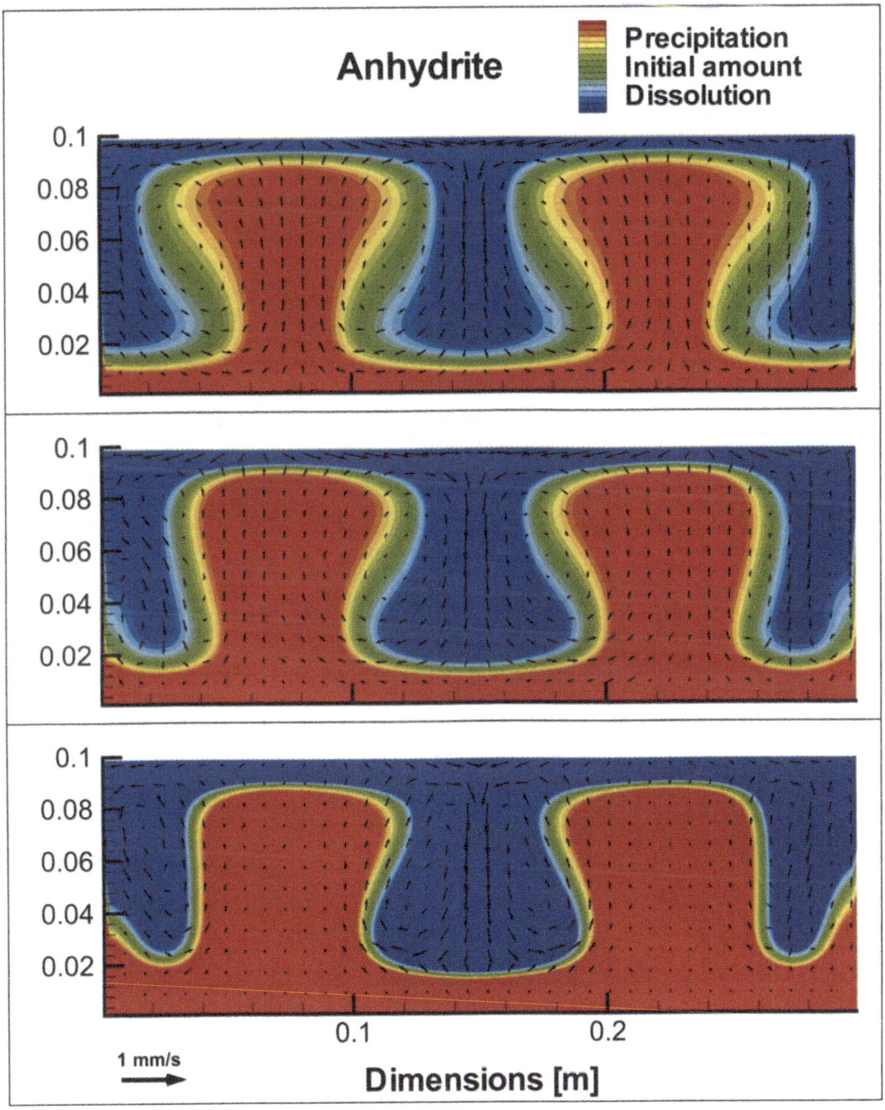

Fig. 4.21. Convection cell in combination with reactive transport processes of anhydrite after 30 (top), 150 (center), and 300 (bottom) minutes; mineral has been evenly distributed at the beginning of the numerical experiment

Finally it can be said that a fully coupled reactive transport model of a hydrothermal system provides new insights and deeper understanding of the processes occurring due to fluid flow, heat transfer, solute transport, and chemical reactions. Reactive transport simulation provides a tool for the detailed investigation of fossil and recent hydrothermal systems (compare following chapters).

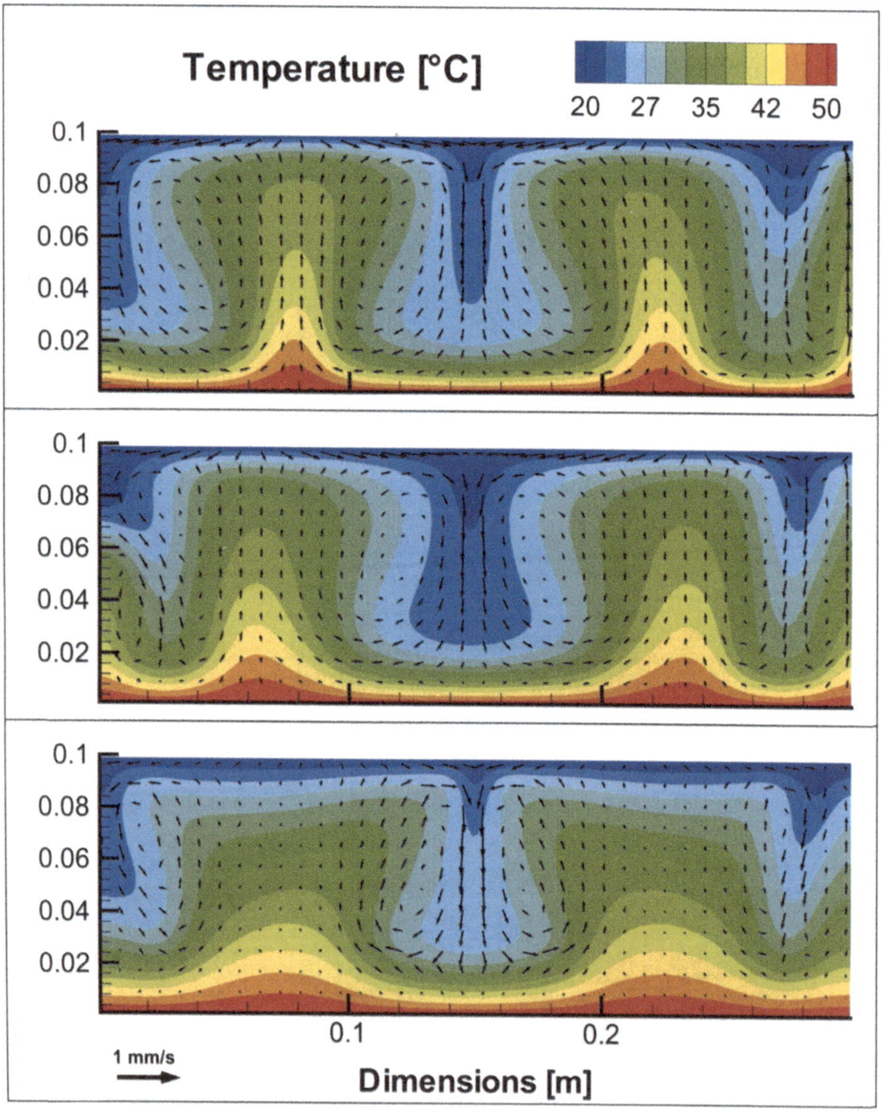

Fig. 4.22. Temperature development within the convection cell due to flow field variation as a result of anhydrite precipitation and dissolution after 30 (top), 150 (center), and 300 (bottom) minutes

5 Fossil Hydrothermal Systems

Reactive transport modeling of the history of fossil hydrothermal systems provides the basis for understanding genesis of ore deposits as well as progress of diagenetic processes. The first part of this chapter presents a brief overview about applications of numerical simulations done by other authors investigating these topics. This is followed by a detailed examination of possible formation scenarios for the observed and considerable anhydrite cementation found at the location Allermöhe in Germany. For that purpose the SHEMAT software has been used in order to numerically determine the thermal-reactive flow-deformational history of the site. In order to explain the observed anhydrite cementation against the background of the entire geologic history of the Allermöhe site the fluid flow, heat transfer, transport, and chemical reaction model has been coupled with a sequence of geologic structures reflecting the stratigraphic development of the site.

5.1 Ore Deposits and Diagenesis

5.1.1 Ore Deposits

Most economically significant ore deposits exist because of the advective transport of solutes and heat by flowing groundwater. Mobilization, transport, and deposition of chemical species are all linked to fluid flow and most often to fossil hydrothermal systems.

Reed (1983) applied a polythermal reaction model (see Chap. 3) to establish a genetic link between massive sulfide deposits resting on metamorphic volcanic rocks. Hereby, the reaction of heated seawater with the rocks of a volcanic pile is assumed to be responsible for regional greenschist grade metamorphism. The solution resulting from rock-water interaction is the source for copper, zinc, and iron sulfide ore deposition in and on the basaltic rocks.

Lu et al. (1992) used a titration model (see Chap. 3) to interpret zinc-lead skarn mineralization at Tin Creek, Alaska. The calculations were carried out using the computer program CHILLER by stepwise titration of reactant rock into the fluid at 300°C. The concept of the model can be regarded as a single reaction front where early-formed minerals can back-react because the fluid does not move away from the reaction site. Lu et al. (1992) concluded that the thermal gradient, progressive fluid reactions, and continuous interaction and dilution of hydrothermal fluids might be responsible for the development of skarn zonation in the Tin Creek area.

Plumlee (1994) used the program CHILLER to study fluid chemistry evolution and mineral deposition in the main-stage Creede epithermal system. The reaction path calculations showed observed mineralogical variations are best accounted for by boiling of the hydrothermal brines, followed by lateral mixing with overlying dilute, steam-heated groundwaters. Conclusion is that epithermal mineral assemblages and zoning patterns can be used to reconstruct the paleohydrology of hydrothermal systems.

The Ozark region of the North American mid-continent hosts a number of Mississippi Valley-Type (MVT) ore deposits. Most of these MVT deposits formed from enormous hydrothermal systems in which fluid flow was driven topographically and tectonically. Beside a comprehensive discussion of host rocks, mineralogy, and alteration processes of several districts from the Ozark region, Plumlee et al. (1995b) performed reaction path modeling with CHILLER providing insights into the ore formation processes applicable to MVT deposits worldwide. They concluded that diverse hydrothermal mineral assemblages could be produced from the same migrating basinal brine by different processes such as boiling, cooling, water-rock interaction, and fluid mixing.

5.1.2 Diagenesis

With the burial of geological layers in sedimentary basins to depths of many kilometers, several diagenetic reactions occur as a result of increasing pressure and temperature. Firstly, this is the process of cementation (precipitation) by silica, calcite, or iron and vice versa their dissolution. A second important group of processes is due to reactions with clay minerals, for example their conversions like smectite to illite or the albitization of potassium feldspars and plagioclase. Prediction of diagenetic changes is important for hydrocarbon or geothermal reservoir exploration due to the associated changes in porosity and permeability.

One of the first studies referring to calculated diagenetic reaction paths is by Harrison and Tempel (1993). They investigated the Gulf Coast Basin with the help of a "loose" coupling between a reaction path model and a groundwater flow model using the program BASIN2 (Bethke et al. 1993). In that way, a history of mineralogical changes can be determined referring to temperature and pressure and varying flow conditions.

Bitzer (1999) presents 2D simulation of clastic and carbonate sedimentation during formation of a basin structure with the developed program BASIN. Processes taken into account in this model are sedimentation, consolidation, subsidence, fluid flow, heat flow, and solute transport. The information gained from such a basin simulation often includes spatial and temporal distribution of petrophysical parameters, which in conjunction can be applied to predict the location of mineral resources.

Lowell and Yao (2002) presented a numerical study, investigating anhydrite precipitation with respect to the extent of hydrothermal recharge zones at ocean ridge crests. They applied a single-pass model driven by fluid buoyancy, schematically shown in Fig. 5.1. Within this model cold seawater enters the recharge zone, penetrates down to 1 km depth, is heated in the vicinity of the magma chamber, enters the discharge zone, and flows out at the surface. Anhydrite precipitation occurs upon heating because the solubility of anhydrite decreases with increasing temperature. The feedback of porosity and resulting permeability reduction on the flow field was taken into account.

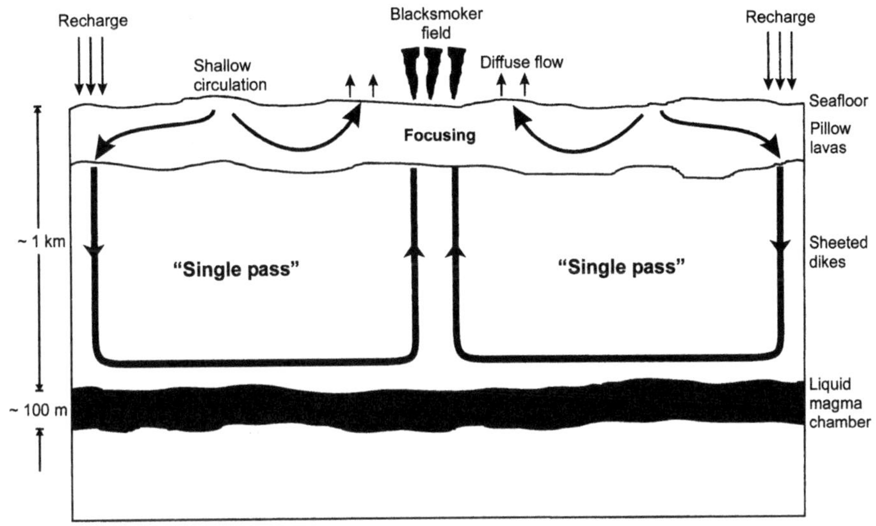

Fig. 5.1. Single-pass model of hydrothermal circulation at mid ocean ridges (adapted from Lowell and Yao 2002)

Aim of the presented calculations was to determine the rate at which anhydrite precipitation can seal permeability in a high-temperature hydrothermal recharge zone and to discover whether this process can provide any constraints on the nature of the recharge zone itself. Lowell and Yao (2002) showed that the rate of precipitation is a strong function of recharge velocity. Generally, precipitation was observed between 150 and 300°C as a consequence of the temperature dependence of anhydrite solubility in seawater. The numerical simulations led to the conclusion that anhydrite precipitation would rapidly seal hydrothermal recharge zones and reduce the heat output of the system unless the recharge zones are 10-100 times larger than the discharge areas.

5.2 Anhydrite Cementation at the Location Allermöhe

It was planned to install a Geothermal Heating Station (GHS) at the location Allermöhe (South-East of Hamburg, Germany) for district heating supply. The target aquifer for water recovery has been the Rhaetkeuper, which feeds the GHS Neustadt-Glewe (100 km east of Hamburg) with up to 120 m^3 per hour since 1995. For that reason, the bore Allermöhe 1 was deepened to a depth of 3,300 m in 1997. It taps a 70 m thick sandstone aquifer with a temperature of 125°C. Although temperature and thickness of the aquifer agree with the conditions needed for geothermal energy use, the pore spaces, originally open with porosities up to 20 %, are filled to a large extent by anhydrite. Mineralogical investigations showed cementation of secondary anhydrite with completely filled pore spaces as well as insular, cloudy, or layered structures. The extractable amount of water, 3 $m^3\ h^{-1}$, determined by a pumping test in 1998, is too low for an economical use of the resource (Baermann et al., 2000a).

The intention of the numerical studies done here with the simulator SHEMAT is to confirm or disprove the hypotheses for anhydrite cementation due to:

- Transport of solutes from neighboring salt structures into the Rhaetian sandstone and subsequent anhydrite precipitation (Lenz et al. 1997) studied here under special consideration of the recent structure of the Allermöhe site and its palaeogeological development, or
- up-flow of brines from deeper stratigraphic units via fault zones and resulting anhydrite precipitation due to changing physical and chemical conditions (Baermann et al. 2000a).

The findings provided by the numerical investigations are finally compared in discussion with two further attempts of explanation for the observed anhydrite cementation:

- Up-flow of brines from the underlying Gipsmergelkeuper formation and precipitation of anhydrite in the Rhaetian sandstone. Christensen et al. (2002) stated that deposition is due to significant pressure differences of the neighboring geologic formations, and
- synsedimentary formation and growth of anhydrite due to capillary evaporation in a highly saline and high temperature sabkha environment due to the arid climate of German Triassic times (Wurster 1965).

5.2.1 Geological Setting and History of the Salt Structures

The investigated site is located SE of Hamburg (Germany) between 53°24'-53°30'N and 10°00'-10°10'E (Fig. 5.2). The study area is concurrent with part no. 2526 of the geological map of Hamburg scale 1:25000 (Sheet Allermöhe, Ehlers 1993). The Allermöhe well has been drilled at the position 53°28'N and 10°06'E in the center of the NE quadrant of the map sheet (Fig. 5.2).

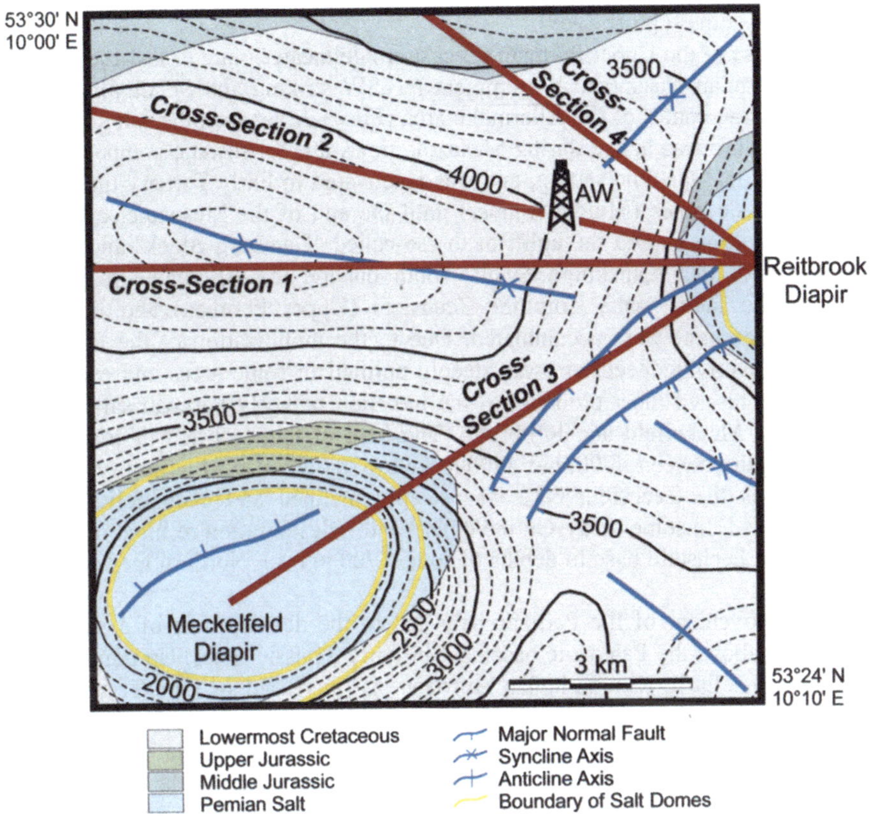

Fig. 5.2. Geological subcrop map (Lower Cretaceous and younger units uncovered) of the study area (TK 2526, sheet Allermöhe, Ehlers 1993, modified after Baldschuhn et al. 2001); isolines: depth contours of base of Keuper horizon (b.s.l.); also shown: locations of cross-sections and position of the Allermöhe well (AW)

More than 100 deep boreholes have been drilled within the area of Allermöhe during hydrocarbon and iron ore exploration. As a result the deeper underground is very well known down to formations of Lower Jurassic times (Lias). Occurrence, thickness, and character of older stratigraphic successions can only be deduced from deep boreholes outside the study area and seismic investigations (Frisch 1993).

The entire region of Hamburg, including the location Allermöhe, is most likely positioned outside the area affected by the Variscan orogenesis during Late Carboniferous times. Conversely, it belonged to the non-deformed molasse basin in front of the Variscan orogen, where probably a complete Upper Carboniferous formation has been deposited. The development of an East-West striking area of subsidence commenced, which provides the contour of a Permian trough called "North German Basin" recognizable until Tertiary times. The area of Allermöhe is situated at the SE margin of the central North German Basin. Subsidence went on until Cenozoic times, just interrupted by a phase of uplift from Jurassic to Early Cretaceous times (Jaritz 1969).

The base of the Upper Permian (Zechstein) delineates more or less the structure of the Permian basement. A prominent NE-SW striking and SE dipping normal fault cuts the center of the Allermöhe area (Fig. 5.2, between the salt structures). This structure was active during Mesozoic times due to isostatic compensation in an extensional tectonic setting, and was reactivated in Early Tertiary times. From Late Triassic times (Middle Keuper) until the end of the Mesozoic, extensional normal faulting caused the uplift of the so-called "Hamburg Block" and the construction of the "Quickborn Swell" (both outside the Allermöhe map sheet). Above the faults, salts from the Zechstein (Upper Permian) and Rotliegend (Lower Permian) were accumulated. Due to the mobilization of the Rotliegend salt, the overlying Zechstein was steeply uplifted in some areas and even partly pierced. This led most probably to the intrusion of Early Permian salts into the diapirs of Meckelfeld and Reitbrook. The base of the Zechstein is located in a mean depth between 4700 and 4900 m below sea level (b.s.l.). However, due to halokinesis the Zechstein base was steeply inclined and even uplifted up to 4300 m b.s.l. in some areas. On the other hand, salt migration resulted in a lowering of the Zechstein base to depths of some 5700 m b.s.l. North of the Meckelfeld salt dome.

The movement of the Permian salts led to the development of a salt pillow structure above the Paleozoic basement in Early Triassic times. Up-doming, initiated by the Early Kimmeridgian epirogenesis, increased in Late Triassic times (Middle Keuper). During Jurassic times the salt pillow developed further into a diapiric stage. In the SE quadrant of the Allermöhe map sheet, the Meckelfeld salt dome evolved during Dogger times and at the eastern margin the Reitbrook salt dome pierced up already in Lias times. Salt diapirism probably continued until Late Jurassic times and uplift gradually ceased in Early Cretaceous times.

The upper surface of the Meckelfeld salt dome is oval and slightly NE-SW striking with an area of approximately 19 km² (Fig. 5.2). Its maximum extension is 7 km in length and 4 km in width and occurs within the surrounding of Upper Cretaceous formations. The actual vertical extent of the Meckelfeld diapir is approximately 3000 m, situated on top of the major Permian saliniferous residuals (Zechstein, Rotliegend).

The western upper surface of the Reitbrook salt dome reaches at the eastern margin into the Allermöhe map sheet. It is also formed oval and slightly stretched in NW-SE direction (Fig. 5.2). The Reitbrook diapir measures 3-4 km in diameter at its maximum extension. The salt dome surface has been uplifted up to a level of 850 m b.s.l. The revealing Cretaceous and Cenozoic layers are not pierced by the Reitbrook diapir.

3D-Structure

The 3D model of Allermöhe has been obtained by digitizing and attributing georeferenced structural contour lines of major stratigraphic units from the "Tectonic Atlas of Northwest Germany", published by Baldschuhn et al. (2001). From these data, triangulated irregular networks (TIN) of the surfaces were constructed and converted to grids with pixel sizes of 250 by 250 m and 50 m in depth. Structural features like traces of faults and folds were incorporated during TIN-constructions. The final 3D GIS model measures 11 km by 11 km and has a depth of 6000 m (Fig. 5.3). The brownish plane displays here the base of the Keuper formation, accommodating the potential Rhaetian reservoir sandstone.

Fig. 5.3. 3D structure of Allermöhe measuring 11 km by 11 km with a depth of 6000 m, obtained by digitizing georeferenced structural contour lines of major stratigraphic units from the "Tectonic Atlas of Northwest Germany", published by Baldschuhn et al. (2001)

The area of Allermöhe is prevailed by the two salt domes Meckelfeld and Reitbrook. They have a huge influence on the stratigraphy and on the shape of the Keuper layer. It is obvious from Fig. 5.3 that especially the deeper formations display a strong relief. In the SW quadrant of the area the salt dome Meckelfeld can be seen almost completely within the study area (Fig. 5.2). The salt dome Reitbrook is situated at the eastern border of the area. The base of the Keuper lies, depending on the location, between 1800 and 4500 m below sea level (b.s.l.).

2D-Cross-Sections

From the GIS-based structural grid data (Fig. 5.3) cross-sections through the 3D model of any desired orientation (Fig. 5.2) can easily be derived using a simple program from Günther (2003a) and used for 2D simulations.

Cross-section 1 from West to East ending in the Reitbrook salt dome is displayed in Fig. 5.4. The main stratigraphic units are the Cenozoic, Upper and Lower Cretaceous, Dogger, Lias, Keuper, Upper and Lower Buntsandstein and the Zechstein layer. The diapir, constituted from Zechstein salt, pierces the overlaying formations from the Buntsandstein to the Dogger, whereas stratigraphic layers from the Lower Cretaceous on are domed up. The pre-Cretaceous layers descend smoothly away from the Reitbrook diapir and start to slightly ascend again in a distance of approximately 6 km to the so-called Hamburg block, which lies outside the Allermöhe map sheet.

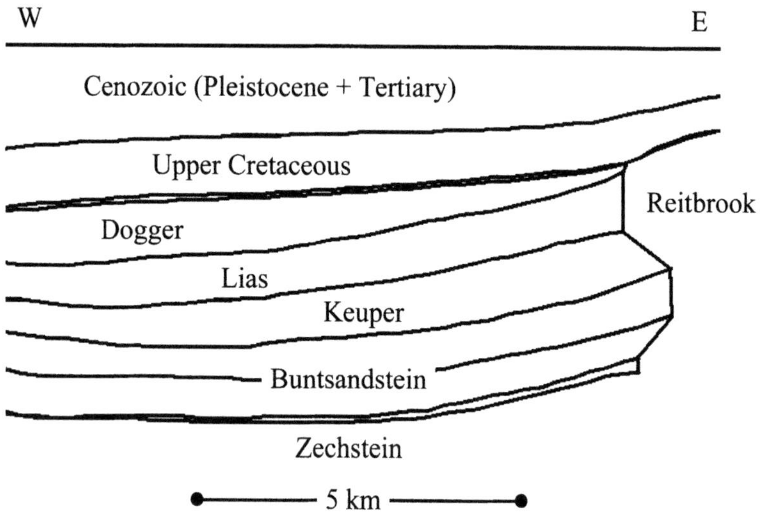

Fig. 5.4. Cross-section 1 (compare Fig. 5.2 for location)

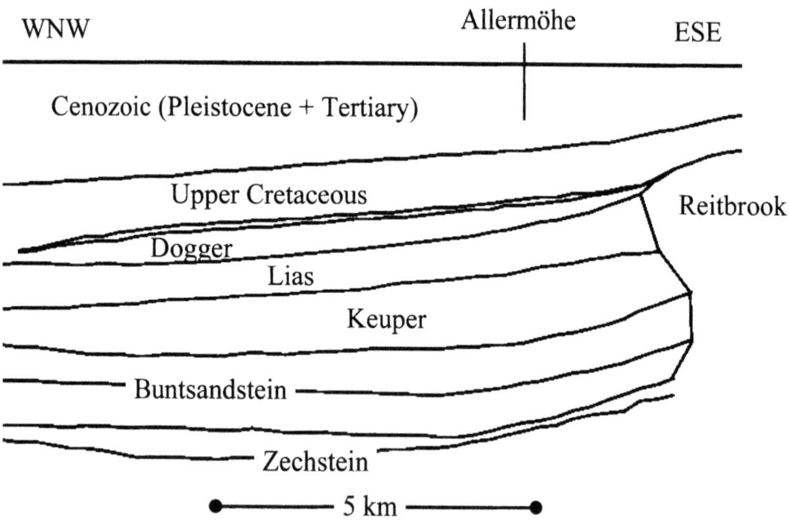

Fig. 5.5. Cross-section 2 (compare Fig. 5.2 for location)

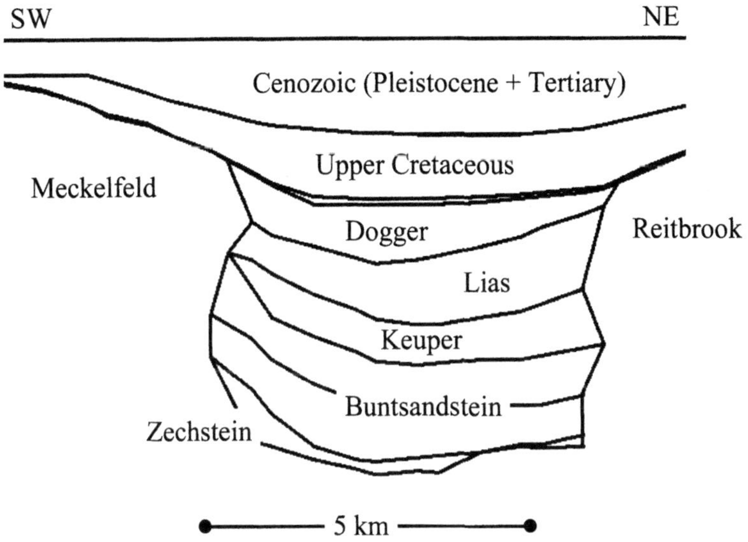

Fig. 5.6. Cross-section 3 (compare Fig. 5.2 for location)

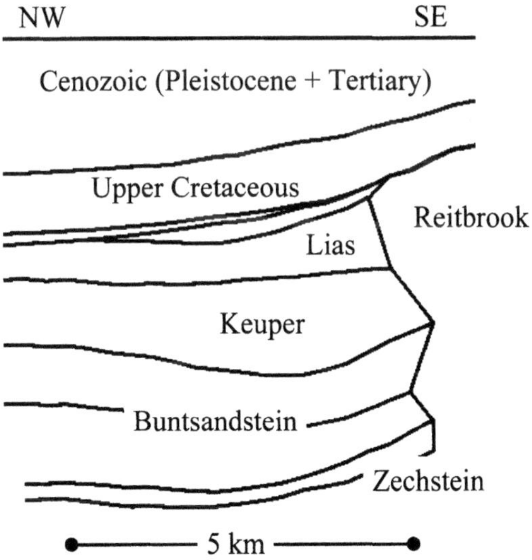

Fig. 5.7. Cross-section 4 (compare Fig. 5.2 for location)

The WNW – ESE cross-section 2 (Fig. 5.5) crosses the well Allermöhe. The stratigraphy is quite similar to the one of cross-section 1, except for the fact that the pre-Cretaceous layers do not ascend again but remain more or less the same level.

Cross-section 3 (Fig. 5.6) is oriented SW to NE and runs through both salt domes. The pre-Cenozoic layers descend away from the diapirs and reach a minimum at the center between them. The formations do ascend steeper to the Meckelfeld salt dome than to the Reitbrook salt dome. The deeper layers of the Buntsandstein reach the Reitbrook diapir almost leveled. The distance between the diapirs is around 5 km.

Cross-section 4 (Fig. 5.7) runs from the Reitbrook diapir in NW direction. A characteristic feature is the border between the Lias and the Keuper, which is level, and also the Buntsandstein layers do not show large depth differences.

The last cross-section, shown here, runs S – N (Fig. 5.8) through the Allermöhe well. This cross-section is specific because it does not cut the Meckelfeld diapir or the Reitbrook salt dome. The Keuper layer slightly descends towards the center of the cross-section and shows at the South as well as at the North border an increased thickness.

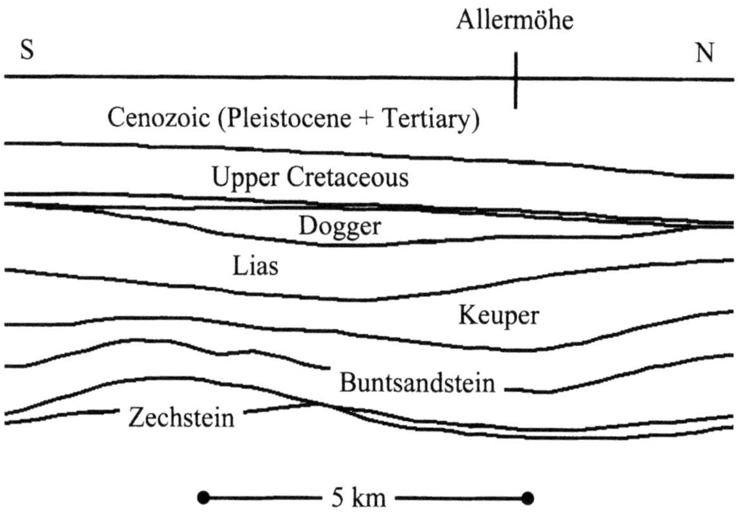

Fig. 5.8. Cross-section 5 (compare Fig. 5.2 for location)

5.2.2 Conceptual Investigation of Reservoirs Near Salt Domes

The salt domes of Meckelfeld and Reitbrook prevail the stratigraphy of the Allermöhe region. This is the reason why, firstly, a conceptual study was conducted on the thermal and reactive flow conditions of an idealized aquifer, representing the Keuper formation, near to one salt dome or flanked by two salt domes.

Six different 2D, vertical, conceptual structures, measuring 3 km by 9 km with a Keuper layer of 500 m thickness each, have been investigated (Fig. 5.9). The salt structure is shown in orange color. The cases A to D include one schematic salt dome and the underlying Zechstein formation as model basement. In concept E and F two salt diapirs flank the permeable Keuper layer, which is displayed in green. In any case the overlying formations are combined in the purple layer and the underlying in the grey layer. Both represent impervious structures.

The question now was, if free thermal convection starts and sodium chloride and calcium sulfate are leached from the salt structures and transported into the Keuper layer maybe subsequently resulting in anhydrite precipitation in the formation. The simulated times have been up to 500.000 years.

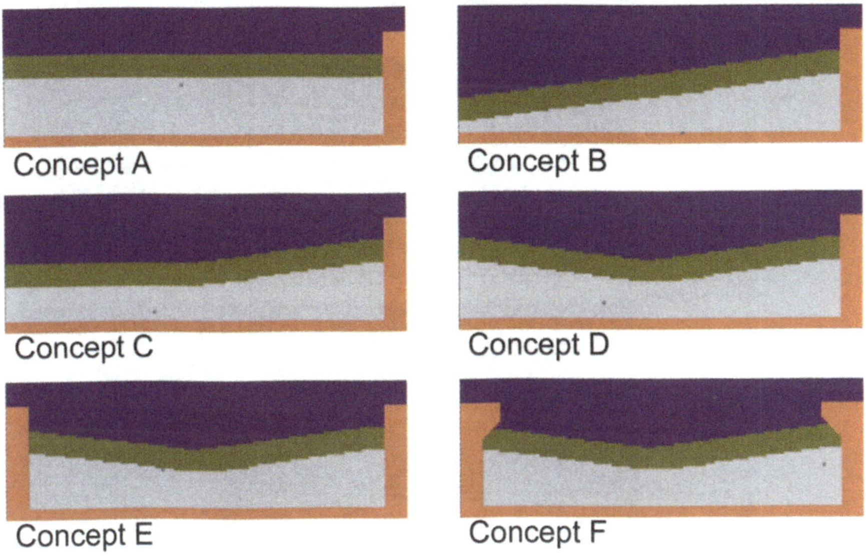

Fig. 5.9. Conceptual structures for 2D reactive transport simulations of a reservoir near one salt dome or flanked by two salt domes

In concept A no anhydrite precipitation occurs in the Keuper layer, although thermal convection starts due to the fact that the critical Rayleigh number is exceeded (Fig. 5.10 A). In concepts B to D thermal convection occurs accompanied by anhydrite precipitation in the permeable Keuper formation. Concepts B to D show quite similar results of anhydrite precipitation. Dependent on the geometric structure of the Keuper layer anhydrite has been precipitated in direction of decreasing levels of the formation (Fig. 5.10 B-D). Additionally, it is observed within the conceptual studies B-D that there is a distinct area in direct vicinity to the salt domes where no anhydrite precipitation occurs or early deposits of anhydrite are re-dissolved with ongoing simulation.

Concept E and F seem to be of identical structure but it can be seen in Fig. 5.11 why such small differences in geometry may prevent anhydrite precipitation near to a salt dome, although free thermal convection occurs.

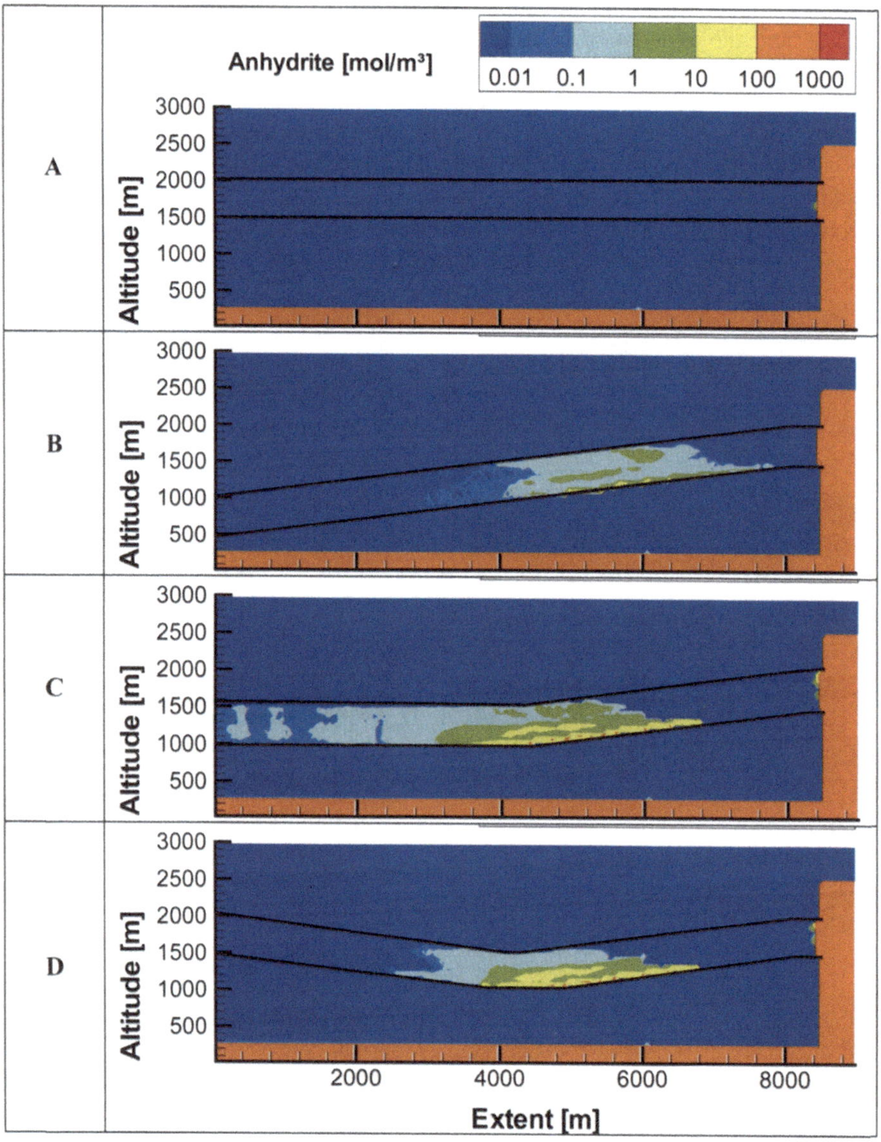

Fig. 5.10. Anhydrite distribution observed in Concepts A, B, C, and D (Fig. 5.9) after a simulated time period of 500.000 years

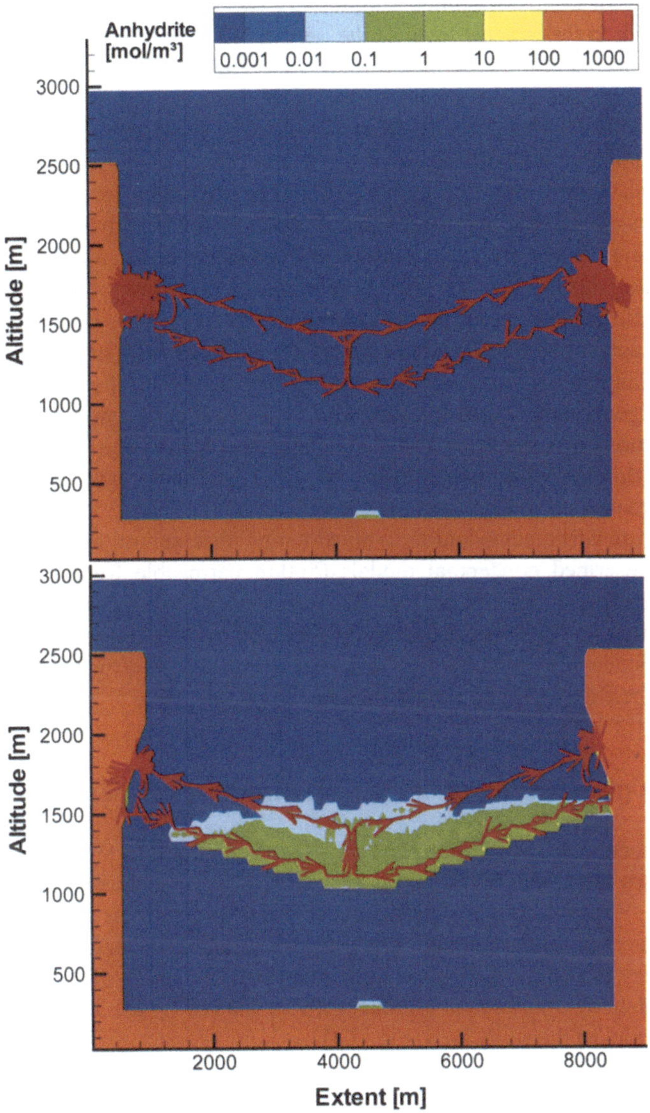

Fig. 5.11. Vertically exaggerated cross sections of Concepts E (top) and F (bottom) show the anhydrite precipitation in mol m^{-3} (contour colors) after 500.000 years; the red arrows display the stream traces calculated from the flow field; varying establishment of convection cells lead to or prevent precipitation within the permeable formation

The contour colors of the vertically exaggerated diagrams depict the anhydrite amount occurring in the simulated area. In both cases free thermal convection occurs, but only when the diapirs are less schematically shaped with an overhanging structure anhydrite precipitation is observed. The stream traces calculated from the velocity field show in the first example that there is an intense convection cell in direct vicinity to each salt dome. Additionally two greater convection cells exist reaching down to the deepest level of the permeable Keuper layer. Away from the diapirs flow occurs at the base of the aquifer and towards the salt structures at the top of the layer. The stream traces of the second example show only the bigger convection cells leading directly from the diapirs down the formation and up again. The small convection cells near to the salt domes do not occur in this case. Detailed investigation revealed higher horizontal flow rates for Concept F compared to Concept E, resulting in mixed convection. Mixed convection takes place when horizontal flows are superimposed on thermally driven flows. Convection cells migrating laterally in the direction of horizontal flow characterize mixed convection (Raffensberger and Vlassopoulus 1999).

It can be concluded that the development of one large flow structure due to laterally migrating convection cells is necessary for the occurrence of solute transport over larger distances into the formation. If a small convection cell exists near to the salt dome all solutes are instantaneously back-transported to the salt structure. The difference in both examples solely consists in a small variation of the diapir's geometry.

To reach anhydrite cementation to an extent as observed in the Allermöhe well within the described conceptual models (500 m permeable Keuper reservoir) a minimum time period of around 150 Mio years would be necessary. This simplified assumption holds for the case that the feedback of reaction on the flow can be neglected. Hence, precipitation or dissolution within the pore space structure does not change the flow field and flow velocities.

5.2.3 Geological History of the Recent Structure of Allermöhe

The importance of the structural geometry, mentioned before, on anhydrite transport processes led to the investigation of historical geological sequences (both in 2D and 3D). Simulations were carried out in order to determine the occurrence of thermal convection and resulting anhydrite precipitation dependent on the structural geometry in certain geological time-intervals. The set up of the 2D sequence is based on data of Schmitz and Flixeder (1993). Source for the 3D sequence is the structural model of the study area (Fig. 5.3).

2D Restoration Sequence

Schmitz and Flixeder (1993) published a roughly WNW-ESE striking 2D vertical cross-section through the Reitbrook diapir. The Reitbrook diapir is of special importance for the Allermöhe well due to its proximity. Their cross section was used to firstly make a sketch of the recent structure (Fig. 5.12). The 2D cross section shows the stratigraphy from Cenozoic down to Rotliegend units of the area around the Allermöhe well and the Reitbrook diapir.

The recent structure of the location Allermöhe (Fig. 5.12) is evolved backwards to Keuper times based on the actual, original, available geologic data. The work has been done applying a structural modeling algorithm to analyze the subsidence history from geological cross-section data (Günther 2003b). The principle here is the step by step (sequential) restoration of the geological horizons to pre-deformational and/or pre-sedimentary conditions, in the way that for each restoration sequence all geological horizons were moved upwards according to vertical shear vectors that were derived from unfolding (straightening) the uppermost horizon of the particular sequence to a fixed elevation datum (0 > m a.s.l.). All lower stratigraphic horizons to be restored are moved upwards accordingly. Paloestructural models both in 2D and 3D can be derived for a specific geological horizon by restoring it incrementally from flattening all higher stratigraphic horizons successively.

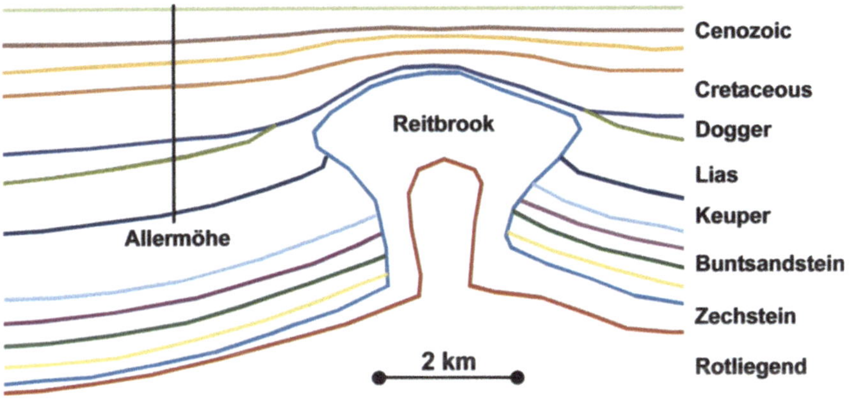

Fig. 5.12. Cross section of the recent structure around the salt diapir of Reitbrook based on data of Schmitz and Flixeder (1993)

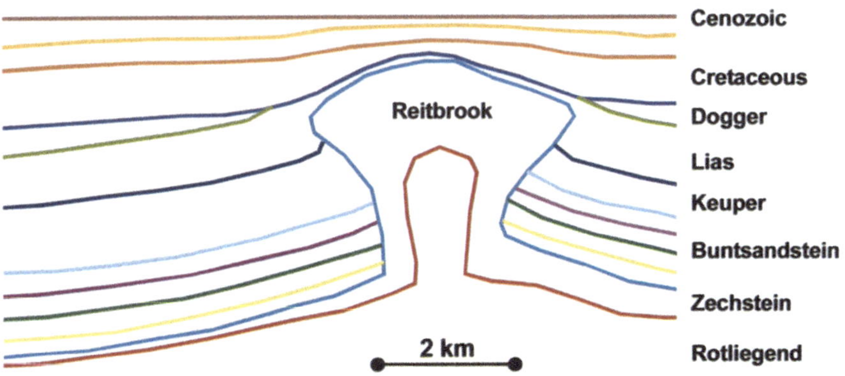

Fig. 5.13. Pre-Pleistocene structure of the Reitbrook diapir representing the stage of late diapirism

The pre-Pleistocene stage in Late Tertiary times is displayed in Fig. 5.13 representing a structure of the area during a stage of late diapirism right after the last movements of the diapir. The stage of early diapirism in Early Jurassic times with the salt piercing the overlying formations is shown in Fig. 5.14. Fig. 5.15 exhibits Late Triassic times with a supposed NE-SW striking and NW dipping normal fault, which cuts the center of the Allermöhe salt structure. This structure must have been present and active during late Triassic times due to different bed-thicknesses of Keuper strata NE and SW of the fault (Fig. 5.15) and probably originated due to isostatic compensation-movements in an extensional tectonic setting.

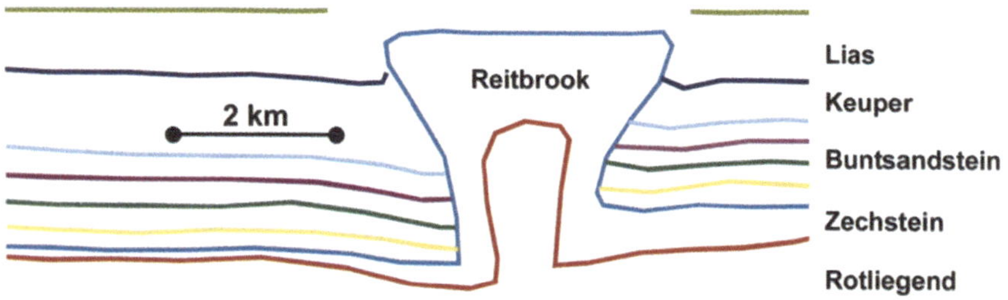

Fig. 5.14. Lias structure during Early Jurassic times representing the stage of early diapirism after the Permian salt just pierced the overlying structures

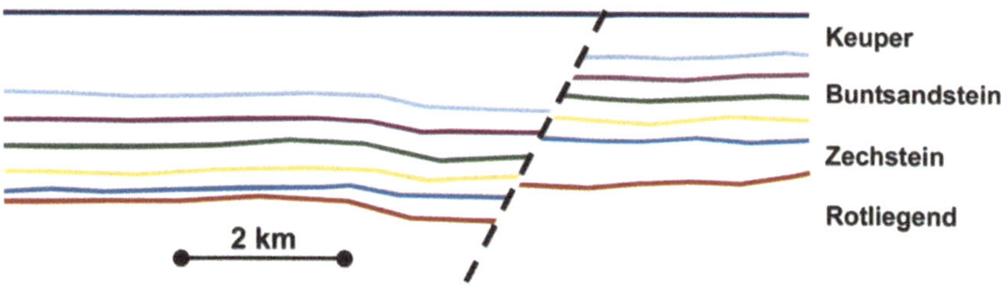

Fig. 5.15. Structure during Keuper times with a supposed NE-SW striking and NW dipping normal fault, which cuts the center of the Allermöhe area; this structure was active during Mesozoic times due to isostatic compensation in an extensional tectonic setting

The presented historic stages (Fig. 5.12 - Fig. 5.15) are shown here as examples of the whole sequence. They represent distinct periods used to determine during which geological times precipitation of anhydrite in the Keuper layer may have occurred.

3D Restoration Sequence

Comparison of 2D and 3D simulations of the Allermöhe site reveal the necessity of a comprehensive investigation in three dimensions (see below). Thus, a second restoration sequence has been derived from the 3D model of the area around the Allermöhe well (Fig. 5.3).

The restoration sequence has been done of the Lias depth base because this is tantamount with the top of the Keuper respectively the Rhaetian layer. Fig. 5.16 displays the shape of the Lias base for different geological times with the Meckelfeld salt dome in the foreground and the Reitbrook diapir in the background. The Lias depth base is shown from bottom to top for Recent, Tertiary, Late Cretaceous, Early Cretaceous, and Dogger times.

The presented 3D historic restoration sequence (Fig. 5.16) is used to investigate if the flow field has significantly varied in different geological times and to determine when anhydrite precipitation has been likely during the geologic history in the vicinity of the Allermöhe well.

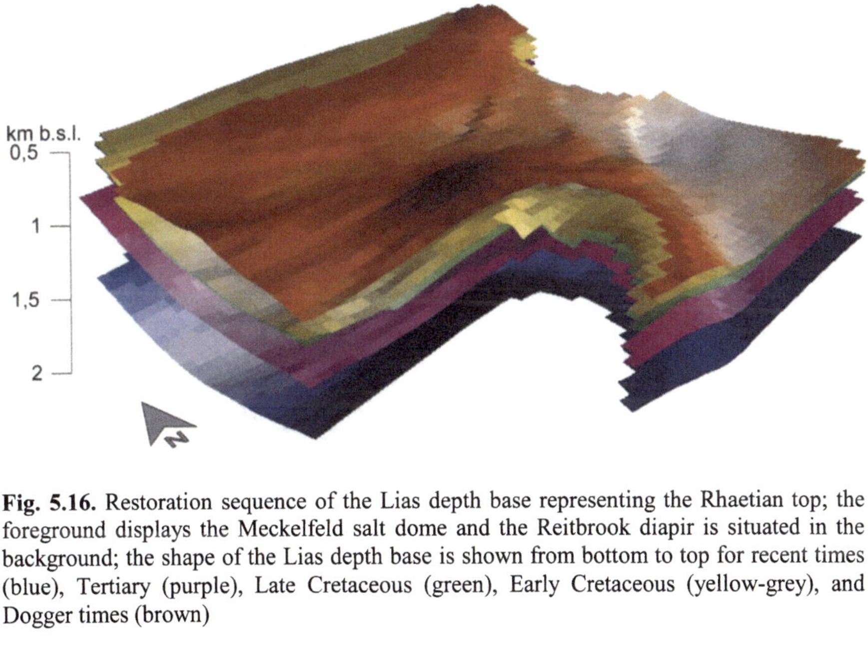

Fig. 5.16. Restoration sequence of the Lias depth base representing the Rhaetian top; the foreground displays the Meckelfeld salt dome and the Reitbrook diapir is situated in the background; the shape of the Lias depth base is shown from bottom to top for recent times (blue), Tertiary (purple), Late Cretaceous (green), Early Cretaceous (yellow-grey), and Dogger times (brown)

5.2.4 Reactive Transport Modeling

The subsurface flow and hydrogeothermal simulation system SHEMAT has been used to investigate processes, which could have led to the observed anhydrite cementation. Numerical investigations were performed to analyze the geologic history in 2D and 3D, the influence of fault zones providing conduits to deeper formations, and exemplary cross sections of the recent structure. The rock alteration index as well as transport of solutes from the diapirs into the Rhaetian sandstone and probably occurring precipitation were finally simulated in the 3D model.

Dependent on the rock composition the model structures have been transferred into reactive transport models under consideration of varying rock properties for thermal capacity ($\rho\ C_{P(r)}$), thermal conductivity (λ_r), porosity (Φ), and permeability (k, Table 5.1). The major rock types have been deduced from the explanations to the geological map of Hamburg (Ehlers 1993). The Keuper layer was thereby divided into a 250 m thick topmost and permeable Rhaetian sandstone layer and the main lower and non-permeable part composed of clay stones. Rock properties were yielded from mineralogical investigations of the original Allermöhe core material and initial water compositions from formation water analysis (Table 5.2). The 2D structures are characterized by a second potential geothermal reservoir in the deeper Buntsandstein formation (Kühn et al. 2002a).

Due to the available computational power it was not possible to simulate the whole 3D model at present. Hence, the Rhaetian layer has been extracted from the 3D structure. The model was then overlain and underlain with distinct geological units in the way that the stratigraphy comprises a depth from 1750 to 4750 m b.s.l. Simulations were conducted for up to 500.000 years.

Table 5.1. Rock properties applied for numerical simulations with data from Schack (1953), Gröber et al. (1963), Fjalov (1959), and Mercer et al. (1982)

Rock Type	$\rho\ C_{P(r)}$ [MJ m^{-3} K^{-1}]	λ_r [W m^{-1} K^{-1}]	Φ [-]	k [m^2]
Sand, Sandstone (dry)	1.6	1.3	0.1-0.3	1.0E-14 – 1.0E-12
Clay stone	2.04	1.28	0.005	1.0E-16
Limestone	2.5	2.2	0.16	1.4E-13
Salt	1.95	6.1	0.0001	0.0
Salt rock	2.02	2.9	0.001	0.0
Rhaetian sandstone	1.6	1.3	0.07	0.4E-12

Table 5.2. Composition of the Allermöhe formation water (Lenz et al. 1997) with total dissolved solids of 218 g L^{-1}, density 1.146 g L^{-1}, and pH 5.4 (depth sample)

Ion	K^+	Na^+	Ca^{2+}	Mg^{2+}	Cl^-	SO_4^{2-}	HCO_3^-
[mg L^{-1}]	1250	75000	6690	1300	132200	465	240

2D Sequence of Geologic History

The situation at the beginning of the 2D historical sequence during Keuper times (Fig. 5.15) was investigated for the sake of completeness only (not shown here). Transport of salt solutions from deeper formations through the fault zone to the topmost Rhaetkeuper is the sole process beside synsedimentary formation (see below), which could have been the reason for anhydrite diagenesis in Late Keuper times. The results reveal that free thermal convection occurs within the Buntsandstein formation but the Rhaetsandstone was completely unaffected. Anhydrite precipitation does not occur.

The second, exemplary, investigated, historic structure represents the stage of development right after the Reitbrook diapir has pierced the overlaying formations in Early Jurassic times (Lias, Fig. 5.14). Free thermal convection still did occur only within the Buntsandstein formation (not shown here). After 500.000 years reactive transport simulation it is obvious that during the stage of early diapirism significant precipitation cannot be observed, neither in the Rhaetkeuper nor in the Buntsandstein formation.

Late diapirism is represented by the Pre-Pleistocene structure of the Allermöhe area (Fig. 5.13). On the contrary to the situation during early diapirism anhydrite precipitation is observed after a time period of 500.000 years, but explicitly within the Buntsandstein and not in the Rhaetkeuper (Fig. 5.17).

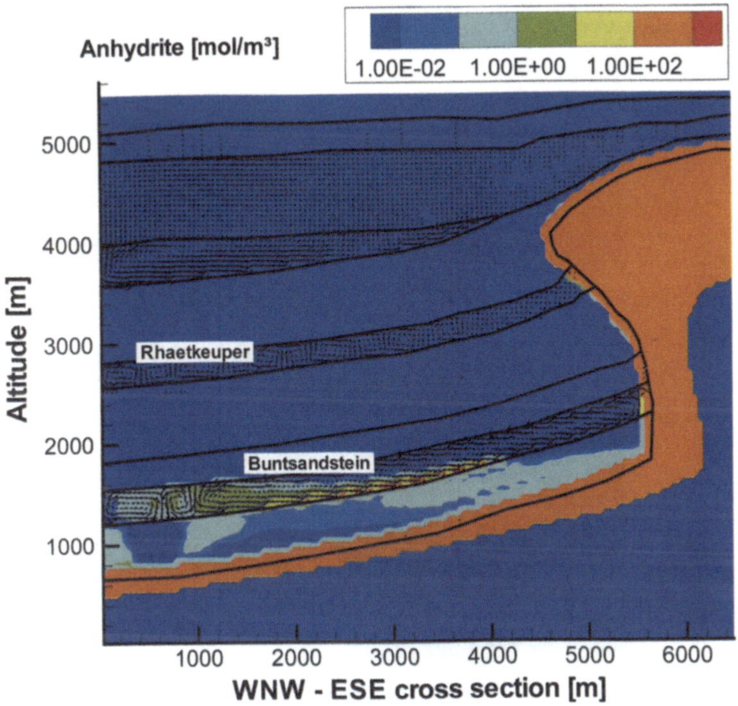

Fig. 5.17. Anhydrite precipitation in the pre-pleistocene structure (Fig. 5.13) in mol m^{-3} within the Buntsandstein formation after a simulated time period of 500.000 years, precipitation in the Rhaetkeuper cannot be observed

From investigation of chemical reactions occurring within the recent structure of the Allermöhe area (Fig. 5.12) it can be seen how halite and anhydrite are leached from the Reitbrook salt dome and are transported through the model. Due to the fact that all other reactions are prevailed by the amount of sodium chloride in the solution this must be the first thing to look at (Fig. 5.18).

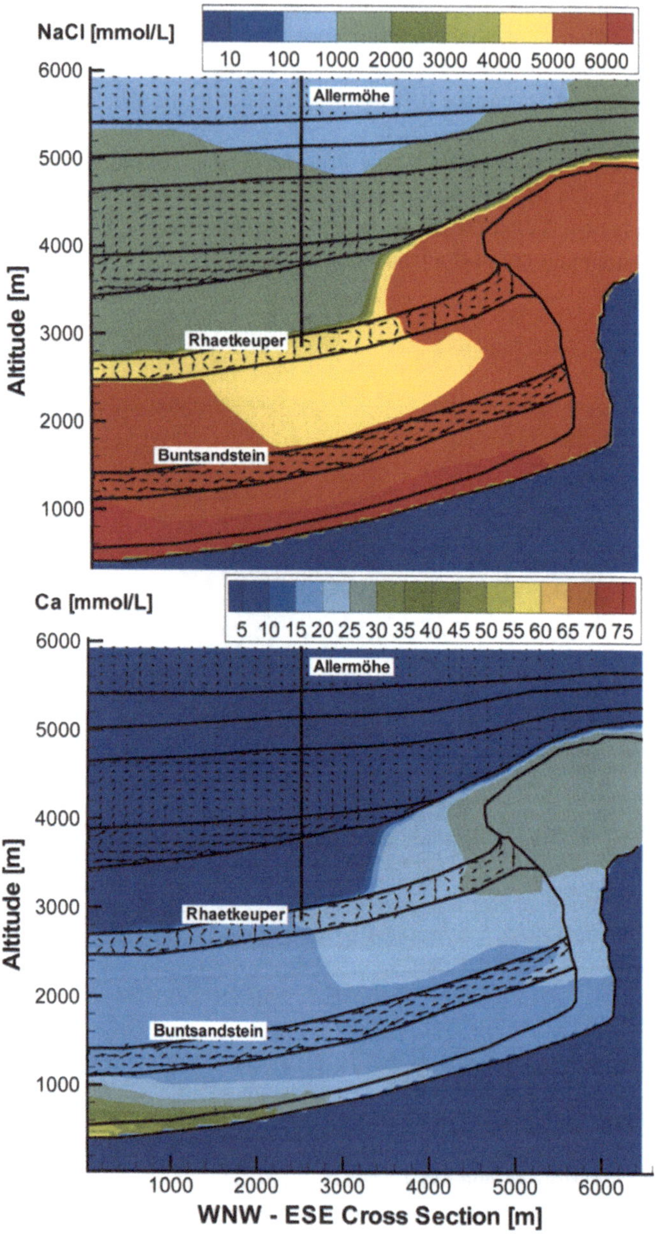

Fig. 5.18. Sodium chloride concentration (top) within the recent structure (Fig. 5.12) after 500.000 simulated years; the resulting concentration within the Rhaetian layer coincides well with the measured value of total dissolved solids; calcium concentration (bottom, sulfate exactly identical) dependent on temperature and sodium chloride concentration (both in mmol L^{-1})

It can be observed how sodium and chloride are distributed in the permeable layers after 500.000 years. The resulting concentration within the Rhaetian sandstone coincides well with measured amounts of total dissolved solids from the water analysis (Table 5.2). The calcium, respectively sulfate concentration is highly dependent on the sodium chloride content. In Fig. 5.18 the resulting Ca distribution is visualized at the end of the investigated time period.

The resulting anhydrite precipitation induced by reactive transport after 500.000 years is shown in Fig. 5.19. From previous timc periods (not shown here) it is observed that precipitation of anhydrite occurs firstly within the Buntsandstein layer and only after 500.000 years also within the Keuper formation.

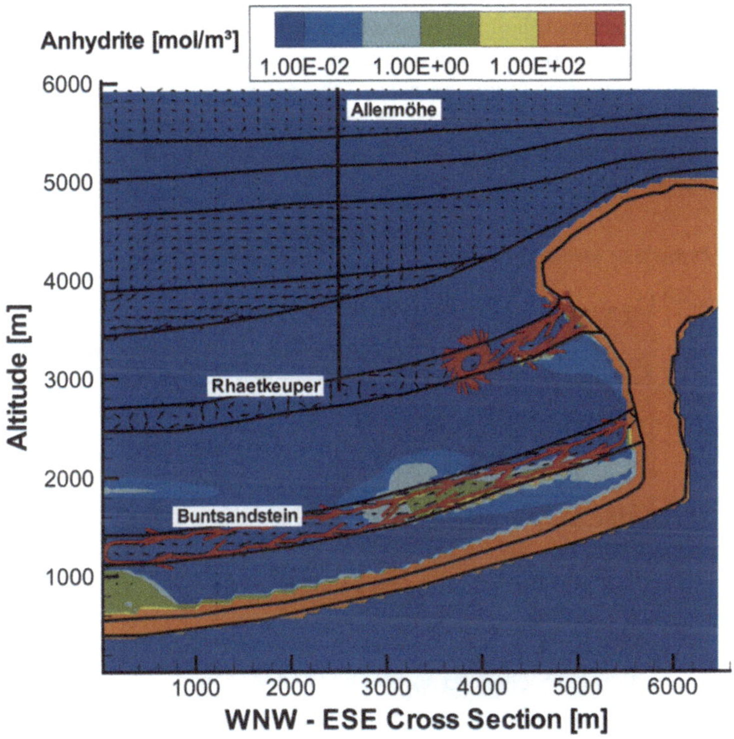

Fig. 5.19. Anhydrite distribution within the recent structure (Fig. 5.12) after 500.000 years; precipitation within the Rhaet formation occurs but at a location different from the Allermöhe well; stream traces (red arrows) deduced from the flow field (black arrows) reveal why the precipitation in the Buntsandstein spreads over a larger region

It can be recognized that anhydrite deposits occur in a larger region within the Buntsandstein than within the Rhaetian layer and that the area affected by precipitation in the Keuper does not spread out to the location of Allermöhe. The reason for this behavior is again the evolution of particular convection structures in the model. The calculated stream traces display a large convection cell in the deeper formation compared again to at least two separated smaller structures within the Keuper layer.

To reach anhydrite cementation up to an extent as observed in the Allermöhe well a minimum time period of 5 to 15 Mio years would be necessary. However, one has to be aware of the fact that the 2D simulation predicts deposition at a location within the Rhaetian formation different from the position of the Allermöhe well.

Influence of Faults in the 2D Recent Model

A second hypothesis for the formation of anhydrite deposits is due to hot water up-flow through vertical fault zones into the reservoir sandstone of Allermöhe. To investigate this scenario the recent structure has been adapted with four arbitrary located faults to favor precipitation near to the Allermöhe well. Thus, the position of one fault zone is close to the location of the well. The fault zones provide conduits from the Rhaetian layer to permeable, deeper formations, in this case the Buntsandstein.

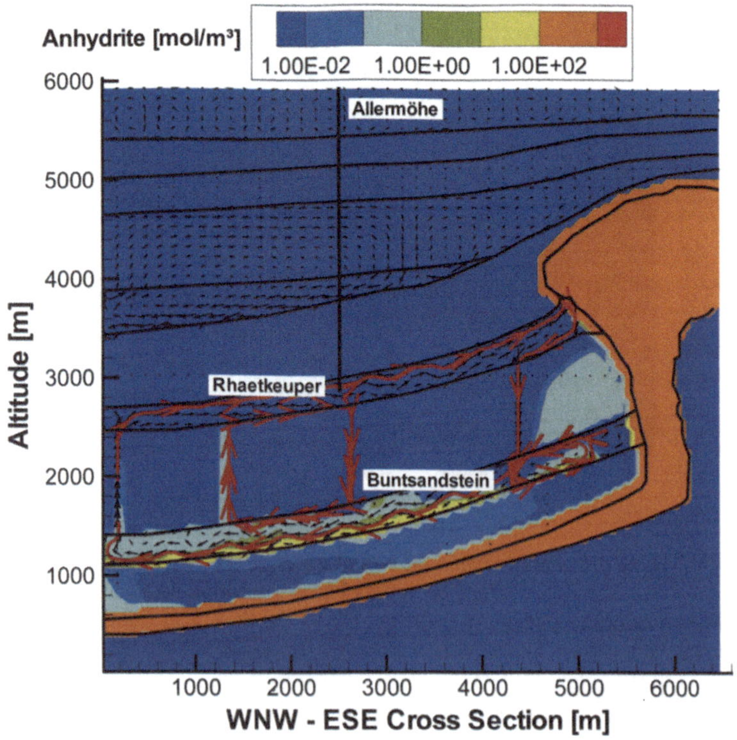

Fig. 5.20. Anhydrite distribution in mol m^{-3} after a simulated time period of 100.000 years in the recent structure of the Allermöhe site adapted with four fault zones; stream traces (red arrows) deduced from the flow field (black arrows) reveal the main flow directions

Fig. 5.20 displays the results after a simulated time period of 100.000 years. It can be seen from the stream traces that the flow field significantly changed compared to the structure without fault zones (Fig. 5.19). The Buntsandstein layer is fed with water from the Rhaetian formation through the two fault zones nearest to the diapir. After flowing down the Buntsandstein layer the brine flows back into the Keuper formation through the other two fault zones.

As a result, the flow field is characterized by down-flow, away from the diapir, in the Buntsandstein layer and up-flow, towards the diapir, in the Keuper layer. The changed flow field leads to precipitation in deeper parts of the Buntsandstein formation, whereas in the Rhaetian sandstone no deposits are observed.

However, the main focus at this point is the question if fault zones lead to a changed structure of precipitation observable in the Keuper. It is obvious from the streamlines that the formation water flows explicitly upwards in the Rhaetian sandstone close to the Allermöhe well. Hence, any precipitation is prevented because temperature decreases in that direction and with it the anhydrite solubility increases.

Rock Alteration Index Determined in 2D and 3D Models

The thermal rock alteration index (RAI, Raffensberger and Vlassopoulos 1999) provides a measure of the amount of thermally driven sediment alteration that might be accomplished as fluid moves through temperature gradients. Calculations of the RAI delineate the patterns of potential diagenesis. In the case presented here the RAI has been investigated for 2D cross sections and the Rhaetian formation extracted from the 3D structure of the Allermöhe site in order to determine areas where precipitation, dissolution, or no reaction occur.

Fig. 5.21 shows the rock alteration index with respect to the mineral phase anhydrite as determined in Cross-section 2 (Fig. 5.5, compare Fig. 5.2). Cross-section 2 has been cut out of the 3D model of Allermöhe. The detail of Cross-section 2 delineates the Rhaetian and Buntsandstein sandstone in the near vicinity to the Reitbrook diapir and the Allermöhe well. The figure exhibits three different colors: (1) red for areas of mineral precipitation, (2) blue for regions of anhydrite dissolution, and (3) green for unchanged regions with initial mineral amounts. It can be seen that most parts of the model show non-reactive conditions. The arrows (purple color) display flow paths calculated from the actual flow field.

With the use of RAI it can be approximated what would happen if anhydrite were available within the entire model. Within the Rhaetian formation anhydrite precipitation is observed at the base of the layer (Fig. 5.21) due to its retrograde solubility, whereas dissolution can be seen at the top. As already observed before, in deeper formations the convection structures are larger (Fig. 5.19). However, if anhydrite is available in the system, not only from the salt domes, relocation of the mineral and resulting depletion and enrichment in distinct areas can be determined. Due to the fact that in the Rhaetian sandstone non-coupled free convection cells act (distinct cells without overlap of stream traces), the transport of solutes from the Reitbrook diapir to the Allermöhe well seems to be unlikely. This would only be the case if mixed convection occurred with coupled convection cells.

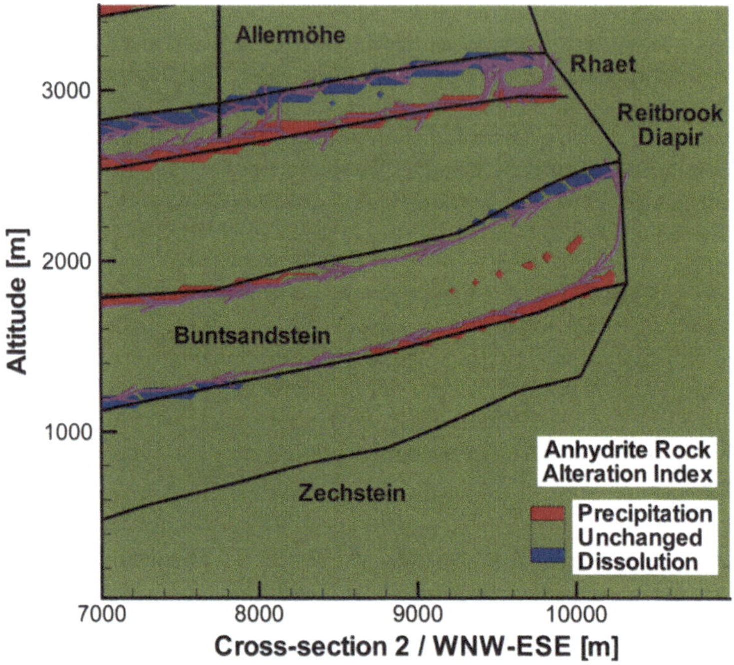

Fig. 5.21. Rock alteration index (RAI) delineating areas of potential anhydrite precipitation (red) or dissolution (blue) in the near vicinity of the Reitbrook diapir and the Allermöhe well from a detail of cross-section 2 (Fig. 5.2)

As mentioned above the 3D model comprises a depth range from 1750 to 4750 m b.s.l. The resulting flow field of the 3D simulation is shown in Fig. 5.22. This plan view of the Allermöhe map sheet exhibits the location of the Allermöhe well and the diapirs of Meckelfeld and Reitbrook. The arrows display the resulting Darcian flow field. It can be seen that the evolved flow field turns clock-wise. Starting from the Reitbrook salt dome, east to the Allermöhe well (x=8000 m, y=6000 m), it can be observed that the main flow is directed south, down the relief of the Rhaetian layer. Approximately 2000 m south of Reitbrook flow turns west towards the Meckelfeld diapir (x=8000, y=4000). North of the Meckelfeld diapir the main flow direction is guided north (x=3000, y=4000) down to the deepest point of the formation (Fig. 5.2, x=2000-4000, y=6000-8000). Coming out of the trough the brine flows in eastern direction upwards to the Allermöhe well and back to the Reitbrook diapir.

The colored arrows represent here the flow direction in combination with the rock alteration index delineating areas of potential anhydrite precipitation in red or dissolution in blue (Fig. 5.22). It is apparent that due to the upward directed flow, leading to increasing temperatures of the fluid, and the retrograde solubility of anhydrite the potential mineral reaction around the Allermöhe well is dissolution. Taking into account the clock-wise flow direction it is most unlikely that enriched amounts of calcium and sulfate, leached from the salt dome, reach the location of the Allermöhe well by advective transport. Under consideration of the recent geometry, the resulting flow field, and the reactive centers it is implausible that the observed anhydrite cementation in the Allermöhe well can be explained by transport of solutes from the salt structures.

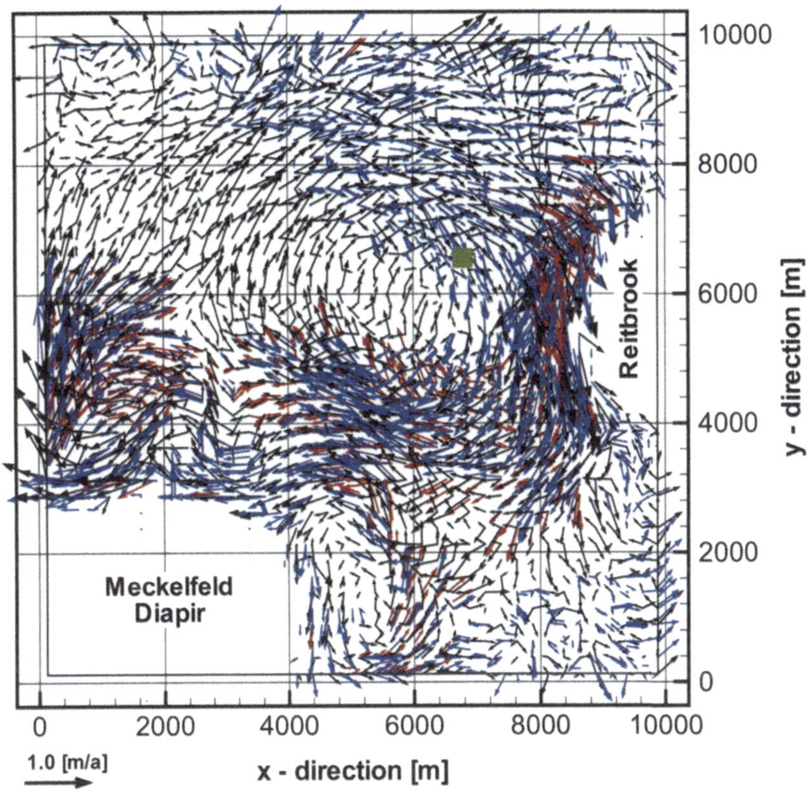

Fig. 5.22. Plan view of the modeled area (Allermöhe map sheet) of the recent structure exhibiting the Allermöhe well (green square) and the diapirs of Meckelfeld (SW) and Reitbrook (E); the arrows display the resulting Darcian flow field from the 3D simulation in direction and magnitude, scaled according to the reference arrow shown; colored arrows represent the rock alteration index (RAI) delineating areas of potential anhydrite precipitation (red) or dissolution (blue) in the Rhaetian sandstone

Temperature Profiles in Comparison to an On-Site Survey

Due to data availability, temperature is the only measure, which can be used for evaluation of the numerical models presented here, according to the simulated scale. This is done by comparison of the temperature survey conducted before the pumping test in 1998 and calculated temperature profiles. For that reason non-reactive simulations of coupled fluid flow and heat transfer as well as heat transfer only were conducted additionally to the reactive flow simulations.

Numerical simulations were performed with a constant basal heat flow of 60 mW m^{-2}, corresponding to published data within the Atlas of Geothermal Resources of Europe (Hurter and Haenel 2002). As a constraint at the upper boundary a constant temperature of 8°C has been assumed in the 2D simulations and of 60°C in 3D, corresponding to the temperature depth of each particular model top.

Fig. 5.23 displays the numerical results based on the recent stratigraphic cross-section published by Schmitz and Flixeder (1993). In general, areas of conductive heat transfer are characterized by high temperature versus depth gradients, naturally 25°C km^{-1}. Conversely, in regions where convective heat transfer predominates the gradients are low, culminating in vertical curves with constant temperatures versus increasing depth. The mere heat transfer simulation of the so-called "recent" structure yields an acceptable fit (Fig. 5.23, blue triangles), although from 500 to 2500 m temperature is overestimated and from 2500 m underestimated.

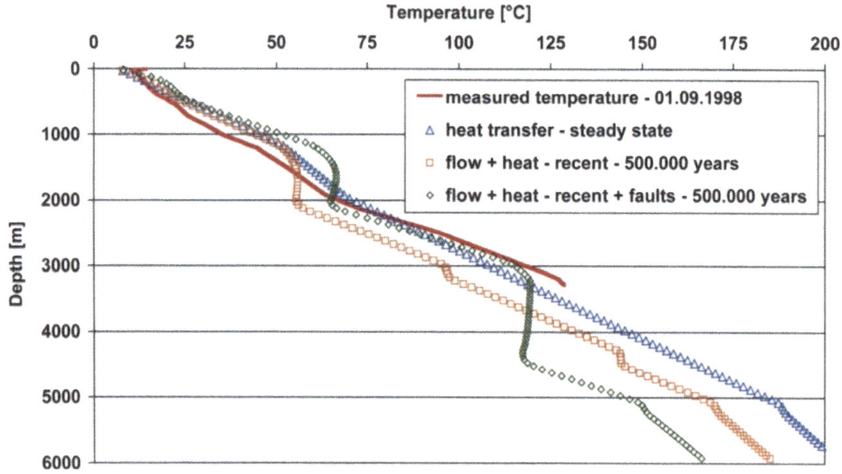

Fig. 5.23. Temperature survey from 1998 (red line) in comparison to results of 2D numerical simulations based on the stratigraphy published by Schmitz and Flixeder (1993) conducted for heat transfer only (blue triangles) or fluid flow and heat transfer (brown squares and green diamonds)

The results gained under consideration of fluid flow and heat transfer processes lead to free convection in the permeable Rhaetian and Buntsandstein stratigraphic units (Fig. 5.24) and due to resulting vertical flow very low temperature gradients (vertical graphs) are observed. Down to a depth of approximately 1000 m the calculated temperature profiles are almost all identical. Whereas the results of the recent structure including fault zones (Fig. 5.23, green diamonds) meet the measured temperature quite well between 2000 and 3000 m depth, the results from the recent structure without fault zones (Fig. 5.23, brown squares) underestimate the on-site temperature survey from 1998. In each of the three described cases the temperature within the Rhaetian sandstone (≈3000-3250 m) is underestimated. This may be due to the assumption of high permeability (higher than recently determined) and resulting fluid flow to determine if the cementation of the pore space in the sandstone could be a result of reactive transport processes.

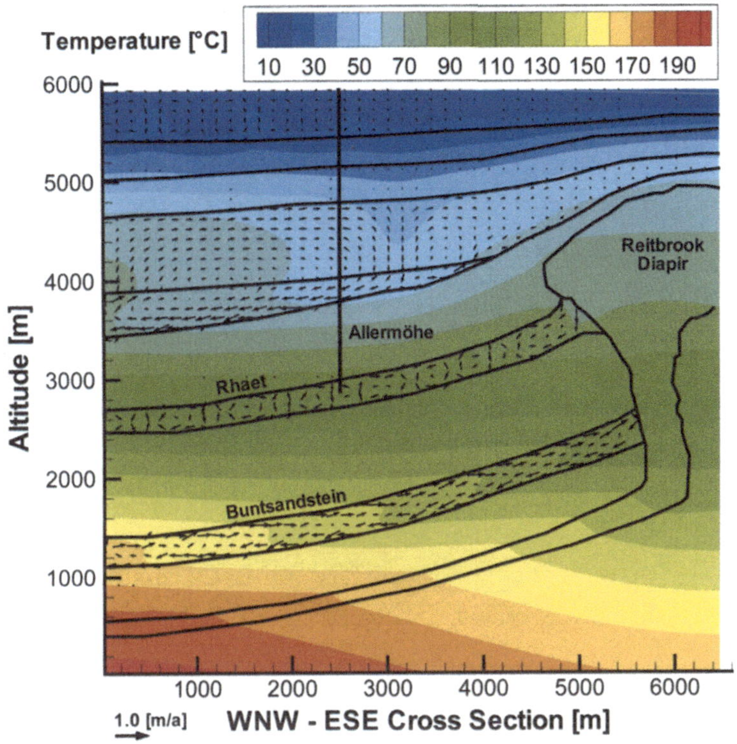

Fig. 5.24. Spatial temperature distribution within the "recent" structure based on the cross-section published by Schmitz and Flixeder (1993), contour colors display temperature and arrows direction and magnitude of flow

Numerical investigation of the cross-sections crossing the Allermöhe well (no. 2 and 5) cut out of the 3D structure of the Allermöhe site (Fig. 5.2) reveals good agreement for Cross-section 2 (Fig. 5.25, blue triangles) and so far the best fit taking fluid flow and heat transfer into account. Cross-section 5 (Fig. 5.25, brown squares) exhibits a temperature profile with overestimation in shallower and underestimation in deeper parts. It can be concluded that the 2D approximation of the 3D natural flow field by Cross-section 2 reflects the flow conditions of the area around the bore far better than Cross-section 5.

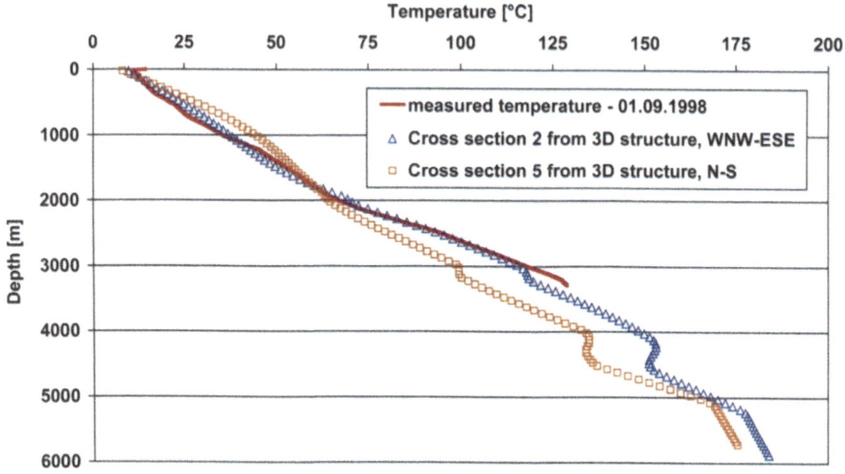

Fig. 5.25. Temperature survey from 1998 in comparison to results of numerical simulations based on 2D cross-sections (Fig. 5.2) cut out of the 3D structure of the Allermöhe map sheet extracted from the Tectonic Atlas of North Germany (Fig. 5.3)

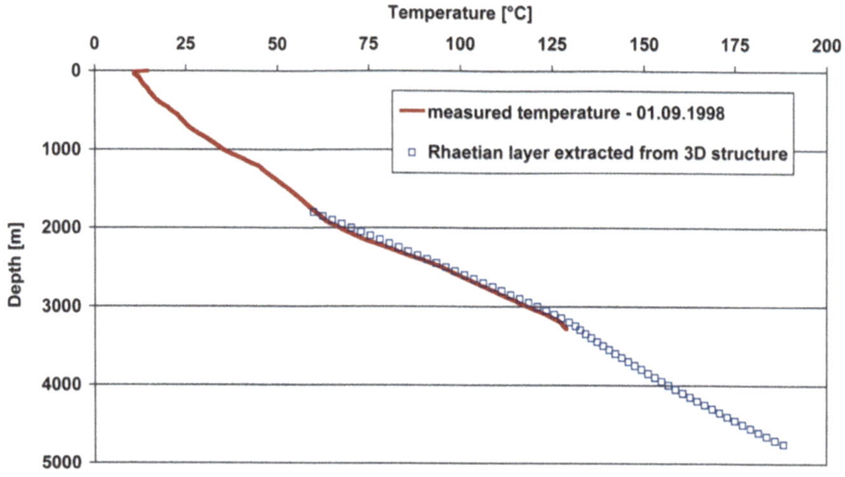

Fig. 5.26. Temperature survey from 1998 in comparison to results of numerical simulations based on the Rhaetian layer extracted from the 3D structure of the Allermöhe site (Fig. 5.3)

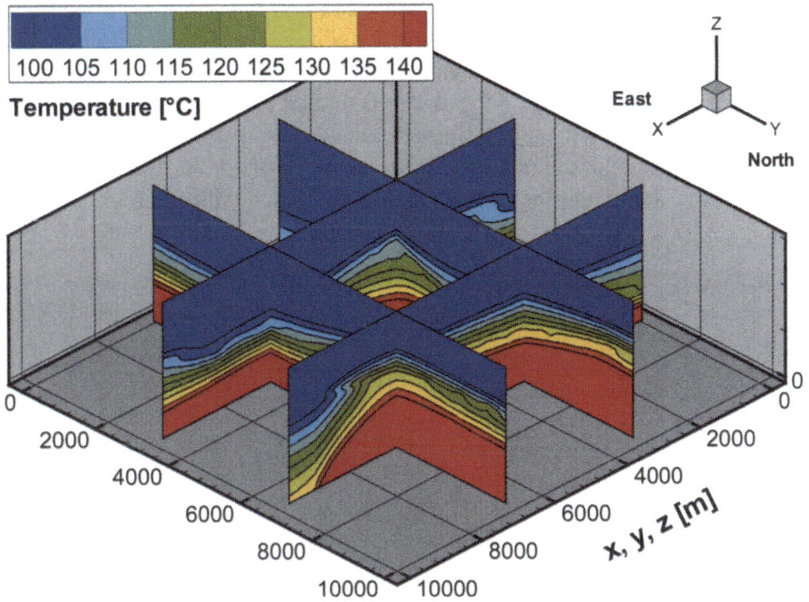

Fig. 5.27. Spatial temperature distribution within the 3D model of the Allermöhe site; intersection of the planes in the foreground displays the location of the Allermöhe well; point of view from NE direction

In Fig. 5.26 the calculated temperature profile from the 3D simulation is compared to the measured on-site survey from 1998. An excellent fit can be observed for the complete intersection between the field investigations and the numerical computations from 1750 to 3250 m depth. Especially within the Rhaetian sandstone (3000-3250 m) very good agreement is determined, on the contrary to all 2D models analyzed so far, because in the 3D model does no vertical convection take place. Vertical convection always leads to low temperature gradients or even constant temperature versus depth in temperature profiles. For the 3D simulation it can be deduced from the numerical results that neither the calculated temperature profile (Fig. 5.26) nor the spatial temperature distribution (Fig. 5.27) exhibit any sign of vertical convection in the permeable topmost Keuper layer. These findings are in accordance with Bjorlykke et al. (1988) who stated that free vertical convection is unlikely to occur in most sedimentary basins. Hence, the horizontal convection process, observed within this 3D numerical case study (Fig. 5.22), is the only transport mechanism able to explain both the measured temperature profile as well as precipitation and dissolution reactions within the vicinity of the Allermöhe well.

3D Sequence of the Geological History

The previously shown results of reactive transport modeling and the comparison of the temperature profiles yielded from the numerical simulations and the on-site survey led to the decision to further derive and investigate a three dimensional sequence of the geological history of the Allermöhe site.

The oldest investigated geological structure is the one representing Dogger times (e.g. the Dogger base lying flat at 0 m a.s.l.) with the Lias layer being the uppermost stratigraphic unit. Diapirism already let to the formation of the salt domes of Meckelfeld and Reitbrook (see above). Fig. 5.28 displays the resulting Darcian flow field from the 3D simulation combined with the rock alteration index.

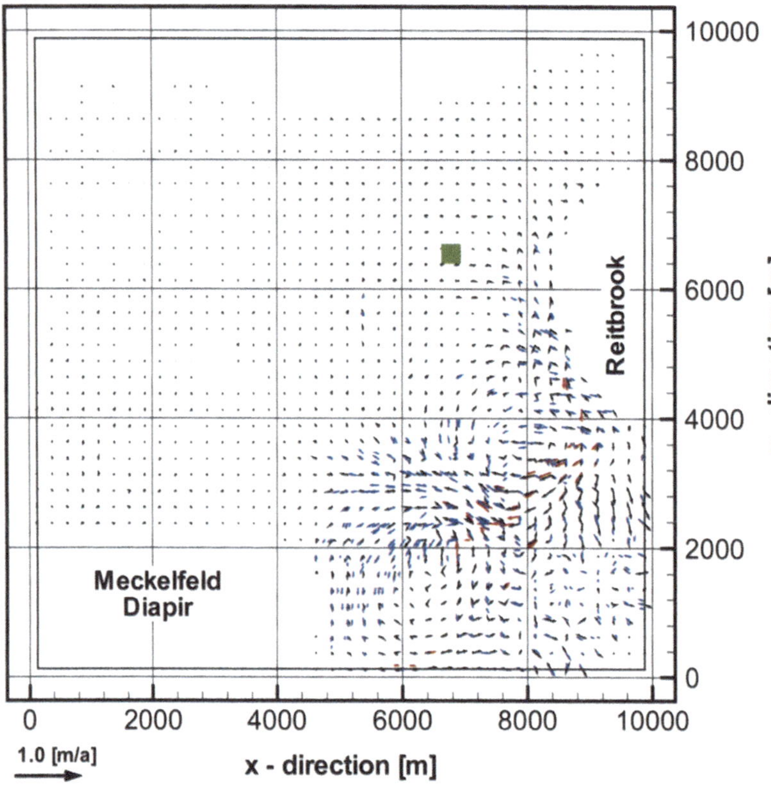

Fig. 5.28. Plan view of the modeled area (Allermöhe map sheet) during Dogger times (Lias = top of stratigraphy) exhibiting the Allermöhe well (green square) and the diapirs of Meckelfeld (SW) and Reitbrook (E); the arrows display the resulting Darcian flow field from the 3D simulation in direction and magnitude, scaled according to the reference arrow shown; colored arrows represent the rock alteration index (RAI) delineating areas of potential anhydrite precipitation (red) or dissolution (blue) in the Rhaetian sandstone

The Rhaetian layer displays only a weak relief compared to the situation within the recent structure. As a result there is hardly any flow in the vicinity of the Allermöhe well. Fluid flow and subsequent potential anhydrite precipitation (red arrows) or dissolution (blue arrows) in the Rhaetian sandstone is restricted to the SE of the modeled area. There is obviously no flow field active, which would be able to transport solutes from the salt domes to the Allermöhe well. Hence, the observed anhydrite cementation cannot be a result from reactive transport processes during Dogger times early after the initial stage of diapirsim.

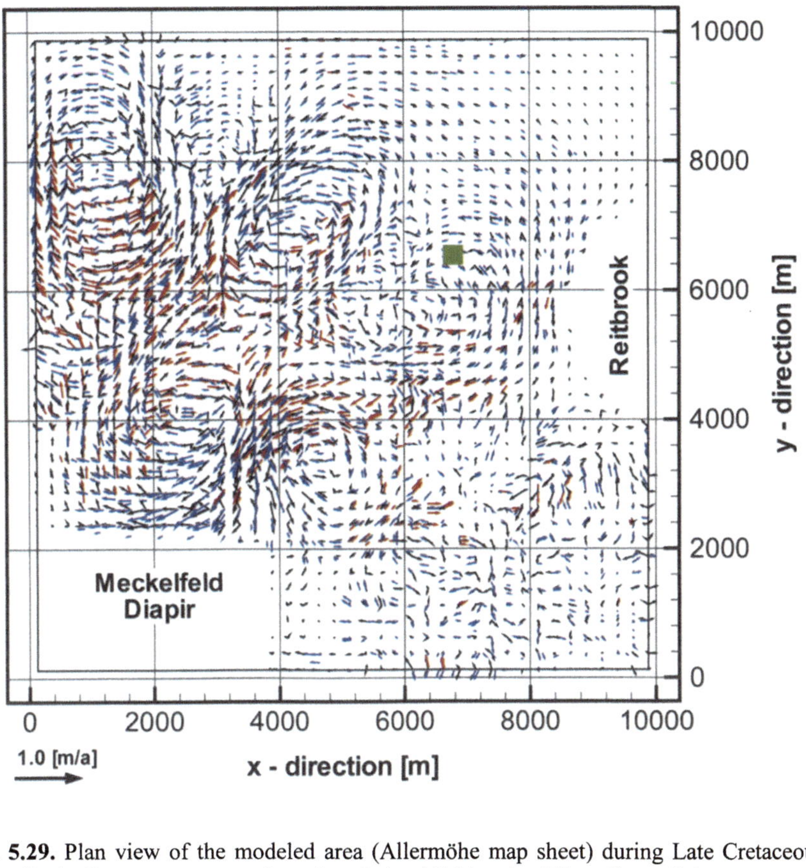

Fig. 5.29. Plan view of the modeled area (Allermöhe map sheet) during Late Cretaceous times (Early Cretaceous = top of stratigraphy) exhibiting the Allermöhe well (green square) and the diapirs of Meckelfeld (SW) and Reitbrook (E); the arrows display the resulting Darcian flow field from the 3D simulation in direction and magnitude, scaled according to the reference arrow shown; colored arrows represent the rock alteration index (RAI) delineating areas of potential anhydrite precipitation (red) or dissolution (blue) in the Rhaetian sandstone

Fig. 5.29 exhibits the situation in Late Cretaceous times with the Early Cretaceous stratigraphic unit being the uppermost layer of the 3D structure. It can be seen that compared to Dogger times (Fig. 5.28) the flow field has changed significantly. The main reactive area due to the flow field is situated in the NW of the map sheet whereas flow in the SE is slightly reduced. The Allermöhe well is situated in the center of a small horizontal convection cell in contact with the Reibrook salt structure. However, the rock alteration index displays an area of potential anhydrite dissolution. The 3D structure of the Allermöhe site during Early Cretaceous times does not supply a reasonable explanation for the observed anhydrite cementation. In general, it can be seen that the entire flow field consists of a number of smaller convection cells compared to the large cell existing within the recent structure (Fig. 5.22).

Fig. 5.30 displays the flow field resulting from the actual geologic structure at the Allermöhe site during Tertiary times with the Late Cretaceous layer being the topmost stratigraphic unit. The arrangement of the flow field has not changed much from the previous historic stage (Fig. 5.29) but the intensity of flow increased significantly.

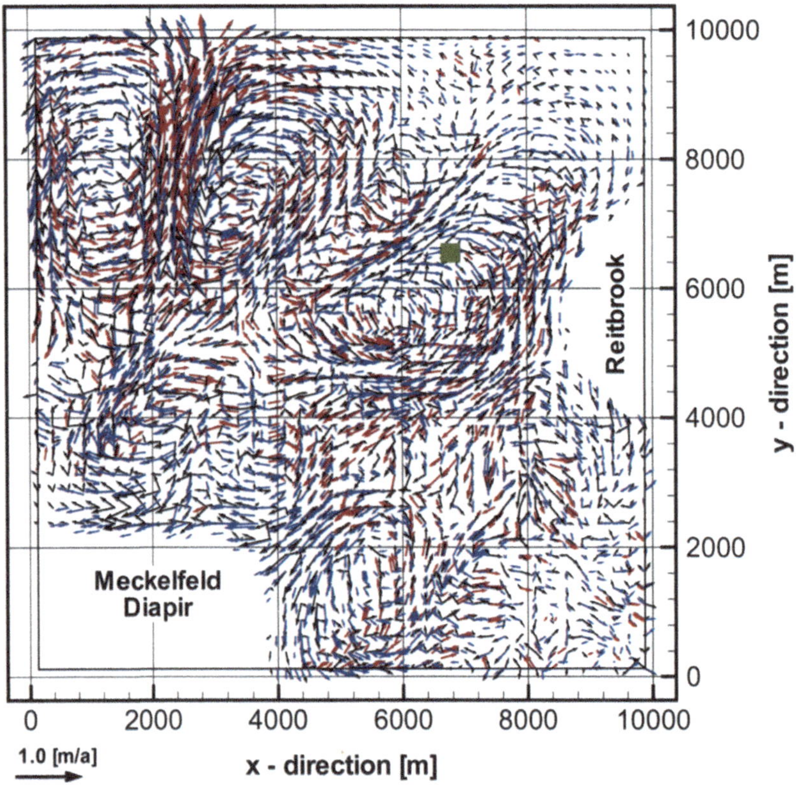

Fig. 5.30. Plan view of the modeled area (Allermöhe map sheet) during Tertiary times (Late Cretaceous = top of stratigraphy) exhibiting the Allermöhe well (green square) and the diapirs of Meckelfeld (SW) and Reitbrook (E); the arrows display the resulting Darcian flow field from the 3D simulation in direction and magnitude, scaled according to the reference arrow shown; colored arrows represent the rock alteration index (RAI) delineating areas of potential anhydrite precipitation (red) or dissolution (blue) in the Rhaetian sandstone

With the increased flow velocities the rock alteration index exhibits high potential anhydrite precipitation and dissolution activity within the entire region of the Allermöhe map sheet. A clockwise convection cell still characterizes the vicinity of the Allermöhe well. Dissolution of anhydrite is most likely around the well during Tertiary times. In the NW quadrant a region with high flow rates and both anhydrite precipitation and dissolution can be seen. Detailed investigation reveals dissolution in shallower parts of the Rhaetian sandstone and precipitation in deeper parts.

From the rock alteration index simulations of the 3D historical sequence, representing a measure for the degree of anhydrite relocation in the investigated area during different geological times, it can be concluded that anhydrite precipitation near to the Allermöhe well is most unlikely even when anhydrite is supplied by a source different from the salt structures. The varying flow fields and rock alteration indices (Fig. 5.22 and Fig. 5.28 - Fig. 5.30) emphasize an area of anhydrite dissolution around the Allermöhe well. However, it is obvious that during all investigated geologic times, the development of horizontal convection cells leads to specific anhydrite precipitation and dissolution pattern assuming that sufficient amounts of anhydrite are available.

Transport of Solutes from the Salt Structures to the Allermöhe Well

Conversely to the investigation of the rock alteration index where anhydrite is supplied from the entire investigated area (previous section), the reactive transport simulation shown here starts with the assumption that anhydrite is available from the salt structures only. The Rhaetian sandstone is initially assumed to be totally free of anhydrite. The aim is to determine anhydrite precipitation patterns for the entire area of the Allermöhe map sheet and to further investigate if precipitation of anhydrite may occur at the Allermöhe well due to anhydrite dissolution from the diapirs and transport of calcium and sulfate through the 3D structures.

Fig. 5.31 displays the resulting flow field within the recent structure of the Allermöhe site after a simulated time period of 50,000 years. The red arrows display areas with anhydrite amounts exceeding 10 mol m^{-3}, due to transport of solutes from the salt structures and subsequent precipitation of anhydrite. It is obvious that significant amounts of anhydrite do occur only in the southern parts of the investigated area but not around the Allermöhe well. This is due to the fact that the large convection cell turns clockwise and the solutes, leached from the salt structures, firstly flow down to deeper and hotter parts of the 3D structure, resulting in anhydrite precipitation. Flow towards the location of the Allermöhe well is directed upwards the Rhaetian sandstone leading to increasing temperatures and lowering the probability of precipitation.

Comparison of the observed precipitation patterns shown in Fig. 5.31 with the RAI determined within the recent 3D structure (Fig. 5.22) reveals agreement of the regions where anhydrite precipitation is simulated. From that point of view it could be supposed that the RAI already provides complete information about regions where precipitation occur and the areas where anhydrite deposits could be excluded. For further examination of this hypothesis the other geological structures of the 3D historical restoration sequence (Fig. 5.28 - Fig. 5.30) were investigated in the same manner. The leaching process from the salt structures followed by transport through the aquifer and subsequent precipitation of anhydrite was simulated. First of all it has been found that anhydrite precipitation does not occur within the vicinity of the Allermöhe well, emphasizing that the RAI is a qualified parameter to delineate areas where anhydrite precipitation is unlikely.

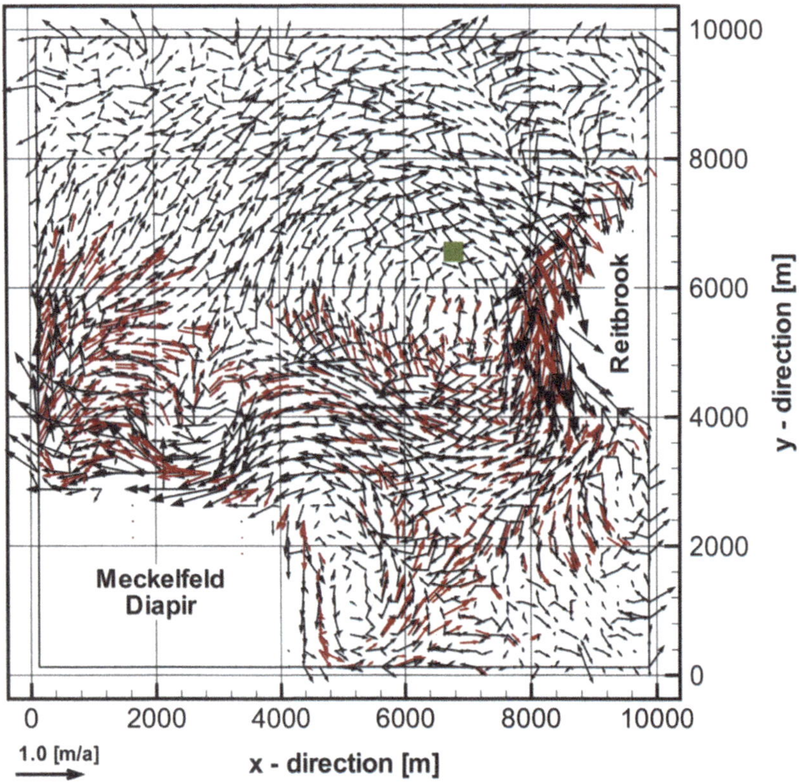

Fig. 5.31. Plan view of the modeled area (Allermöhe map sheet) of the recent structure exhibiting the Allermöhe well (green square) and the diapirs of Meckelfeld (SW) and Reitbrook (E); the arrows display the resulting Darcian flow field from the 3D simulation in direction and magnitude, scaled according to the reference arrow shown; red arrows represent anhydrite precipitation in the Rhaetian sandstone with amounts above 10 mol m^{-3} after a simulated time period of 50,000 years; in this case anhydrite is supplied from the salt structures only

The numerical simulations of the 3D structures representing Dogger times and Late Cretaceous times (compare Fig. 5.16), both not shown here, exhibit hardly any precipitation at all as result of the leaching process over the entire region of the Allermöhe map sheet. Whereas this has been observed already with the RAI for the structure representing Dogger times (Fig. 5.28), significant anhydrite precipitation is supposed by the RAI within the region north of the Meckelfeld diapir in the structure representing Late Cretaceous times (Fig. 5.29). Fig. 5.32 displays the resulting flow field of the structure representing Tertiary times. Significant amounts of anhydrite (> 10 mol m^{-3}, red arrows) occur only in the southern part of the Allermöhe map sheet, north of the Meckelfeld diapir and between both salt structures. These results are in contradiction to the ones gained from simulations of the RAI (Fig. 5.30). Especially the area of precipitation, predicted by the RAI, in the NW quadrant of the Allermöhe map sheet cannot be observed during the leaching process.

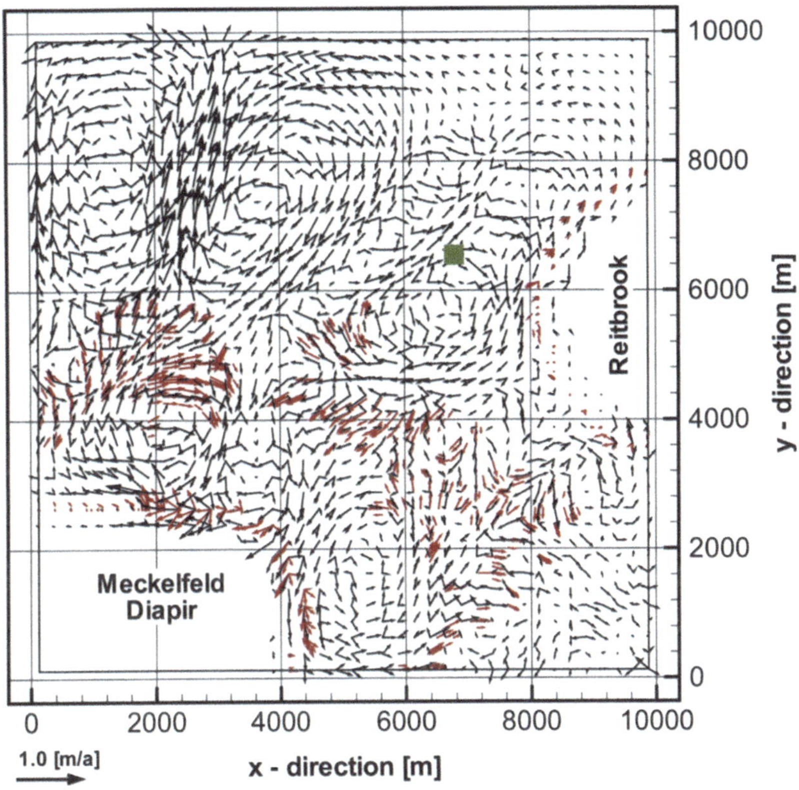

Fig. 5.32. Plan view of the modeled area (Allermöhe map sheet) during Tertiary times (Late Cretaceous = top of stratigraphy) exhibiting the Allermöhe well (green square) and the diapirs of Meckelfeld (SW) and Reitbrook (E); the arrows display the resulting Darcian flow field from the 3D simulation in direction and magnitude, scaled according to the reference arrow shown; red arrows represent anhydrite precipitation in the Rhaetian sandstone with amounts above 10 mol m^{-3} after a simulated time period of 50,000 years; in this case anhydrite is supplied from the salt structures only

It can be concluded that the observed flow field of a particular model and the RAI determined within this structure (Fig. 5.22, Fig. 5.28 - Fig. 5.30) already delineates the regions where anhydrite deposition can be excluded. Hence, during the simulation of a leaching scenario (transport from the salt structure) precipitation does not occur where the RAI predicts dissolution. But the information from the RAI with respect to precipitation has to be further checked by simulations investigating the leaching process to determine more precisely probable anhydrite distribution patterns.

The plot of the temporal development during the simulated time span of the temperature, the sodium chloride and calcium concentrations at the Allermöhe site based on the recent structure is shown in Fig. 5.33. The displayed parameters reveal that the system develops from initial conditions (see above) to equilibrium within approximately 20,000 years corresponding to flow velocities around 1 m a^{-1} and the investigated area of around 10 km by 10 km. Whereas temperature approaches the value of 125°C measured at depth and the calcium amount remains almost constant, the sodium chloride concentration increases from the observed amounts of around 3000 mmol L^{-1} to values above 5000 mmol L^{-1}. These results can be interpreted in two different ways: (1) A horizontal convection cell exists, emphasized by the agreement of measured and simulated temperature profile and Ca concentration. But the processes controlling the sodium chloride contents of the formation water are still not completely understood and precisely enough implemented in the numerical model. (2) A horizontal convection cell can be excluded, if sodium chloride concentrations as high as simulated would actually result from such convection within the investigated geologic structure. This assumption holds if the processes of halite dissolution and precipitation are adequately reproduced in the model. However, in this case the fluid flow conditions within the investigated area are only poorly understood.

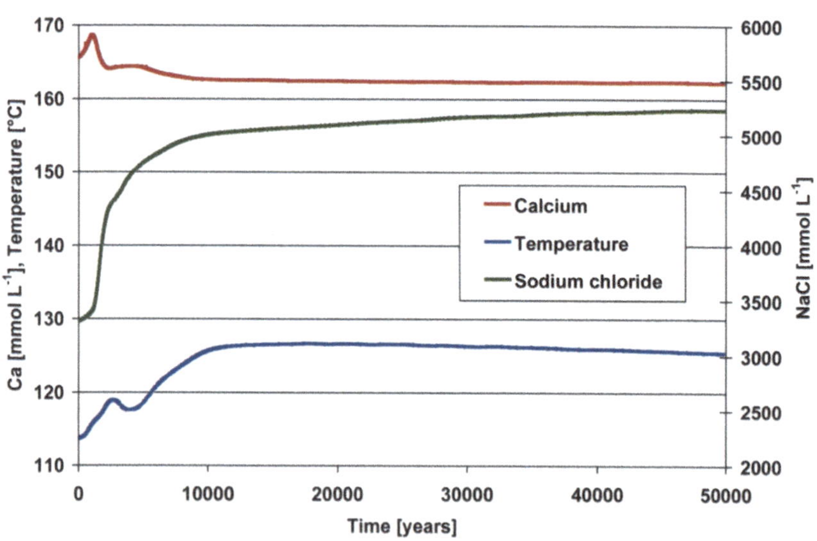

Fig. 5.33. Variation of temperature, sodium chloride and calcium content at the Allermöhe well from initial conditions to a simulated time period of 50,000 years within the recent 3D structure

5.2.5 Summary and Conclusions of the Allermöhe Case Study

In sufficiently thick and permeable formations free thermal convection is a potential process for mass and heat transport and diagenesis in sedimentary basins (Bjorlykke et al. 1988, Raffensberger and Garven 1995). A conceptual study was performed under the assumption of a 500 m thick and permeable reservoir in the vicinity to salt domes (Fig. 5.9). Free convection led to transport of solutes leached from the salt diapirs and resulting precipitation of anhydrite in the formation. The 2D numerical simulations revealed that the occurrence of anhydrite deposits strongly depends on the geometry of the stratigraphic units (Fig. 5.11).

The relevance of the structural geometry for the Allermöhe site was investigated by a 2D historical sequence (Fig. 5.12 - Fig. 5.15). However, the simulations showed that only the geometry of the recent structure might cause anhydrite precipitation in the Rhaetian sandstone (Fig. 5.19). A minimum time period of 5 to 15 Mio years would be necessary to produce anhydrite cementation to an extent as observed in the Allermöhe well. Nevertheless, the 2D simulation of the recent structure predicts anhydrite to precipitate particularly outside the region where the Allermöhe well is situated (Fig. 5.19). Even fault zones, providing conduits to deeper formations, are not suitable to explain the reduction of porosity at the Allermöhe site. On the contrary, the anhydrite precipitation observed in the Rhaetian aquifer in the recent structure is prevented due to the flow field evolving as a result of the fault zones (Fig. 5.19 and Fig. 5.20).

The temperature survey from 1998 has been used to evaluate the results of the numerical studies. Simulation of the 3D structure of the Rhaetian formation provides an excellent fit for the temperature profile with the on-site survey (Fig. 5.26) conversely to the 2D calculations, which all exhibit especially within the Rhaetian layer (3000-3250 m) disagreement with the measurements (Fig. 5.23, Fig. 5.25, and Fig. 5.26). Flow field and temperature distribution emphasize that free horizontal convection occurs, whereas vertical convection, as predicted within the 2D models is most unlikely. Three-dimensional models are a prerequisite for an adequate reproduction of fluid flow and heat transfer processes at the Allermöhe site.

The thermal rock alteration index (RAI, Raffensberger and Vlassopoulos 1999) provides a measure of the amount of thermally driven sediment alteration due to fluid movement. The simulations of the 3D historical restoration sequence of the Allermöhe site reveal that the area around the Allermöhe well is only affected by dissolution (Fig. 5.22, Fig. 5.28 - Fig. 5.30). The observed flow field of a particular model and the RAI determined within this structure already delineate the regions where anhydrite deposition can be excluded. Hence, during the simulation of a leaching scenario (transport from the salt structure, Fig. 5.31 and Fig. 5.32) precipitation does not occur where the RAI predicts dissolution. But the information from the RAI with respect to precipitation has to be further checked by simulations investigating the leaching process to determine more precisely probable anhydrite distribution patterns.

The conclusions which can be drawn from the presented numerical simulations, studied here under special consideration of the geological structure of the Allermöhe site and its historical development falsifies the hypothesis of Lenz et al.

(1997). Their assumption that the observed anhydrite cementation at the location of the Allermöhe well may be due to solutes, leached from the salt structures, transported through the system, and subsequently precipitated, is most unlikely. Firstly, it has been shown that the geometry of areas in vicinity to salt structures is highly important. Secondly, this led to the conclusion that 2D simulations of the Allermöhe site are not able to represent flow field and temperature distribution accurately. Distinct investigation of various palaeo 3D structures emphasized that the vicinity of the Allermöhe well is a potential area of anhydrite dissolution and that precipitation occurs predominantly in southern parts of the area. But even the horizontal convection cells, observed within the 3D simulations, are questionable, because the simulated temporal development of the sodium chloride concentration at the Allermöhe well significantly deviates from chemical measurements (Fig. 5.33). Hence, transport of solutes from the diapirs to the Allermöhe well seems to be implausible.

The hypothesis of Baermann et al. (2000a) that up-flow of brines from deeper stratigraphic units via fault zones and resulting anhydrite precipitation due to changing temperature and chemical conditions does not hold either. Although significant changes of the flow field has been shown within the 2D simulations investigating the influence of fault zones, the up-flow always leads to decreasing temperatures and a decreasing potential of anhydrite precipitation.

Christensen et al. (2002) also state that up-flow of brines may be the reason for the observed anhydrite cementation at the Allermöhe site. They assume that solutions from the Gipsmergelkeuper, underlying the Rhaetian sandstone formation, led to precipitation of anhydrite due to significant pressure differences of the neighboring geologic formations. Fact is, that the close neighborhood of Rhaet and Gipsmergelkeuper excludes significant pressure differences for the case that fault zones hydraulically connect both formations. Hence, the solubility difference of anhydrite within the formation waters is small. Additionally, the analyzed sulfate isotopic signature of Allermöhe core samples is significantly different from Keuper samples as published by Baermann et al. (2000b). If dissolution, transport, and subsequent precipitation are the reason for an observed cementation the sulfate isotopic signature of source and target must be identical. Both arguments disprove the theory of Christensen et al. (2002).

Major arguments have been presented that falsifies the hypotheses attributing the anhydrite cementation observed at the Allermöhe site to the leaching of solutes from salt diapirs or deeper stratigraphic units, their transport into the Rhaetian sandstone, and subsequent precipitation of anhydrite. However, a mechanism suitable to explain the anhydrite formation within the Rhaetian sandstone at Allermöhe must be compatible on the one hand with the "regional scale" of the anhydrite cementation (meters to kilometers) and on the other hand with the structures observed at the "well scale" (decimeter). Another hypothesis, the formation and growth of anhydrite due to capillary evaporation in a highly saline and high temperature sabkha environment (sabkha is an Arabic word for salt flat), not discussed so far, may provide a mechanism suitable to combine anhydrite distribution patterns on all scales.

Sabkhas are low-lying salt-encrusted marine or continental mudflats where displacive and replacive evaporite minerals are forming in the capillary zone above a saline water table (Warren 1989). Evaporites form where a potential exists for more water to leave the basin by evaporation than to enter the basin by a combination of rainfall, surface, and subsurface inflow. Thus, evaporite deposits are most often found in arid and semiarid deserts (ancient or recent). This is in agreement with the arid climate stated by Wurster (1965) for the German Triassic times. Characteristic environments thought to be important for the development of sabkhas exhibit a continuous spectrum from marine coastal to fluviolacustrine and eolian dominated types (Handford 1981). Sandstones of the Rhaetian in the area of Hannover (gasfield Thönse, ≈150 km south of Allermöhe) were deposited in a westward prograding fluvial-dominated delta. Lateral shifts of the delta position are assumed to be responsible for sudden cessation of sand sedimentation. The basin wide sea level rise during the Rhaetian is expected to be the major reason for the landward retrogradation of deltaic facies belts to the east (Gaupp 1991). It has generally been observed that many ancient sabkhas evolved during a single progradational or retrogradational episode in the coastal plain from marine coastal to eolian to fluviolacustrine-dominated systems or vice versa (Warren and Kendall 1985).

Most of the evaporites in a sabkha are deposited in the capillary zone as intrastratal nodules and crystals. Sabkhas form shoaling sequences in both continental and coastal mud flats. Idealized sabkha sequences are cycles with displacive evaporites in the upper part of the salt flat. Each cycle is capped by an erosion surface, the result of displacive crystal growth raising the surface of the wet mud flat into the vadose zone. Prograding marine sabkhas exhibit gradation with finer material to the top, while continental sabkhas tend to coarsen upward. Sandy reservoirs are found in marine sabkhas in the subtidal and intertidal sand bodies sealed by evaporites. Sandy reservoirs in continental sabkhas reflect infilled stream channels and dune sand sheets (Warren 1989). Mineralogical investigations of the Rhaetian sandstone samples performed by Baermann et al. (2000a) showed comparable sequences of alternating layers of shale and sandstone, sandstone with interbedded shale fillings, and massive bedded sandstones.

Many ancient evaporites are composed of salts originally formed by the concentration of seawater-derived brines. When the brine has about five times the concentration of seawater (175 g L^{-1} total dissolved solids compared to 218 g L^{-1} of the Allermöhe formation water) gypsum precipitates. As the system is concentrated to a level slightly before halite saturation (eleven times seawater, 385 g L^{-1}), anhydrite replaces gypsum. The calcium sulfate minerals gypsum and anhydrite are important phases and their occurrences depend on prevailing temperature, pressure, and salinities. Anhydrite can precipitate and grow by capillary evaporation due to high salinities or temperature, or it can replace gypsum already present in the sabkha. Hence, when bedded gypsum is buried and the temperature rises above 60°C it is transformed to nodular anhydrite (Warren and Kendall 1985). Baermann et al. (2000a) found similar patterns of anhydrite cementation in the Allermöhe core samples: beside sandstone layers with pore spaces totally filled with anhydrite, insular or cloudy, and layered anhydrite cementation were observed in cyclic sequences. Evaporites are susceptible to diagenesis from the time they are first laid down, and this often destroys much of the early depositional textures. For example, subsurface movement of pore waters can leach evaporites in siliclastic matrices creating secondary porosity.

From the known facts, synsedimentary formation of the observed anhydrite cementation is the most likely of all discussed hypotheses. Growth of anhydrite due to capillary evaporation in a highly saline and high temperature sabkha environment provides an opportunity to explain a high variety of cementation structures. Ward et al. (1986) presented characteristics of various settings of evaporite deposits. They described the coastal mud flats of shoaling cycles of evaporites and quartz sands and silts sealed with gypsum and/or anhydrite ("well scale"). The extent of the described sabkhas is around 40 km ("regional scale"). That means areas with such high anhydrite cementation, as observed at Allermöhe, should be restricted to a limited area. For example, the Rhaetian sandstone of the location Neustadt-Glewe (≈100 km east of Allermöhe) does not show any anhydrite cementation. If it is possible, with the help of geological investigations, to precisely reproduce the shoreline with respect to a particular formation and to delineate sabkha environments, the risk of geothermal reservoir exploration might be significantly decreased.

6 Recent Hydrothermal Systems

The investigation of recent hydrothermal systems, especially the understanding of their development and structure, is one of the main fields of application of reactive transport simulation models. The aim of numerical studies of recent hydrothermal systems is to set-up or evaluate conceptual models of geothermal areas which are able to describe the processes of fluid flow and heat transfer as well as to explain the formation of observed alteration products. This is the preliminary stage to the application of reactive transport simulation for reservoir management (compare Chap. 7) where evaluated models are used for parameter estimation in response to the exploitation of a hydrothermal system.

Within the first part of this chapter typical, currently published numerical studies of recent hydrothermal systems are summarized. The published case studies describe sophisticated numerical simulations contributing new insights to the understanding of the structure and development of hydrothermal systems. The following second part is a detailed case study of the shallow hydrothermal system of Waiwera (New Zealand). The case study evaluates the proposed conceptual model of the geothermal field and the derived natural state is used for history matching of the exploitation since 1863. Under consideration of the current conditions reservoir development is estimated until the year 2018.

6.1 Investigating Geothermal Field Development and Structures

6.1.1 Generic Model of the Taupo Volcanic Zone (New Zealand)

White and Mroczek (1998) performed a conceptual study investigating the Taupo Volcanic Zone of New Zealand (TVZ) using CHEM-TOUGH, a version of TOUGH2 modified to include transport of reacting chemicals. This study was performed in order to deepen the understanding of the structure and development of hydrothermal circulation in regional geothermal systems. These simulations to investigate silica transport in both subcritical and supercritical flow regimes are based on results of Kissling (1997) with regard to model dimension, heat source, and magmatic water composition.

Kissling (1997) has argued that a single magmatic intrusion provides insufficient heat to drive a geothermal field like the TVZ. It is estimated that a geothermal field requires 1000 km^3 of magma implying several hundred intrusions of 1-

10 km^3 distributed spatially and in time. These intrusions are represented as a single source in the model of White and Mroczek (1998). The TVZ conceptual system, 13 km wide and 10 km deep, has been modeled for a time period of 500,000 years. The heat source was modeled with constant temperature of 550°C at a depth of 10 km together with an inflow of fluid at a rate of 10 kg s^{-1} over this area (Fig. 6.1). This mass flow results in approximately 5 % magmatic water in the reservoir near the surface.

White and Mroczek (1998) investigated that groundwater circulation from the top down to the base of the model only occurs for a short time (< 50,000 years, Fig. 6.1). At the end of the simulated period of 500,000 years permeability has increased by a factor of 100 at the base of the model in the up-flow zone due to dissolution of significant amounts of quartz in the 500°C hot fluid. Above this, between about 6 km and 8 km depth, a region with decreased permeability evolved caused by quartz precipitation. Little change in permeability has been observed above 6 km. This model of the TVZ geothermal reservoir shows that the lower reservoir becomes impermeable after a short time and a hydrothermal circulation system is only present in the upper part of the model.

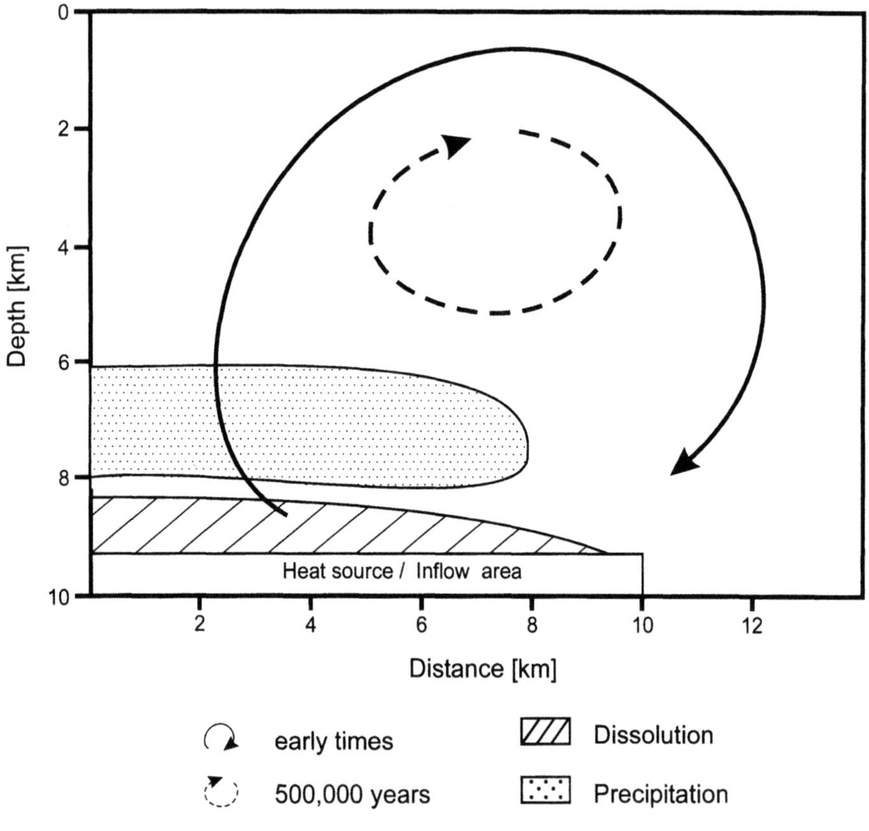

Fig. 6.1. Conceptual model of the Taupo Volcanic Zone (after White and Mroczek 1998)

6.1.2 Mineral Alteration in the Broadlands-Ohaaki Geothermal System (New Zealand)

The Broadlands-Ohaaki geothermal field is located at the east side of the Taupo Volcanic Zone (New Zealand). It is a liquid dominated, boiling hydrothermal system hosted by a sequence of Quaternary felsic volcanic rocks and Mesozoic sediments. A comprehensive description of the geochemical and thermal structure and the hydrogeology of the epithermal environment is given by Hedenquist (1990). Simmons and Browne (2000) investigated the relationship between mineral distribution patterns and alteration processes occurring in this active geothermal system. Among other things, they performed reaction path modeling using the program code CHILLER. The distribution of aqueous and gaseous species and the amounts of minerals deposited were calculated along both an isenthalpic boiling and a mixing path.

Simmons and Browne (2000) emphasized that the distribution of hydrothermal minerals at Broadlands-Ohaaki can be related to ascend of a hydrothermal fluid (Fig. 6.2), containing high amounts of dissolved gases (CO_2, H_2S, etc.).

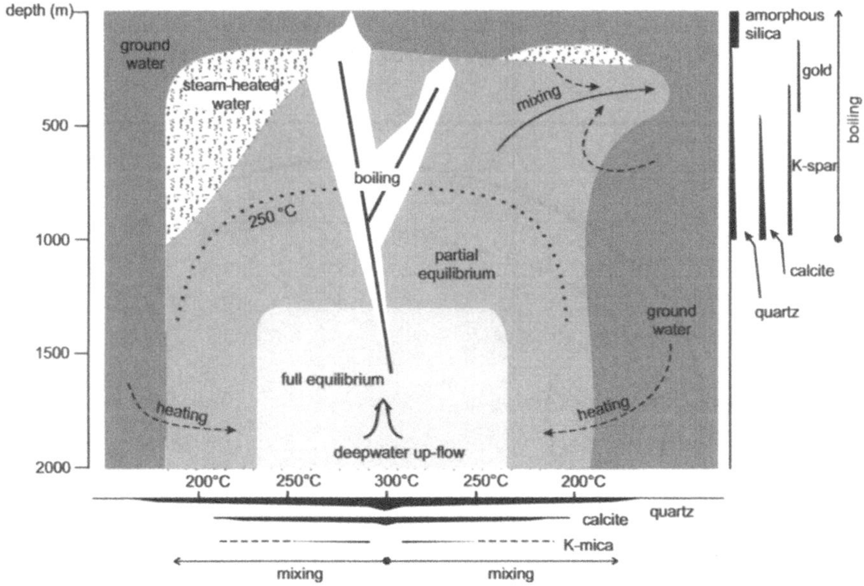

Fig. 6.2. Conceptual model of the Broadlands-Ohaaki geothermal system (adapted from Simmons and Browne 2000)

The hydrothermal fluid undergoes water-rock interactions by boiling and/or mixing processes. Within the deep central up-flow zone "full equilibrium" (Giggenbach 1984, compare Chap. 2) is attained. At shallow depth and on the periphery hydrolytic (H-metasomatism) alteration takes place due to CO_2-rich steam-heated water resulting in argillic alteration assemblages. Outward from the margin of the system, alteration intensity diminishes and grades into fresh rocks. Final conclusion of Simmons and Browne (2000) deduced from field observations and reaction path modeling is that boiling is the main process influencing deposition of precious minerals (gold) and that mineral distribution patterns of silica, quartz, calcite, and feldspars associated with different water types delineate the hydrogeological features of the hydrothermal system (Fig. 6.2).

6.1.3 Deep Circulation System at Kakkonda (Japan)

The Kakkonda geothermal field (≈ 600 km NE of Tokyo) is one of the most active liquid dominated fields in Japan. McGuinness et al. (1995) described the production history and hydrology of the production zone. A less detailed model of Hanano and Seth (1995), 18 km wide and 6 km deep, provides the basis for the reactive transport simulations of Kakkonda done by White and Mroczek (1998). The Kakkonda geothermal field is characterized by a large neo-granitic pluton at a depth 2000-2800 m (Fig. 6.3). Simulations were conducted to determine the development of hydrothermal circulation systems during a geological time span of 100,000 years, due to and right after the emplacement of an 800°C hot pluton. The occurring circulation processes associated with the cooling granite intrusion led to permeability changes in the reservoir caused by dissolution and precipitation of quartz.

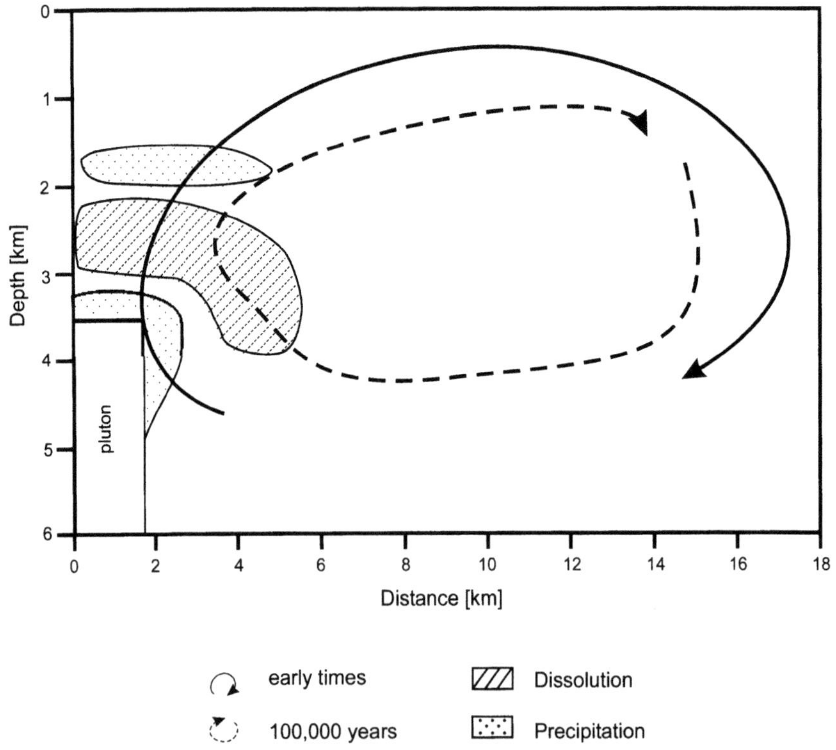

Fig. 6.3. Conceptual model of Kakkonda geothermal field (after White and Mroczek 1998)

White and Mroczek (1998) observed permeability reduction above the granitic intrusion as a result of hydrothermal convection and reactive transport during early times (Fig. 6.3). Permeability reduction occurred during the whole modeled time period caused by the reduced silica solubility near the critical point of water. Above the region with reduced permeability follows a region of enhanced permeability and a second area of reduced permeability (Fig. 6.3). Enhancement of permeability is due to cool water being drawn into the up-flow zone and becoming undersaturated in silica when heated. Ascending within the up-flow zone the water cools and becomes finally supersaturated in silica and precipitation begins. The final convection cell after 100,000 years results from silica precipitation and accompanying permeability reduction.

6.1.4 Alteration Halo of a Diorite Intrusion

White and Christenson (2000, 2002) investigated the alteration halo of an idealized diorite intrusion, because magmatic intrusions are often the heat sources of geothermal fields (for example Kakkonda, Sasada et al. 1998). The idealized simulations of an intrusion, here according to the diorite body intersected during drilling in the Ngatamariki field (New Zealand, Christenson et al. 1997), were performed with CHEM-TOUGH. Extensive hydrothermal alteration within the plutonic rock and in a halo surrounding it suggests that it acted as a heat source for a convective hydrothermal system. The simulations focused on the influence of magmatic vapor intruding into the base of a geothermal reservoir and resulting alteration around the magmatic intrusion as well as water compositions in shallow parts of the field, due to fluid flow, heat transfer, transport, and reaction processes.

The model of White and Christenson (2002) takes into consideration thermal and chemical effects right after emplacement of the diorite intrusion. The hydrothermal reservoir contains a reactive fluid dominated by CO_2. White and Christenson (2000) demonstrated the possibility to simulate the effect of a pulse of magmatic vapour (CO_2, SO_2, H_2S, HCl) into the base of a reservoir and to calculate changing chemical and physical conditions over a time period of 15,000 years. They observed convective upwelling above the intrusion (comparable to TVZ, Fig. 6.1) with highly oxidizing fluids near the heat and vapor source. The addition of the volatiles resulted in hydrolysis reactions and led to dissolution of albite and anorthite close to the intrusion. Elemental sulfur and alunite form in the low pH and oxidizing environment. Similarly, Ca^{2+} released from anorthite is taken up in anhydrite.

The reactive transport approach of White and Christenson (2002) shows a way to set up concepts for reactions occurring in the environment of unexplored heat sources at depth and to associate with it fluid chemistry and alteration products observed in shallow parts of the reservoirs.

6.2 Waiwera – New Zealand

Waiwera is a small east coastal township, north of Auckland, New Zealand, which has developed mainly as a resort area. It is a unique and important asset to the people of the region, mainly due to the thermal waters. The main attraction is a public thermal swimming pool complex, used throughout the year. A low temperature geothermal reservoir is located underneath the township of Waiwera. The proximity to the sea is an important feature of the Waiwera geothermal aquifer. Geothermal fluid of approximately 50 °C is feeding into the well fractured Waiwera reservoir. Fig. 6.4 shows the surroundings of Waiwera.

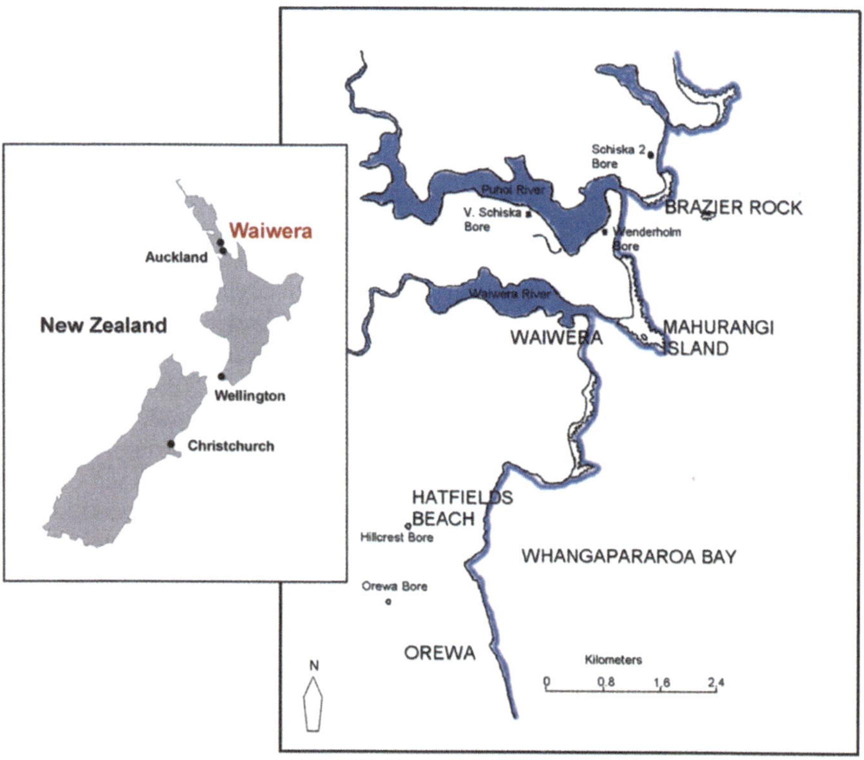

Fig. 6.4. The surroundings of the study area Waiwera (New Zealand)

The main landscape features are: the Waiwera River, which extends 10 km westward; the river estuary with tidal mud-flats; the flat peninsula on which Waiwera is situated at the mouth of the Waiwera river; and the rolling hills between Waiwera and Hatfield's Beach, which rise to 171 m at Te Whau Hill 1 km west of the township (ARC 1999).

6.2.1 History

Geothermal springs had long been a source of pleasure for Maori, the native inhabitants of New Zealand. Since the 19th century the New Zealand Government developed with the help of European investors these springs into spas. The first spa was at Waiwera (Weeber et al. 2001).

The Scottish businessman Robert Graham purchased the Waiwera area in 1842. It is told that the sight of up to 3000 Maoris assembled on Waiwera's beach and bathing in holes in the sand had fascinated Graham. "Waiwera" is a Maori word and means "hot water". The springs, and the Maoris use of them, were described as early as 1841:

"At the mouth of a creek ... the main spring gushes out from a high cliff, about two feet from its base; and successive jets, apparently from the same source, bubble up through the sand, along a line of about a hundred yards, from south to north, all covered by high water. ... The natives have recourse to these springs for the cure of different cutaneous disorders with which they are commonly affected. ... When any person wishes to bathe he digs himself a pool in the sand ... [and] he may then enjoy a comfortable bath" (Rockel 1986).

European utilization of the thermal water began in 1863 with the construction of hotels and was extended with bathhouses in 1872. At that time, boreholes discharged naturally by artesian flow. During the 1950s, the proliferation of the geothermal water utilization began to affect the thermal water supply. Deeper wells had to be drilled, nearly to the full depth of the warm water aquifer. During the 1960s exploitation led to periodically discontinuance of the artesian flow and pumping was necessary for the first time. The last reported natural artesian flow from boreholes occurred in 1969 and the hot springs on the beach apparently ceased to flow between 1975 and 1976 (ARWB 1980).

In 1975, residents informed the Auckland Regional Water Board (ARWB), now Environmental Management Department of the Auckland Regional Council (ARC), of their concern about declining water levels. The Water Board initiated a study designed to assist in the protection, allocation, and management of the resource. A first Waiwera Thermal Groundwater Allocation and Management Plan was adopted by ARWB in 1987 (ARWB 1987)

6.2.2 Geological Setting

The area of interest covers the township of Waiwera and the estuary of the Waiwera River (Fig. 6.5). The dominant rock type at Waiwera is the Waitemata Group Sandstone of the Miocene Pakiri Formation, interlayered with siltstone (Fig. 6.6). The stratified rocks have been tilted, folded, faulted, and fractured by tectonic movement, providing pathways for the groundwater. The fractured Waitemata rock forms the aquifer from which the boreholes at Waiwera extract thermal water. Borehole logs indicate that 400-425 m of Waitemata Group Sandstone is overlying very compact, indurated greywacke of the Jurassic Age Waiheke Group (Fig. 6.6).

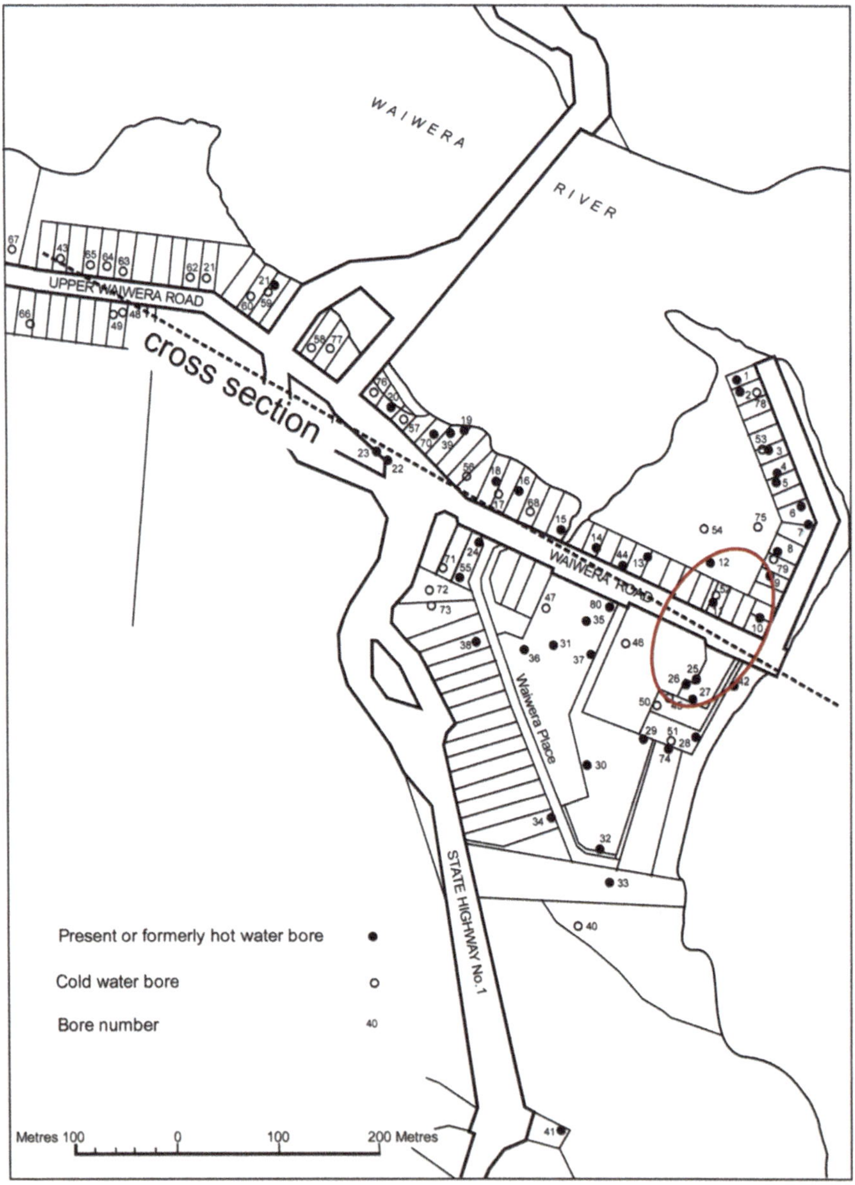

Fig. 6.5. The Waiwera Township with borehole locations; center of the geothermal area is marked with the red ellipse

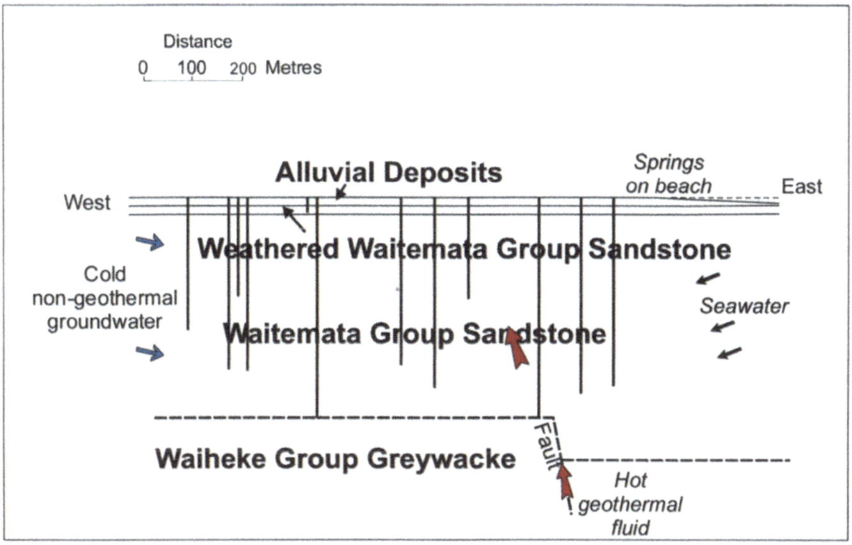

Fig. 6.6. Conceptual model of the Waiwera geothermal aquifer

The flat peninsula at the Waiwera river estuary mouth is composed of unconsolidated alluvial and marine sands, silts, and clays of Holocene age up to 13 m thick. The clays and silts, which overlie the Waitemata Group rock of Waiwera, together with the weathered surface of the Waitemata Group rock, are believed to form a confining layer overlying the aquifer. The weathered layer in this area is commonly 3 to 20 m thick (ARWB/ARC 1980, 1987, 1991, 1999).

The assumed hydrogeological model of the study area is shown in Fig. 6.6 in cross section together with the assumed flow paths of the different types of water (geothermal fluid, groundwater, seawater) entering the aquifer. Geothermal water rises via a fault zone into the Waitemata Group Sandstone forming the low temperature reservoir. The center of the geothermal area is believed to be 100 m from the beachfront approximately on a level with Waiwera Road (Fig. 6.5). At the western margin of the geothermal aquifer the geothermal fluid is cooled by conductive heat loss and dilution with cold, non-geothermal groundwater. At the eastern, seaward margin of the aquifer there is a seawater-freshwater interface.

The composition of the water and the associated gases preclude a volcanic origin. The chemistry of the thermal waters suggests that this water is most likely meteoric. The temperature of the water, about 50°C at 150 m depth, is inconsistent with the temperature gradient of approximately 30 K km^{-1} observed in the region (ARWB 1980). If the unmixed thermal groundwater has been heated from ambient 15°C by the natural geothermal gradient, a source depth of 1,200-2,300 m can be inferred (ARWB 1987). Thus, water from deep down is finding its way to the Waitemata strata via a leakage path such as a fault in the basement greywacke rock (ARWB 1980). Application of the Na-K-Mg geothermometer after Giggenbach (1988) to the Waiwera geothermal waters emphasizes this hypothesis (Fig. 6.7).

Fig. 6.7 displays water samples, which can be classified unambiguously as a strictly geothermal water type, in green triangles. The water samples are from boreholes no. 11, 31, 35, and 37 (Fig. 6.5). The Garden Bore (no. 37) from the public pool is situated almost right above the center of the geothermal field. The relative water compositions plot close to the full equilibration line indicating that the water may have been in mineral-rock equilibrium at a temperature around 80°C. With the ambient temperature and the regional geothermal gradient this corresponds to a source depth of approximately 2,200 m. It is significant to note from Giggenbach (1988) that the relative composition of seawater influenced water samples approaches the full equilibration line from the Na-corner. The samples from the bores no. 1, 7, and 8 (Fig. 6.5), certainly influenced by seawater, are displayed with orange diamonds in Fig. 6.7. Additionally shown are two undoubtedly fresh water samples with blue squares, from the bores no. 40 and 59. They are characterized by its non-attainment of full water rock equilibrium, mainly caused by the slowness of the reaction processes supplying the comparatively large equilibrium contents of Na, especially at lower temperatures. The remaining samples are left unassigned.

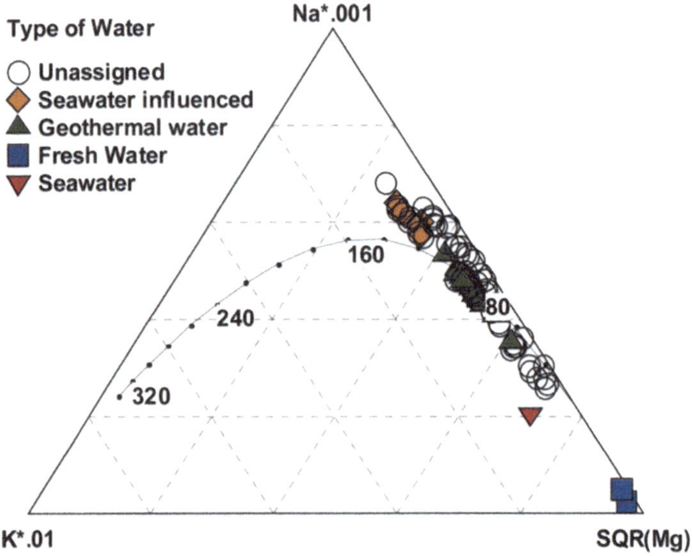

Fig. 6.7. Relative Na-K-Mg contents of the Waiwera water samples (SQR = square root); geothermometer after Giggenbach (1988) given with the full equilibration line; colored samples represent a certain water type

6.2.3 Observation Data

Hydrological Data

In 1979 a pumping test has been carried out and the main conclusions from the pump test according to ARWB/ARC reports are (1) a calculated transmissivity of 320 m^2 d^{-1} which is a high value in comparison with cold water aquifers in similar Waitemata sandstone elsewhere in that region, (2) that all geothermal bores appear to be hydraulically connected, and (3) that the aquifer has a small storativity. The assumptions underlying for calculation of the transmissivity, a homogeneous isotropic aquifer of infinite extent, a fully penetrating well, a fluid of uniform density, temperature, and salinity, are only partially met. Therefore the value has to be used with great care (ARWB 1980).

Measured piezometric head levels in the Waiwera aquifer are always depth dependent due to a vertical flow component (compare conceptual model, Fig. 6.6) and varying density of the fluid with depth, due to temperature and salinity changes. Hence, contour maps of piezometric heads should only be produced for specific depth and are due to the lack of sufficient amounts of data not provided here.

Over 70 boreholes have been used to produce groundwater from the Waiwera aquifer, of which approximately 40 were for geothermal water (Fig. 6.5). Most boreholes are tapping the geothermal aquifer at 200-400 m depth and are cased for 10–40 m through alluvial sediments and weathered rock. Geothermal water in Waiwera is used for recreational purposes, i.e. swimming pools, the main user being the Waiwera Thermal Pool complex (ARC 1999).

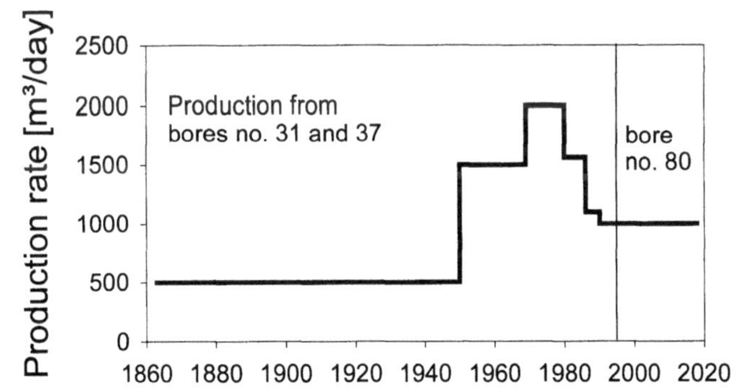

Fig. 6.8. Approximate production rates for the Waiwera geothermal aquifer for the years from 1863 to 2018; major amounts of water were extracted from bores no. 31 and 37 before 1998 and after that time from bore no. 80

Approximate production rates are shown in Fig. 6.8 since 1863. In 1955, the estimated production of geothermal water was 500 m^3 d^{-1}. The peak production from the reservoir took place between 1970 and 1980 and has been approximately 2000 m^3 d^{-1}. Since 1991 the production rate is constant at around 1000 m^3 d^{-1} due to the Water Allocation and Management Plan. Production rates are shown until

2018 because predictive modeling has been performed until that time. The major amounts of water were extracted from boreholes no. 31 and 37 before 1998 and after that time from borehole no. 80.

Geothermal water levels in the ARC monitoring borehole (no. 74, see Fig. 6.5) have been automatically recorded without interruption since 1977 (Fig. 6.9). The water levels from 1977 to 2002 show a decline from approximately sea level in 1977 to 1 m below sea level in 1980. A stabilized level was recorded from 1981 to 1984 and a general increase since 1984 to approximately 1 m above mean sea level in 1990. There was a decline to sea level in 1993, followed by an increase to 0.5 m above mean sea level during 1996 and 1997, and to 1.5 m above mean sea level in 1999. Since 2000 the water level decreased just to 1.0 m above mean sea level (ARC 1999).

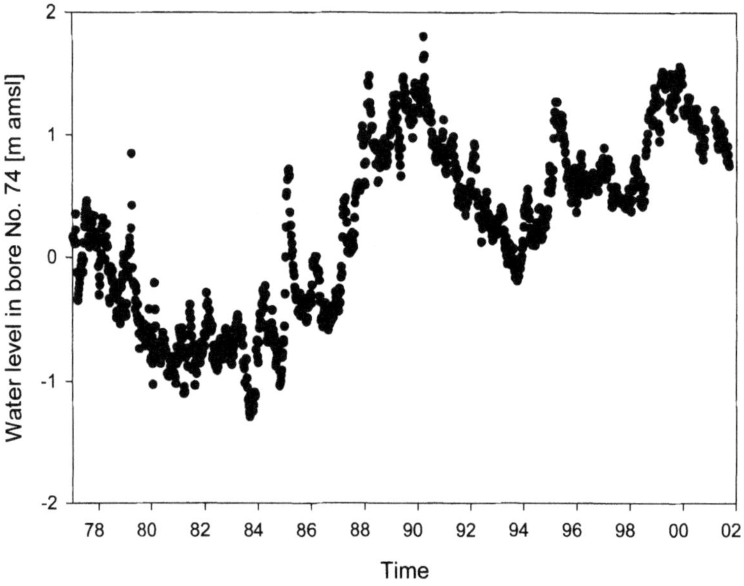

Fig. 6.9. Water levels (weekly mean) in ARC monitoring bore no. 74

All geothermal bore water levels are affected by marine tidal fluctuations up to a maximum of 0.6 m between high and low tide on the seafront and decreasing inland to 0.2 m near State Highway 1. The effect is compressional, caused by changing weight of seawater by the seaward extension of the aquifer with each tidal cycle (ARC 1999).

Temperature Data

Production temperatures of deep groundwater boreholes at Waiwera range from 24°C in bore no. 12 to 53°C in bore no. 80. Boreholes in the center of the geothermal aquifer yield the highest, those at the western edge of the aquifer the lowest production temperatures. Borehole production temperatures have been measured since 1979. The long term monitoring bores are no. 8, 12, 21, 22, 29, and 31 (Fig. 6.10). Bores that are pumped regularly with high abstraction rates show most consistent production temperatures (e.g. bore no. 21 and 31). The maximum production temperatures of boreholes at the edge of the geothermal aquifer have not changed significantly from 1979 to 1997 (e.g. bore no. 21, 22). The maximum temperatures of boreholes near the center of the geothermal aquifer have increased slightly (e.g. bore no. 31, ARC 1999).

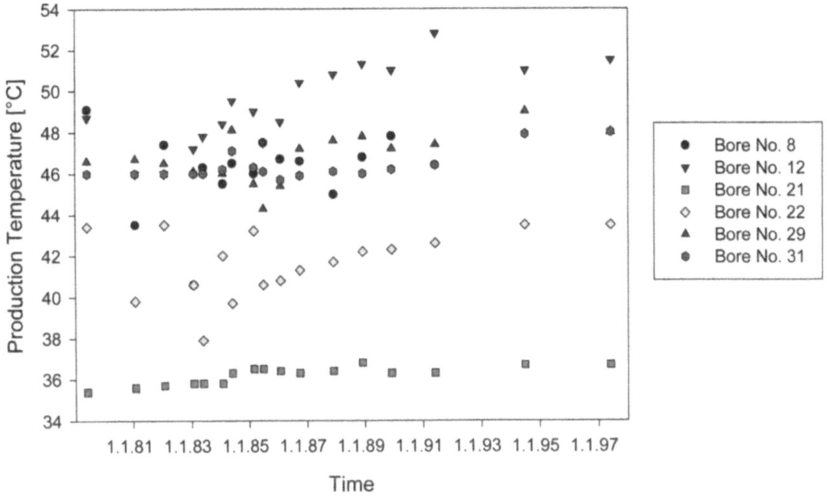

Fig. 6.10. Development of production temperatures between 1979 and 1998 for the long term monitoring bores

According to ARC (1999) the consistent increase in temperature since 1984 may indicate a warming of the Waitemata sandstone aquifer caused by an increase in pressure in the aquifer as a result of reduced extraction rates due to the Allocation Management Plan. On the contrary, an increase in the temperature of the deep geothermal fluid emerging from the basement greywacke rock can be considered unlikely.

In February 1980, down hole temperature profiles were measured on thirteen boreholes located within 250 m of the center of the geothermal reservoir by the Auckland Regional Water Board (ARWB 1980). Additionally Pandey (1982, measured 1980) carried out a temperature survey on bore no. 2. Gonzalez (1986) measured twelve boreholes, five boreholes in the center of the geothermal reservoir and seven on the extremity or outside the geothermal area of Waiwera. In the scope of the study presented here, in all fourteen bores, accessible in November and December 2001, down hole temperature surveys were carried out.

As until recently, most of the boreholes at Waiwera were neither screened, nor cased to any great depth. It can be assumed that temperatures measured in the boreholes are representative of the temperatures of the aquifer (ARWB 1980).

Measured temperature profiles of the deep geothermal groundwater boreholes are compiled in Fig. 6.11 - Fig. 6.15. Within 250 m of the center of the geothermal aquifer, bores show a steady temperature gradient of approximately 0.2 K m^{-1} within the upper 60–120 m depth (bores no. 2, 3, 7, and 8 - Fig. 6.11; bore no. 19, 27, and 29 - Fig. 6.12; bores no. 30, 34, and 36 - Fig. 6.13; bore no. 74 - Fig. 6.14). This is in excess of the natural geothermal gradient of the Auckland region. Below 100–140 m and down to the bottom, temperatures remain constant within a range of 40–50 °C, indicating thermal convection (bores no. 2, 3, and 8 - Fig. 6.11; bore no. 27 - Fig. 6.12; bores no. 34 and 36 - Fig. 6.13; bore no. 74 - Fig. 6.14). Near surface water temperatures are above 15°C ambient, indicating that some upward movement of hot geothermal water occurs.

Temperature profiles of the bores no. 48/49 (Fig. 6.13) and 66 (Fig. 6.14) show an increased geothermal gradient pointing out the influence of the geothermal water. These bores are located towards the western extremity. At the western boundary the non-geothermal groundwater region is located (Fig. 6.6).

Several temperature profiles were measured from bores outside the center of the Waiwera geothermal reservoir. Temperature profiles Wenderholme, and Wenderholme DW (Fig. 6.15) show that there has to be flow of tepid geothermal groundwater in a permeable fracture zone which extends northward at least 1.5 km to Wenderholme, but not as far as 2.5 km north across the Puhoi River (Schiska 2, Fig. 6.14). Both the Hillcrest and the Orewa bore are not influenced by the geothermal water, they show the naturally occurring geothermal gradient (Fig. 6.15). The temperature profile measured in the bore V Schiska (Fig. 6.14) displays apparently convective heat transfer. But, due to the fact that the bore is pumped regularly and the unattended times before the survey were too short in 1986 as well as in 2001, a convective temperature profile exhibits an artifact.

Performed temperature surveys from bores no. 8 (Fig. 6.11), 11, 27 (Fig. 6.12), and 34 (Fig. 6.13) exhibit, additionally to the different years displayed, measurements of four days back-to-back (a, b, c, and d refer to the 19th, 20th, 21st, and 22nd of February, respectively). This comparison reflects the accuracy of the measured temperature profiles and provides a range of reliable temperatures. However, the monitored temperatures emphasize a relatively characteristic, constant and stable temperature distribution within the Waiwera aquifer over the years.

Bore no. 36 is the only one exhibiting great deviation in the temperature profile (below 200 m depth) between the surveys of the ARWB (1980) and the one done here. Whereas the profile from 1980 provides strictly convective heat transfer in deeper parts, the measurements done in 2001 suppose a colder water inflow between 200 and 300 m depth. This could be a reason for the relatively low production temperature observed in this bore (ARWB 1980), although it is situated within the center of the Waiwera reservoir in the near vicinity of bores with highest production temperatures.

It has to be noted that bore production temperatures are usually, significantly less than maximum profile temperature.

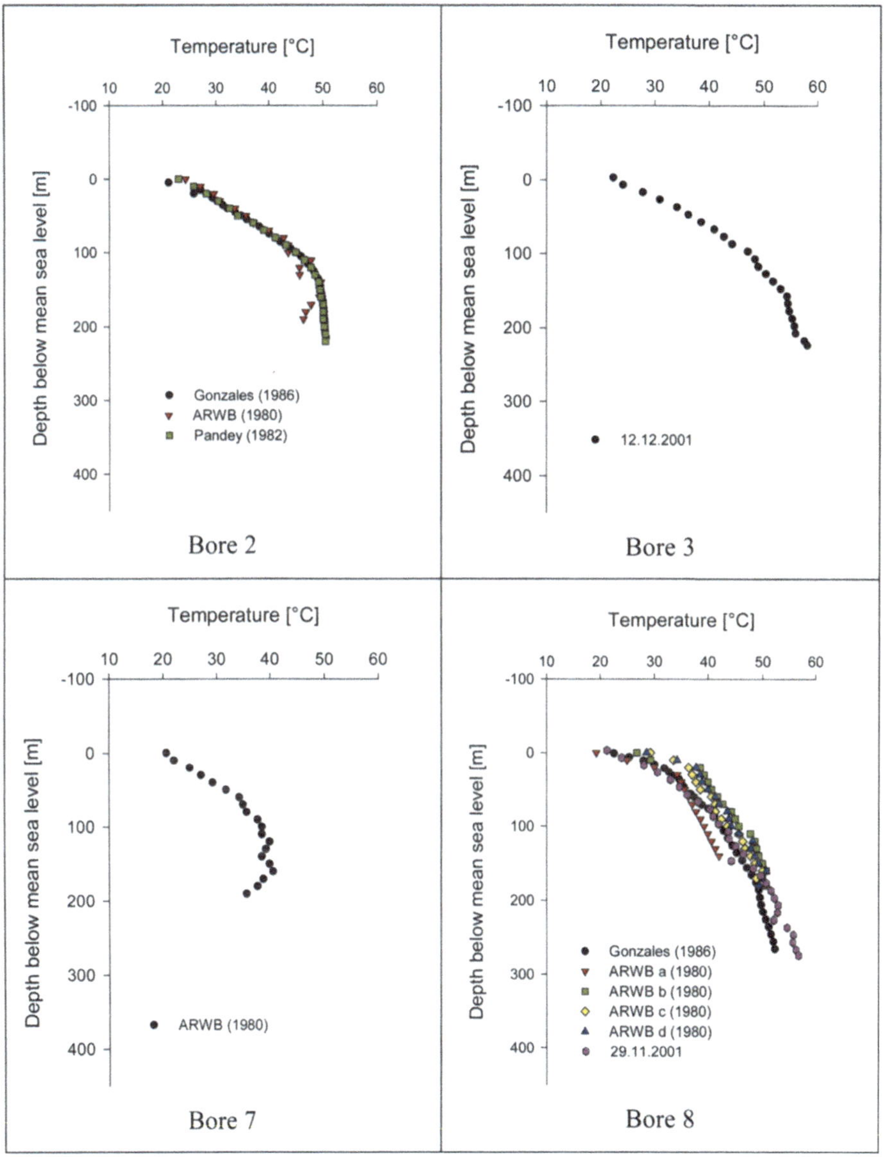

Fig. 6.11. Temperature profiles of bores no. 2, 3, 7, and 8 (Fig. 6.5)

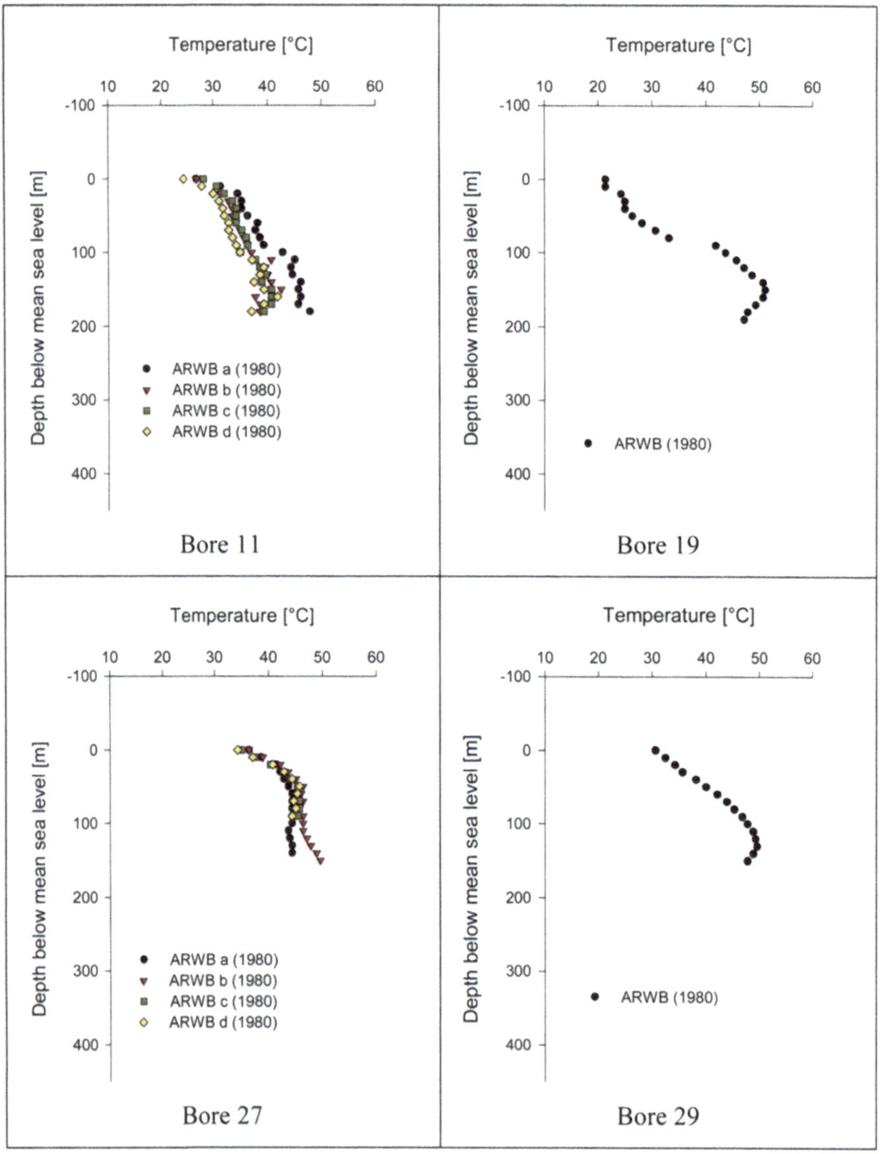

Fig. 6.12. Temperature profiles of bores no. 11, 19, 27, and 29 (Fig. 6.5)

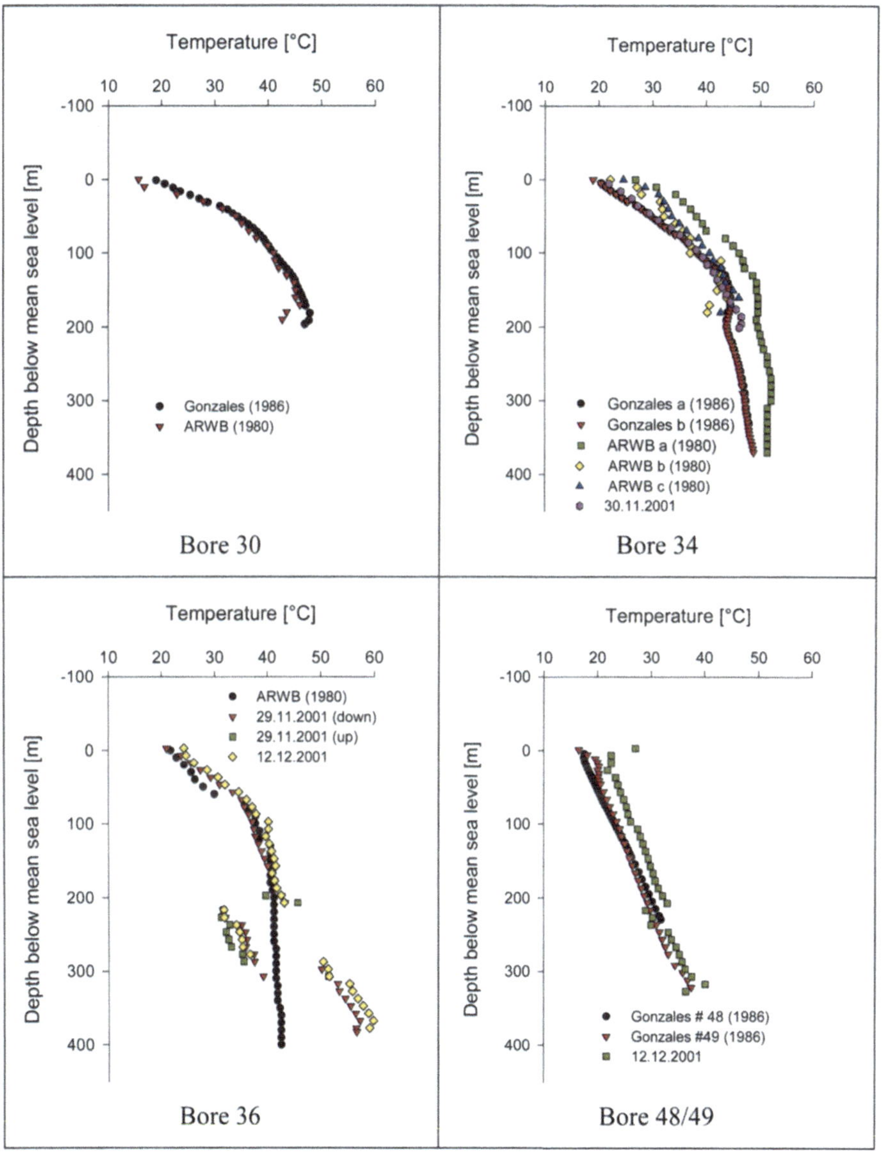

Fig. 6.13. Temperature profiles of bores no. 30, 34, 36, and 48/49 (Fig. 6.5)

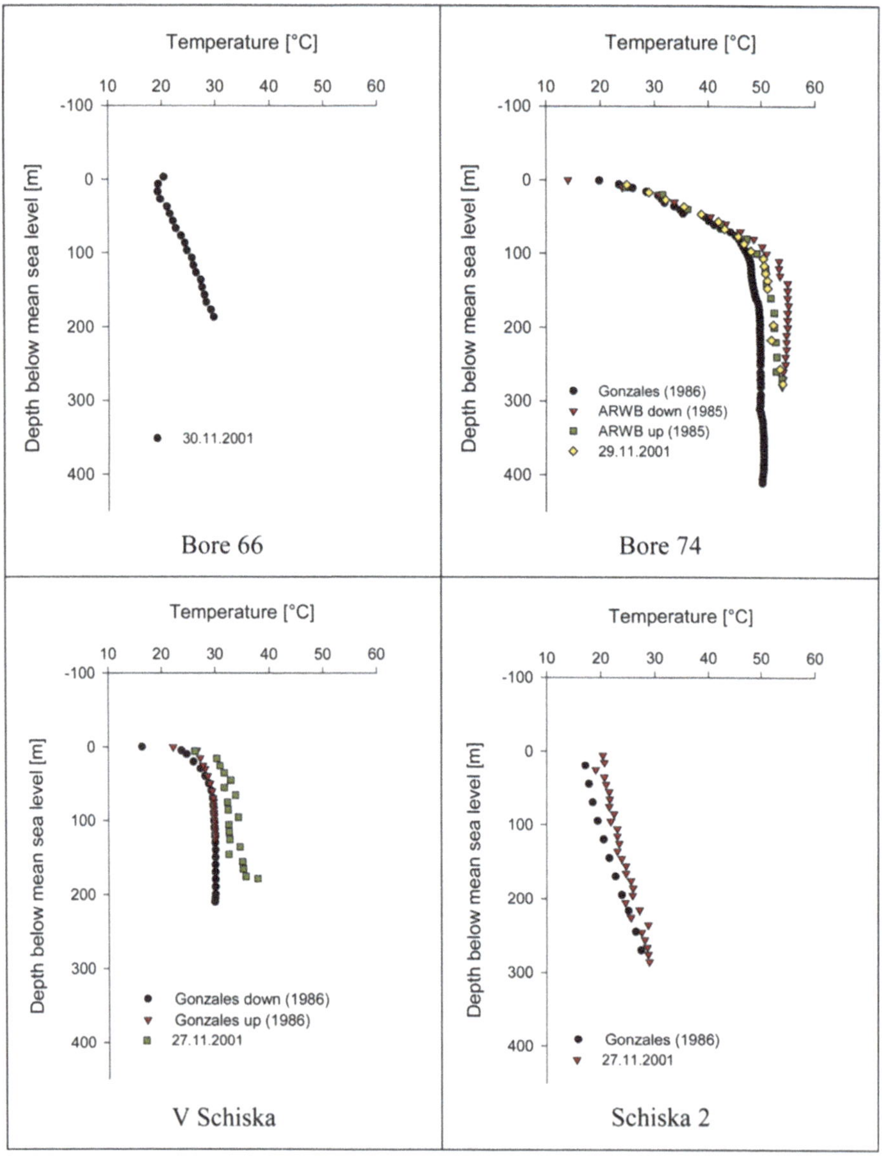

Fig. 6.14. Temperature profiles of bores no. 66 and 74 as well as bores V Schiska, and Schiska 2 (Fig. 6.5)

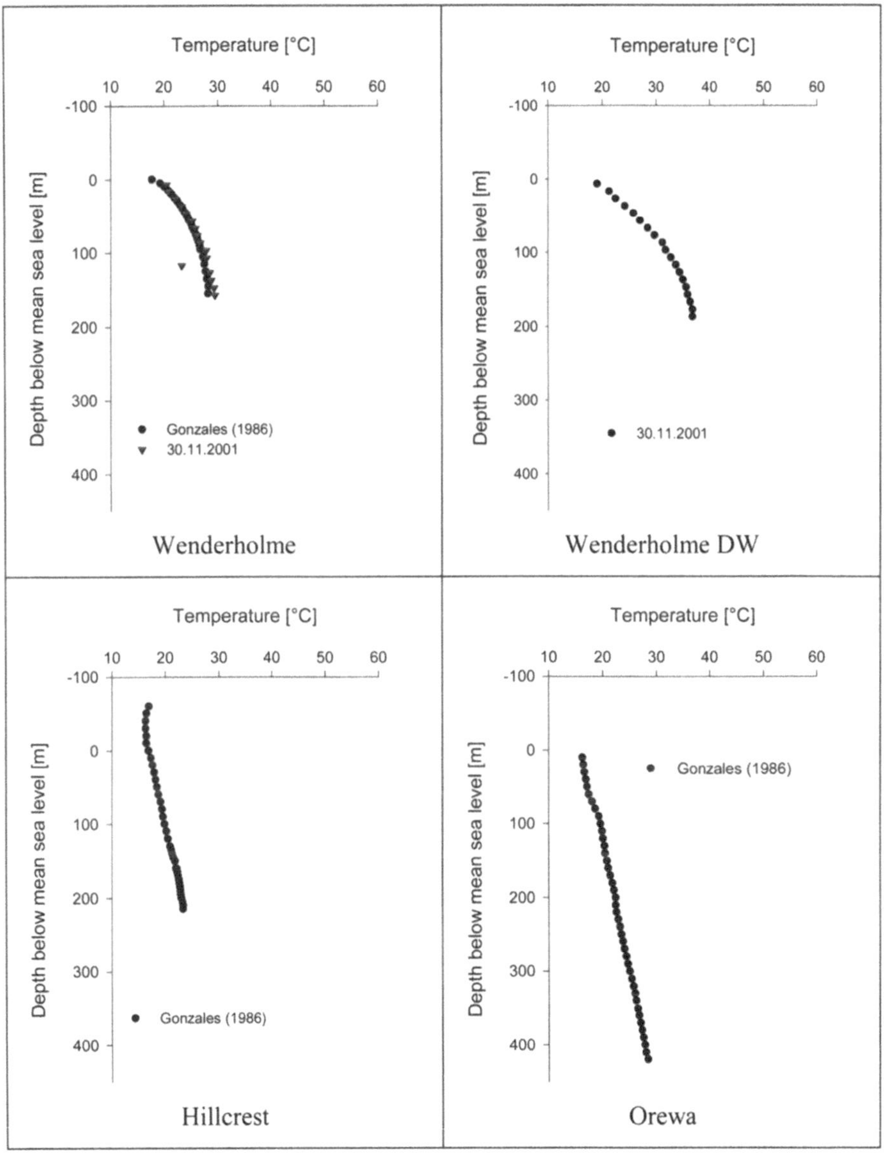

Fig. 6.15. Temperature profiles of bores Wenderholme, Wenderholme DW (drive way), Hillcrest, and Orewa (Fig. 6.5)

Chemistry

Analysis of chemical constituents can provide information about the origin of thermal, cold non-thermal, and seawater within the studied area. Geothermal groundwater from Waiwera displays a considerable variation in chemical composition, particularly in chloride concentrations, but can be classified into five types. These are fresh groundwater, seawater, unmixed geothermal water, geothermal water mixed with seawater, and geothermal water mixed with cold fresh groundwater. Table 6.1 lists mean values of the main water types of fresh, geothermal, and seawater as they are analyzed and used in the simulations.

Table 6.1. Chemical composition of fresh, geothermal, and seawater at Waiwera

Species	Freshwater [mmol L^{-1}]	Geothermal water [mmol L^{-1}]	Seawater [mmol L^{-1}]
Na^+	8.40	28.30	435.00
K^+	0.10	0.20	10.20
Ca^{2+}	0.30	1.20	9.30
Mg^{2+}	0.50	0.00	51.00
Cl^-	4.20	31.90	508.00
HCO_3^-	5.50	0.30	1.10
pH	7.04	8.06	8.05
Temperature [°C]	18.50	46.00	10.00
Sat. Index Calcite	0.02	0.09	-0.03

A more detailed presentation of the water characteristics from the Waiwera geothermal field is given in a Schoeller diagram (Fig. 6.16). The highest amount of total dissolved solids has been found in the estuary water sample (seawater, red), which can be seen above the main array of curves. The fresh water (blue) is characterized by a more or less balanced composition of all constituents, whereas all other samples are predominated by the sodium chloride content. Compared with each other, the seawater influenced and geothermal water samples can be distinguished by their low and high magnesium and hydrogen carbonate concentration, respectively. A difference between geothermal water and fresh water samples is obvious in their magnesium, sodium chloride, and hydrogen carbonate contents. While the geothermal water contains less amounts of magnesium and hydrogen carbonate than the fresh water its sodium chloride concentration is higher.

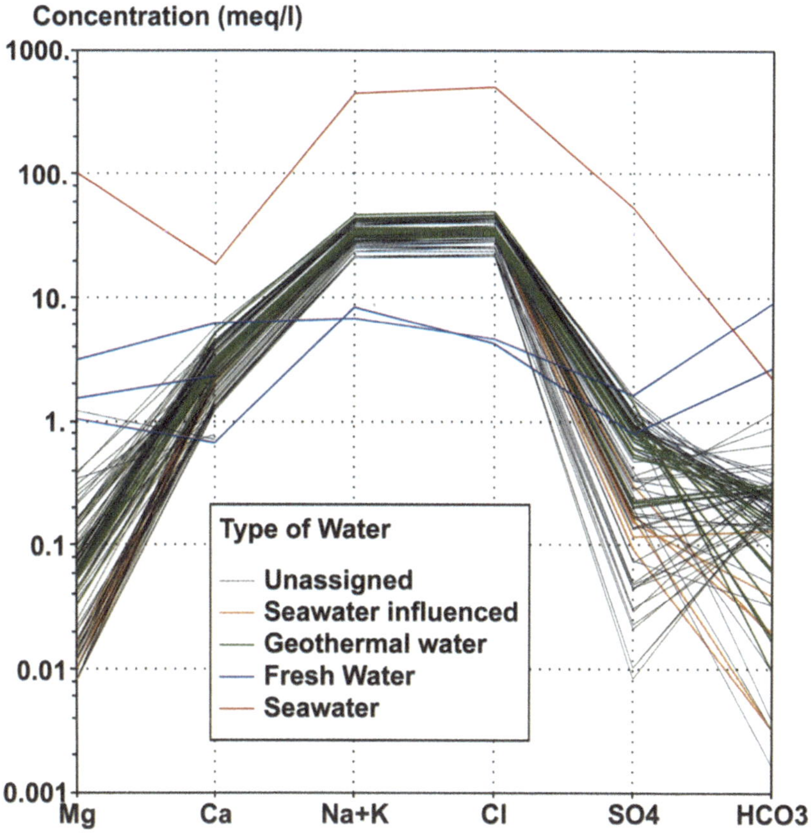

Fig. 6.16. Schoeller diagram of Waiwera water samples

All bores yielding unmixed geothermal water are located in the eastern part of the geothermal aquifer, within 300 m of the center (100 m from the beach front, Fig. 6.5). All bores yielding geothermal water mixed with seawater are located within 200 m of the beach. Bores yielding geothermal water mixed with groundwater are all located on the western edge of the geothermal aquifer, farthest away from the geothermal source. It has to be noted that the chloride concentrations in the Waiwera aquifer are very low, considering the proximity to the sea and the thickness of the aquifer. Historical data suggest that there has been little change in composition of the deep geothermal fluid over the past 60 – and even probably 110 – years.

Five representative long term monitoring bores have been sampled for water chemistry since 1979 (Fig. 6.17). The two bores no. 12 and 31 yield unmixed geothermal water. Bores no. 21 and 22 are believed to show chloride concentrations of a mixture between geothermal water and fresh groundwater. Fig. 6.17 shows the development of the chloride concentrations from 1979 to 1997. There has been a significant but irregular long term decrease in chloride in bore no. 29 near the beach in the southern part of the geothermal aquifer (Fig. 6.5). Bores no. 12, 21,

22, and 31 show no long term consistent change. The chloride concentration in bore no. 8 increased from 1979 to 1990 followed by a decrease to the initial value in 1992. No long-term trend in chloride concentrations can be detected for bores at the western edge of the geothermal aquifer (ARC 1999).

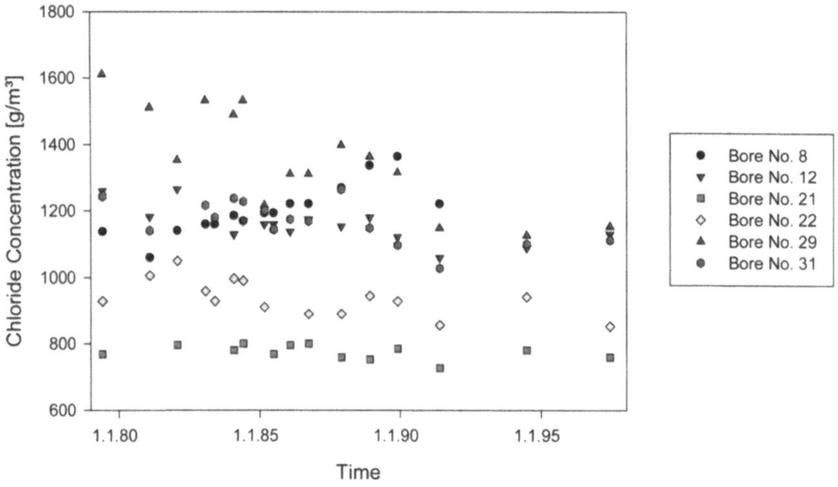

Fig. 6.17. Long term development of chloride concentrations in monitoring bores

6.2.4 Numerical Simulations

In previous manuscripts the set up of a 3D coupled fluid flow, heat transfer, and solute transport model of the Waiwera area is described in detail (Stöfen 2000, Stöfen et al. 2000, Stöfen and Kühn 2003), which is used here to carry out several numerical investigations using the code SHEMAT. This model defines a natural state of the Waiwera reservoir and is used to simulate varying scenarios to answer following questions:

1. Is the conceptual model (Fig. 6.6) of the Waiwera area reasonable?
2. Does seawater intrusion into the aquifer occur?
3. Is the present heat exploitation sustainable?
4. Do chemical reactions alter the reservoir?

Conceptual Model and Boundary Conditions

The following section briefly describes the set up of the numerical model (for details the reader is referred to Stöfen 2000, Stöfen et al. 2000, Stöfen and Kühn 2003) drawn from the previously discussed concept of the geothermal field of Waiwera (Fig. 6.6) and the observation data. An equivalent porous medium is assumed to represent the Waiwera aquifer. The applied boundary conditions and reservoir properties are shown in cross section in Fig. 6.18. Since temperature profiles are the most reliable and complete data at Waiwera, they are used to define the thermal boundary conditions for the numerical model. Recharge across the upper boundary is specified for the alluvial deposits. The rest of the upper boundary is treated as impervious. Groundwater flux is defined at the western boundary to simulate the groundwater-geothermal water mixing zone. The eastern boundary under the sea is set so that a seawater-freshwater interface can build up. The greywacke forms an impermeable boundary at the bottom of the model with a defined influx of geothermal water in the center of the reservoir.

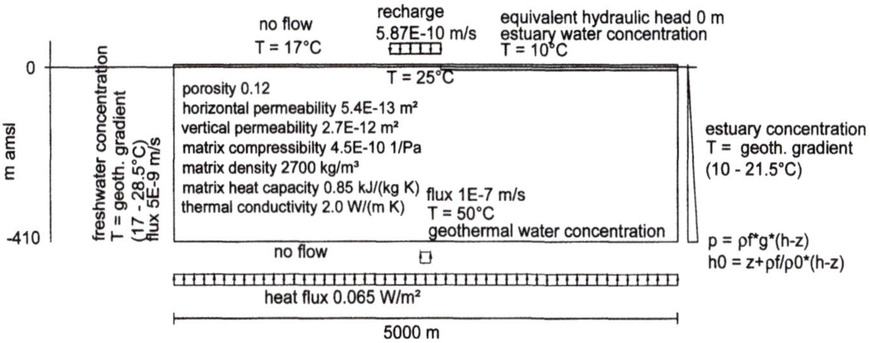

Fig. 6.18. Boundary conditions and rock properties of the Waiwera model

Natural State

Results based on a total of 6000 simulation years show that the steady state is reached within about 2400 years, starting from artificial initial conditions.

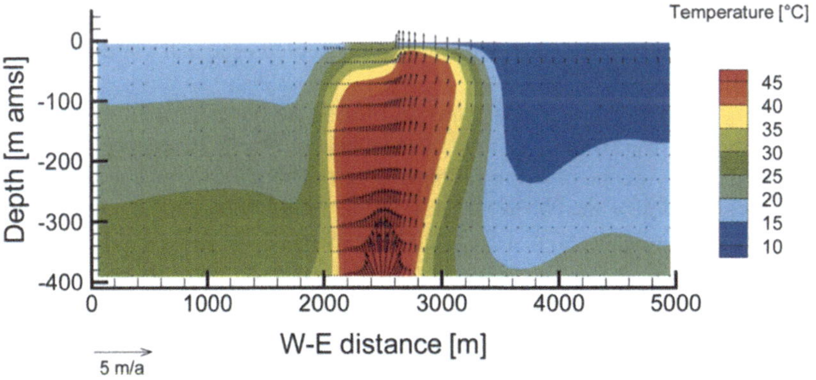

Fig. 6.19. Temperature distribution in the natural state; W-E cross section through the center of the aquifer (Fig. 6.5); arrows show the Darcy velocity

Fig. 6.19 shows the resulting temperature distribution in a cross section through the center of the aquifer for the natural state. There is a strong up-flow of geothermal water underneath the Waiwera Township, where the 45°C isotherm is located 100 m below mean sea level. Underneath the sea it is located 50 m below mean sea level.

Fig. 6.20 shows the corresponding chloride concentrations, which reflects the mixing of fresh, geothermal, and seawater. A freshwater–geothermal water boundary evolves to the east and a geothermal water–seawater boundary to the west. The inflowing geothermal water prevents the seawater from entering into the aquifer.

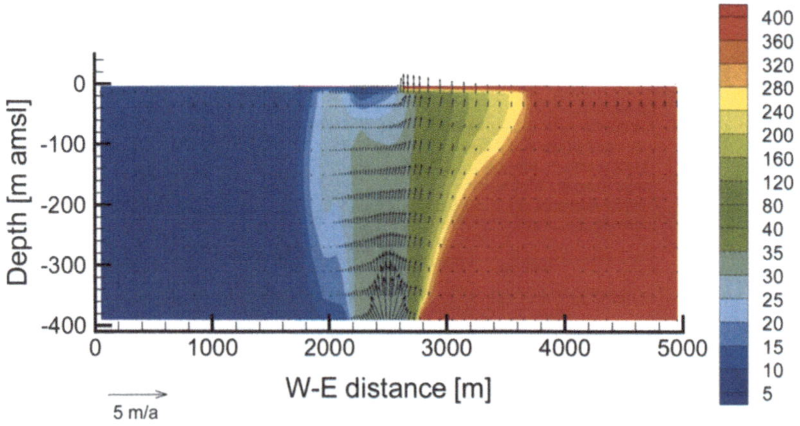

Fig. 6.20. Chloride concentration in the natural state; W-E cross section through the center of the aquifer (Fig. 6.5); arrows show the Darcy velocity

Temperature profiles are used to compare the modeling results with measured values (Fig. 6.21 - Fig. 6.23). In Fig. 6.21 the simulated and measured temperature profiles of the bores situated on the flat peninsula at the beachfront (Fig. 6.2) are shown. For the bores no. 2 and 3 the results are in very good agreement, whereas for the bores no. 3 and 7 deviations can be observed especially in deeper parts of the profiles.

Fig. 6.22 displays the temperature profiles of the bores situated east to the center of the Waiwera reservoir towards the beachfront. With respect to the mentioned accuracy of the measured temperature profiles it can be concluded that the numerical model of the Waiwera geothermal reservoir is appropriate to describe temperature gradients and distribution at the seaward side of the model.

Bores no. 19, 48/49, and 66 are situated at the extremity of the simulated area towards the non-geothermal groundwater region (Fig. 6.5). Their simulated and measured temperature profiles are shown in Fig. 6.23. Whereas the numerical results for bore no. 19, which is still nearby the geothermal center, are generally in quite good agreement with the measurements, bores no. 48/49 and 66 exhibit a simulated geothermal gradient too low compared to the observations. However, the process by which heat is transferred within nature and the numerical model is the same. Calculated as well as observed values reveal that both bores are in an area where conductive heat transfer prevails.

The Hillcrest bore is situated outside the region covering the Waiwera geothermal area. Thus, the temperature profiles display the naturally occurring geothermal gradient in perfect coincidence between simulation and measurement (Fig. 6.23).

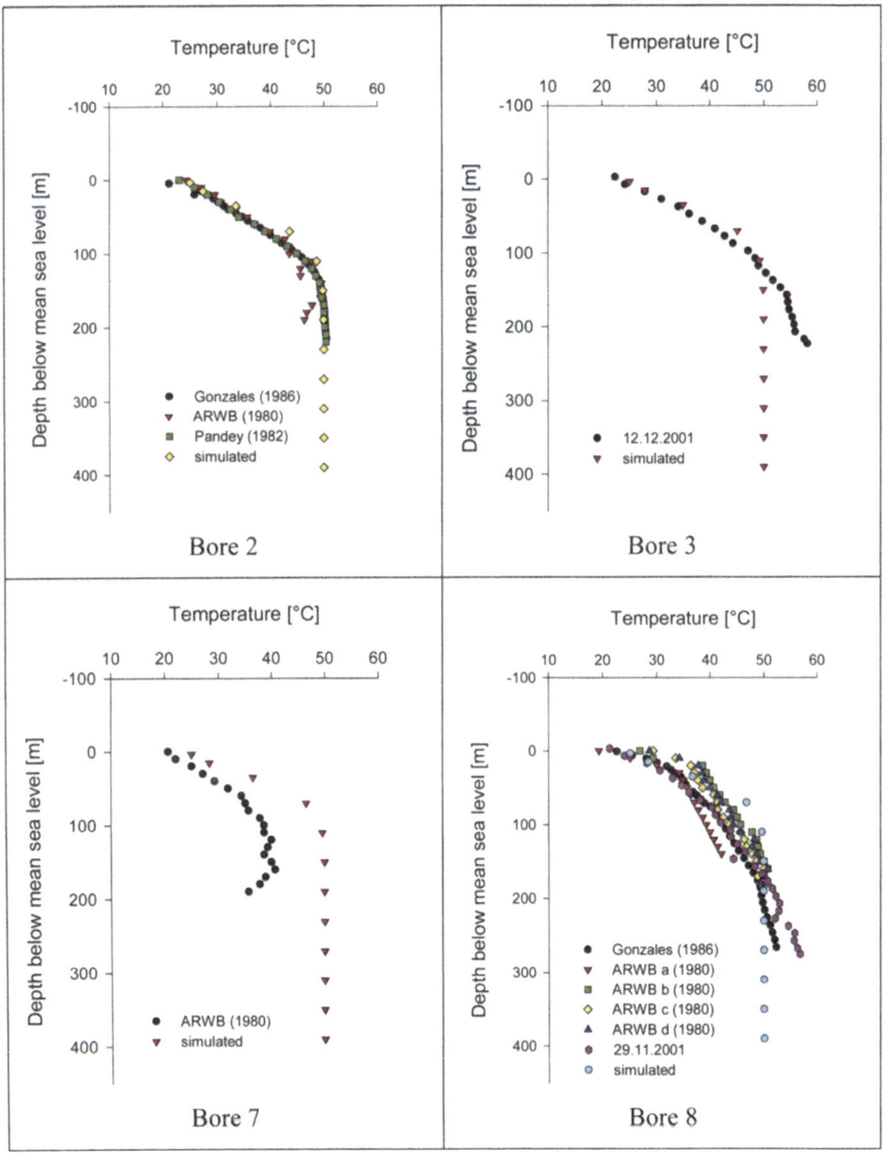

Fig. 6.21. Comparison of measured and simulated temperature profiles of bores no. 2, 3, 7, and 8, all situated on the flat peninsula at the beachfront (Fig. 6.5)

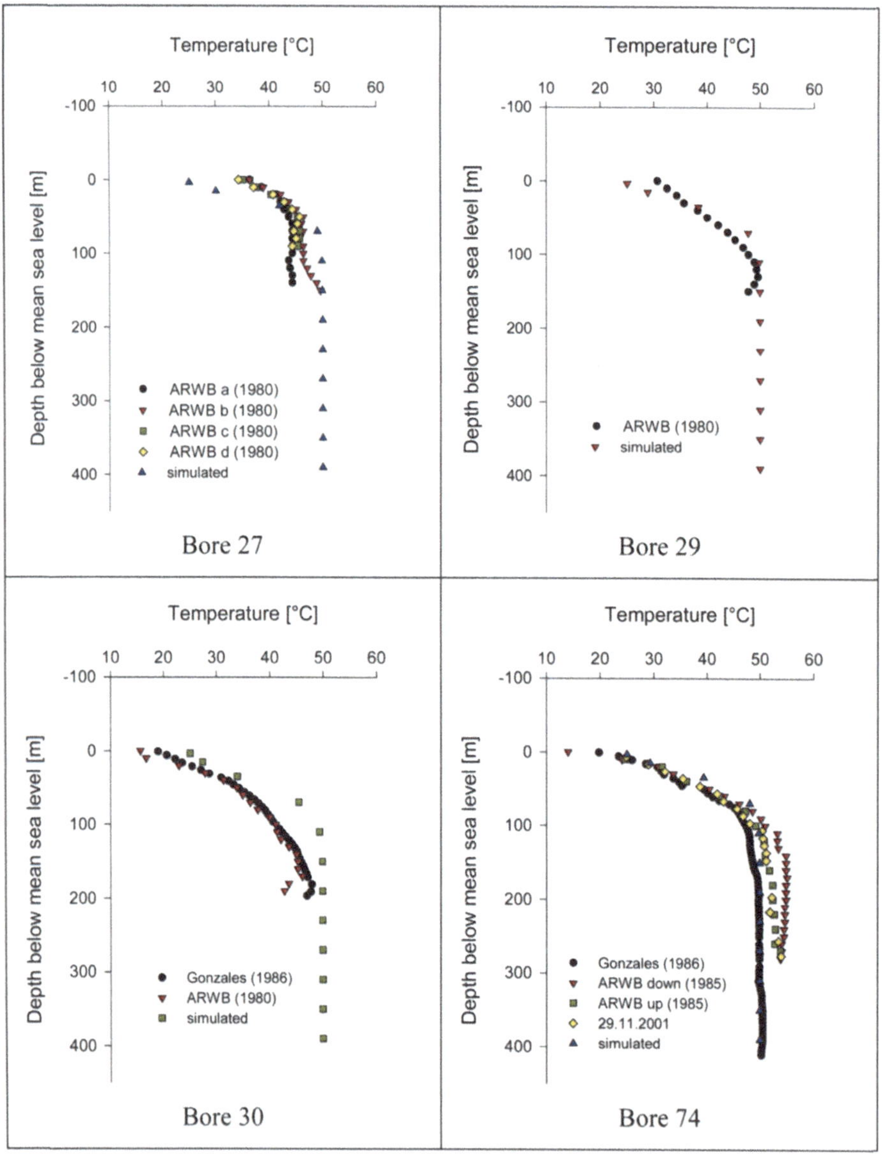

Fig. 6.22. Comparison of measured and simulated temperature profiles of bores no. 27, 29, 30, and 74, all situated east to the central part of the Waiwera reservoir towards the beachfront (Fig. 6.5)

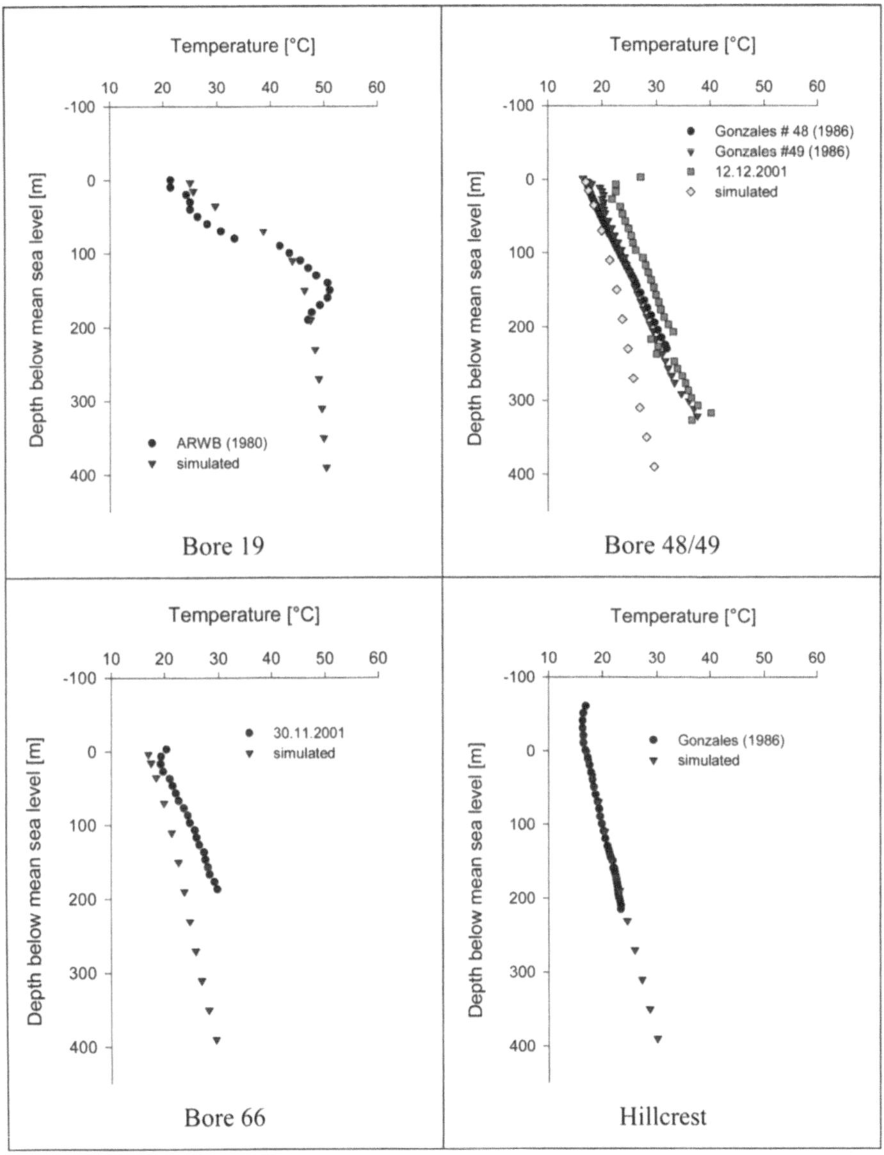

Fig. 6.23. Comparison of measured and simulated temperature profiles of bores no. 19, 48/49, and 66, all situated at the extremity of the simulated area towards the non-geothermal groundwater region, as well as the Hillcrest bore, which lies outside the central region of the reservoir (Fig. 6.5)

Exploitation

From what is known about the development of the exploitation of the Waiwera geothermal aquifer a rough approximation is used for the exploitation case of the 3D model of Waiwera. The pumping scheme implemented in this case, is displayed in Fig. 6.8. The results of the simulated natural state served as initial conditions for the exploitation test case.

The temperature distribution in Fig. 6.24 shows a detail of the cross section in Fig. 6.19 from 1800 m to 3600 m W-E distance and between the model top and a depth of 100 m. Solid black, gray, and white lines represent the natural state, and the exploitation in the years 1980 and 2018, respectively. Due to exploitation the hottest region is shrinking. The two production wells operating in the upper part of the reservoir during the 1980s, resulted in a rise of the 45°C isotherm. With the newly constructed production well operating in a lower part of the reservoir the 45°C isotherm is again moving downward between 1998 and 2018.

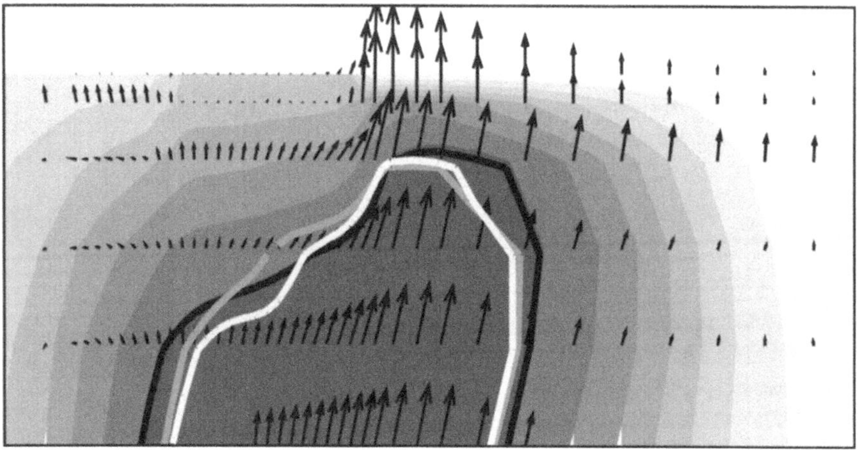

Fig. 6.24. Detail of the temperature distribution of Fig. 6.19 from 1800 m to 3600 m W-E distance and the top of the model to a depth of 100 m; shrinking of the hottest zone during exploitation: 45°C isotherms for the natural state (black line), the situation in 1980 (gray line) and the situation predicted for 2018 (white line)

Fig. 6.25 shows the development of the chloride concentrations at the two boreholes no. 29 and 34, which are mostly influenced by the intrusion of seawater. Obviously, seawater intrudes into the upper parts of the aquifer. Even with declining abstraction rates after 1980 and the use of the deeper bore no. 80 (Fig. 6.8), it takes about 30 to 40 years within the numerical simulation before the natural state concentration profile is achieved again.

The observed effect of saltwater intrusion implies the question if this has an influence on the temperature profiles. But, because the water has to cool the rock material before a temperature decrease in the profile can be observed, the influence of seawater intrusion is not significant for the simulated temperature profile of the bores no. 29 and 34 (not shown).

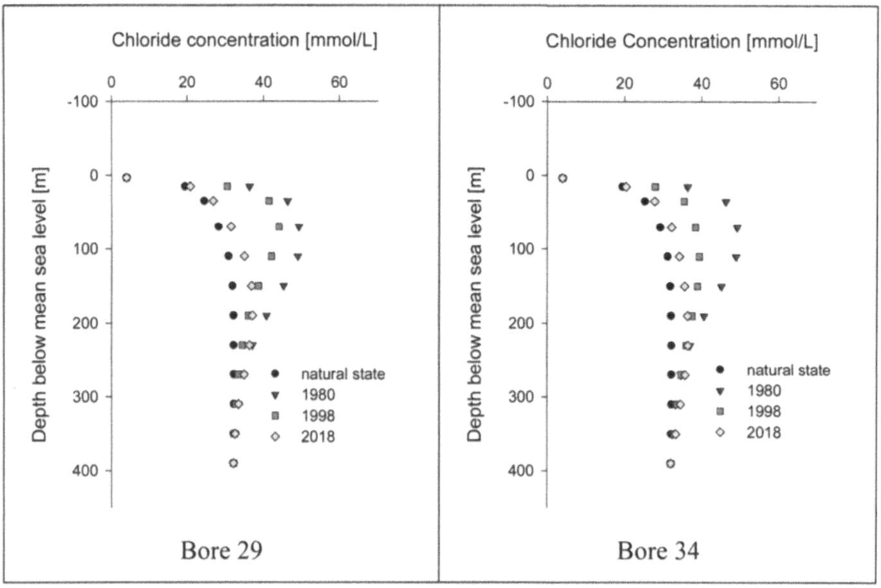

Fig. 6.25. Simulated chloride concentration profiles for borehole no. 29 and 34: natural state and during exploitation

Fig. 6.26 displays the simulated distribution of the chloride concentrations at a depth of 35 m below mean sea level in the center of the geothermal field. Gray shading and black solid lines represent the natural state. Bold solid and dotted lines represent the concentrations during the exploitation phase in the years 1980 and 2018, respectively. The pumping regime, present in the 1980s, triggers intrusion of seawater into the southern parts of the geothermal aquifer. In fact, elevated chloride concentrations have been observed in the production water from boreholes in this area. After a reduction of the pumping rate, the simulated chloride distribution in 2018 shows a retreat of the geothermal–seawater boundary, but the initial conditions are still not reached again.

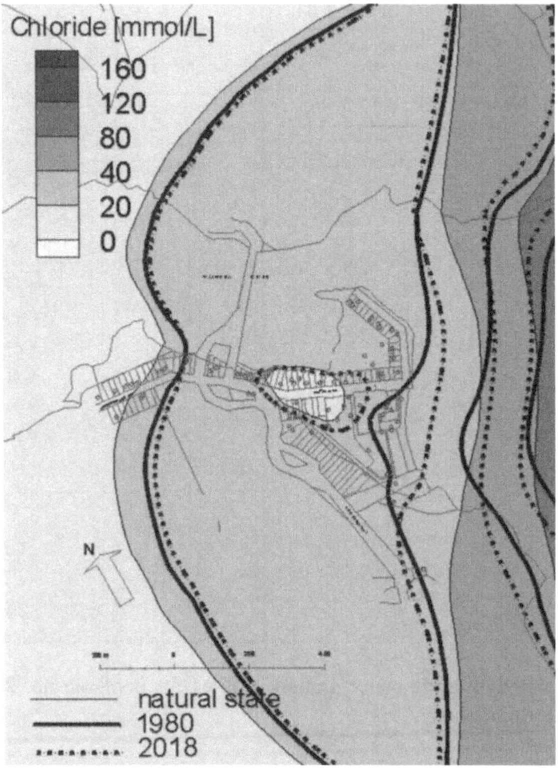

Fig. 6.26. Change with time of the chloride distributions at a depth of 35 m resulting from the intrusion of seawater into the southern part of the Waiwera geothermal aquifer due to exploitation; a partial retreat of this intrusion is predicted until the year 2018

Alteration

Geochemical simulations of the fresh water, the geothermal water, and the estuary water, as they are analyzed (Table 6.1), with the chemical module of the numerical code of SHEMAT yield, that they are almost in thermodynamic equilibrium with calcite as the mineral phase (saturation indices between 0.09 and -0.03).

Due to mixing of the different water types and changes in temperature with time, it has to be expected that calcite precipitates and therefore possibly alters the reservoir properties porosity and permeability. Under the recent physical and chemical conditions calcite is the only mineral phase potentially precipitating from the waters. It has been found that the Waitemata sandstone is partly cemented by calcite (ARWB 1980, ARC 1997).

Calcite precipitation and dissolution takes place in both the natural state and during the exploitation phase. Fig. 6.27 illustrates for the year 2018 during exploitation, how precipitation occurs within the center of the reservoir in the hot up-flow region of the geothermal water and within the mixing zone of freshwater and geothermal water. A zone of dissolution surrounds the central area of precipitation. Neither precipitation nor dissolution alters significantly the reservoir porosity and permeability, because the quantities of mineral, which precipitate or dissolve, are too small (Kühn et al. 2001).

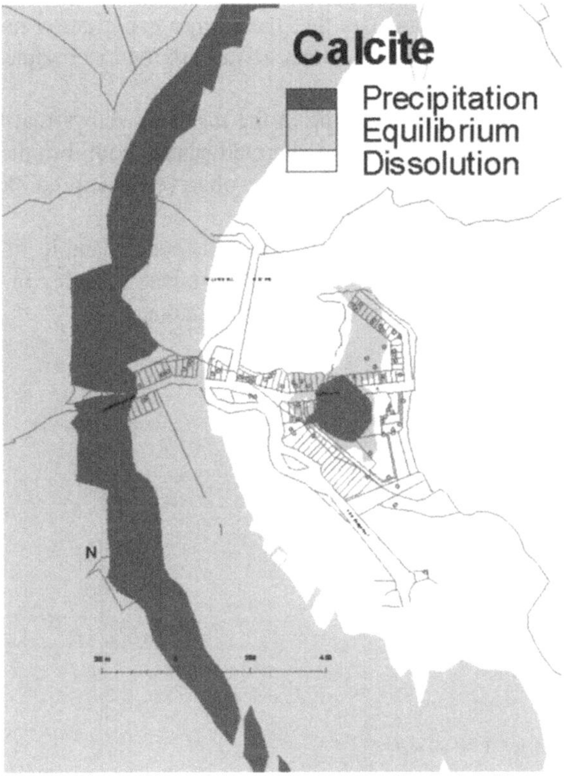

Fig. 6.27. Calcite precipitation and dissolution at a depth of 35 m below mean sea level in the center of the Waiwera geothermal reservoir; calcite dissolution (white) occurs around the hot water up-flow zone with calcite precipitation (dark gray) and is limited to the west by an area of calcite precipitation within the freshwater–geothermal water mixing zone

6.2.5 Waiwera Case Study Conclusion

The conceptual model of the Waiwera geothermal field has been evaluated by the presented numerical investigation. The simulated temperature profiles agree well with data measured at Waiwera. A qualitative agreement between observed and simulated data can be shown for the chloride concentrations.

The results exhibit that the inflow of geothermal water at the bottom of the aquifer prevents seawater from entering the Waiwera aquifer. If seawater intrusion does occur, it is due to over-exploitation. While the aquifer is over-exploited, seawater intrudes into the upper parts of the geothermal aquifer (between 20 m and 200 m below mean sea level) in contrast to the situation at an undisturbed seawater-freshwater interface where seawater intrudes at the bottom of an aquifer.

Comparing the natural state with the predicted distributions of temperature and chloride concentration, it appears that the former exploitation had not been sustainable. However, after modifications of the production regime the geothermal system is recovering again.

The study of the chemical regime in the reservoir shows that freshwater, geothermal water, and seawater are in thermodynamic equilibrium with calcite. In spite of mineral reactions involving calcite, observed precipitation and dissolution do not alter the hydraulic aquifer properties.

The Waiwera geothermal aquifer is an excellent example of how numerical simulation serves to deepen the understanding of the complex interaction of density driven flow, heat transfer, and chemical reactions.

7 Reservoir Management

Geothermal power generation affects chemical processes within reservoirs and in turn chemical reactions affect geothermal power generation. Reactive transport modeling is a technique that provides opportunities to help reduce costs and environmental impact due to geothermal power generation. Additionally numerical simulation is a means to investigate and approximate the long term performance of installed wells of geothermal plants. The following set of practical chemical problems, which should be and could be studied in detail by reactive transport modeling, arise from industrial experience:

- Chemical brine rock interaction due to the injection of undersaturated, supersaturated or acidic brine in wells.
- Reservoir management aided by modeling chemically reactive tracers.
- Recovery of precious minerals from geothermal brines.
- Minimizing gas production and probable resulting scaling products through optimized water injection and or production.
- Effect of exploitation on CO_2 flux from geothermal systems.

The first part of this chapter gives a short overview of numerical simulations performed by other authors investigating the mentioned topics. In the second part a detailed investigation is presented of the long term performance of the geothermal potential Stralsund (Germany).

7.1 Brine Rock Interaction, Reactive Tracer, Mineral Recovery, and Gas Contents

7.1.1 Brine Rock Interaction

Fluids to be re-injected in geothermal power plants are often silica supersaturated. The injection of supersaturated brines may lead to precipitation around the wellbore. This results in a decrease of porosity and in turn in reduction of permeability. Predicting the rate of silica scale in geothermal wells and aquifers can extend their life by optimizing fluid flows and temperature to achieve minimum precipitation compatible with hydrogeological properties of the aquifer. For lifetime prediction Mroczek et al. (2000) developed a combined test of fluidized bed experiments and reaction rate calculations. Results of further simulations (Mroczek et al. 2002) showed that negligible silica is precipitated greater than 40 m from the well.

Applying the program package SHEMAT, Kühn et al. (2002b) and Kühn and Schneider (2003) studied numerically the injection of a mineralized fluid into a reservoir as a function of temperature and chemical reactions of the minerals anhydrite and barite. Baermann et al. (2000a) identified anhydrite as the reason for substantial reduction of permeability in a north German reservoir (see above), and there is also a potential risk of precipitation of barite during re-injection (Kühn et al. 1997). The temperature-controlled dissolution of anhydrite around the cool well increases the permeability, while the precipitation of anhydrite at the warm temperature front reduces it. The negative effect of the injection temperature on the injectivity is in part compensated by the relocation of anhydrite, as the permeability increase has a larger effect than the decrease (see below). In contrast to anhydrite, barite possesses prograde solubility. The re-injection of cold brine leads to supersaturation of the solution in respect to barite. Although barite precipitates around the injection well no significant permeability change or hydraulic effect is observed during the simulation period.

Brines are sometimes acidified prior to re-injection to prevent silica or iron scaling or to improve the injectivity of already damaged wells. The trend in concentration change with time is highly dependent on the minerals present in the system, their amounts, and the brine-rock reactions rates. Pham et al. (2001) presented simulations using TOUGHREACT where they envisaged calcite dissolution and simultaneous kaolinite precipitation in the volcanic rock of a granite reservoir due to low pH fluid injection.

7.1.2 Modeling Chemically Reactive Tracers

Kissling et al. (1996) modeled the CO_2 chemistry of the Wairakei geothermal field in New Zealand using the program CHEM-TOUGH. They investigated chemical reactions of pH, CO_2, and H_4SiO_4, while Cl was treated as conservative, non-reactive tracer, to calibrate the hydrogeothermal model of the Wairakei reservoir. Hereby, some of the most important chemical processes within a geothermal reservoir are incorporated into an established model of a geothermal field.

With their simulations, Kissling et al. (1996) were able to reproduce observed chemical changes in the reservoir. This kind of calibration process, using reactive chemicals, provided further confidence in an already successful model, but also highlighted possible discrepancies between physical processes taking place in the model and those occurring at Wairakei. For the first time, the advantage of including equilibrium reactions in a reservoir model has been shown.

7.1.3 Mineral Recovery

Vast reserves of dissolved minerals can be found in saline geothermal waters around the world and have been investigated in manifold ways (Brown 1986, Brown et al. 1996, Dorrington and Brown 2000, Gammons and Barnes 1989, Kühn et al. 1998, Schenberger and Barnes 1989, Spycher and Reed 1989). Usage of these reserves is an opportunity to significantly reduce the cost of geothermal energy production.

The commercial recovery of zinc from the brine at the geothermal fields in Imperial Valley of California is already carried out. The production of silica and manganese will soon follow and extraction of silver and lead is under consideration. Pham et al. (2001) modeled the recovery process using TOUGHREACT. The calculated zinc concentration distribution is in agreement with the actually found zinc within the reservoir under similar temperature and pressure conditions.

7.1.4 Gas Contents

The forecast of long term trends of the gas content in geothermal steam and brine affects the cost and environmental impact of geothermal energy production. For example, with increasing gas content in steam the efficiency of power generation in a plant decreases and the discharge of greenhouse gases from the plant increases at the same time. Pham et al. (2001) examined with the numerical program TOUGHREACT how the gas content can be affected by the injection process.

As a consequence of the concerns about greenhouse gas accumulation in the atmosphere (Kyoto protocol) the assessment of CO_2 flows from geothermal systems has become important. Reactive transport modeling may help to decide if the discharged amounts of greenhouse gases are minimal, natural, or significant (Sheppard and Mroczek 2002). Concerning the CO_2 only discharged from the geothermal power plant, minimal or significant in this case means compared to conventional thermal plants. If the amount of CO_2 is produced anyway and no new quantity is added to the environment the flow is called natural. Main question hereby is if the produced CO_2 is a result of the exploitation process.

The solubility of CO_2 highly depends on the pressure. Hence, the pressure gradient around the production well may lead to degassing and resulting calcite precipitation. Satman et al. (1999) investigated the effect of calcite deposition in a formation concerning the reduction of inflow performance of geothermal wells producing brine with a significant CO_2 content. Through the derivation of analytical expressions for the rate of calcite precipitation, they identified the key operational and reservoir parameters influencing the magnitude of impairment by calcite deposition and its effect on the flow rate. Final conclusion has been that decreasing the pressure gradient near the well may significantly reduce the degree of calcite precipitation around the well.

7.2 Long Term Performance at Stralsund (Germany)

Due to the geological situation the exploitation of geothermal energy for space and district heating in the North German basin is mainly provided from deep sandstone aquifers. The common arrangement of bore holes is the well doublet, consisting of one well for hot water production and one well for cooled water re-injection. One reason for water re-injection is to maintain the reservoir pressure, but the more important one is to avoid contamination of shallow aquifers and surface streams by geothermal brines. The prediction of the long-term evolution of the hydraulic and thermal reservoir parameters is necessary with respect to the economically required operation period of a geothermal heating plant of at least 30 years.

Re-injection of cooled brine into deep aquifers strongly affects the mass and energy flows in the reservoir. Temperature and pressure conditions within the aquifer are significantly changed due to a running geothermal plant. This results in a shift of the chemical equilibrium states between different minerals in the hosted rock and the formation fluid. Hence, there is strong interaction between flow, heat transfer, transport, and chemical reactions in the aquifer. Understanding the pore space changes caused by thermally induced precipitation and dissolution reactions and their effect on the flow field is the key to predict changes in most of the involved parameters.

In preliminary numerical studies (Kühn et al. 1999, Kühn and Schneider 2003) permeability changes due to chemical water-rock interaction were investigated referring to the near vicinity of an idealized injection well penetrating a sandstone aquifer moderately cemented with anhydrite. Due to retrograde solubility the injection of cold water leads to dissolution of anhydrite in a growing region around the well. The dissolved species are transported through the aquifer and it turns out that subsequent precipitation of anhydrite occurs at the thermal front. Nevertheless, the associated permeability increase around the well predominates the permeability decrease at the thermal front. It is concluded that retrograde dissolving cement minerals (like anhydrite or calcite) affect such a system in a positive way as shown, for example, by Bartels et al. (2002). They investigated anyhdrite relocation due to dissolution and subsequent precipitation in more detail in a comparing study where a laboratory experiment has been used to check the corresponding numerical simulations.

At the Stralsund location a geothermal resource has been investigated and confirmed in previous studies (Bartels and Iffland 2000) in Buntsandstein layers at a depth of about 1520 meters. For this modeling study of the long-term behavior of the reservoir, additional data were considered and re-examined to have a complete data set of the rock formation properties and the composition of the highly saline water.

Numerical simulations of a typical production regime of heat exploitation for district heating, applying the program SHEMAT, referring to the Stralsund location focus on:

- the simultaneous temporal and spatial evolution of hydraulic, thermal, and chemical parameters and their contribution to injectivity trends,
- permeability changes due to anhydrite and calcite mineral reactions,
- the sensitivity of permeability changes in respect to the assumed initial pore space structure.

7.2.1 Geological Setting of the Geothermal Potential

The city of Stralsund is situated at the Baltic Sea in North East Germany at the northern edge of the North German Basin. Three wells are already drilled and within the depth range of 1500 to 1600 m they reached the Detfurth sandstone with a thickness between 33 and 36 m. This aquifer, suitable for geothermal exploitation, belongs to the Buntsandstein formation (Fig. 7.1).

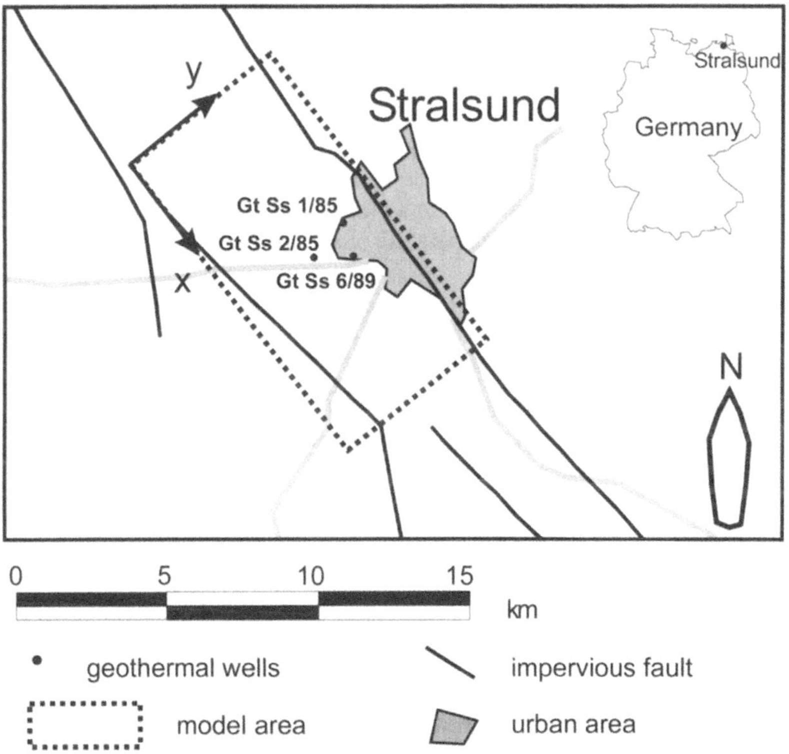

Fig. 7.1. Stralsund geothermal site with the wells Gt Ss 1/85, Gt Ss 2/85, and Gt Ss 6/89 (black dots); the reservoir is partly delineated by impervious faults (black lines); the model area (dotted rectangle) measures about 12 km x 6 km

Within the mesozoic stratigraphic sequence, the study area is affected over a wide range by distinctive tectonic faults. They evolved during the Keuper in the old Kimmeridgian stage. The faults are bound to and run along old Rheinisch striking elements of the deep basement. The displacements of the faults in the Buntsandstein, between 100 and 500 meters, are affected by salt tectonics. Thus, hydrodynamic separation of neighboring blocks can be assumed.

The project area considered is part of an approximately 5 km wide faulted block which trends north west - south east in relatively undisturbed stratigraphic conditions. Reflection seismometer measurements within the faulted block do not exclude smaller zones subjected to tectonic stress with resulting cleavage. But interference tests do not reveal any hydraulic obstruction between the wells. The block is bounded by fault zones, which can be assumed as impervious (Fig. 7.1). Regional aquifer flow is negligible.

From bore profiles and core samples it is yielded that the Detfurth sandstone is a well sorted, weakly consolidated fine to medium sandstone interlayered by silt and coarse sandstone within an alternated stratification. It is feldspatic quartz sandstone, low-graded with clay (< 2 %) and cement minerals (4-5 %). The cementation mainly consists of calcite and a minor amount of anhydrite (Bartels and Iffland 2000). For the following numerical simulations it is assumed, that no other mineral phases but calcite and anhydrite react. Porosity varies between 15 and 32 % and permeability between 0.1×10^{-12} and 2×10^{-12} m^2 with mean values of 23.9 % and 0.54×10^{-12} m^2, respectively.

The high saline formation water is of the Na-(Ca-Mg)-Cl type with a solute content of 280 g L^{-1} corresponding to a mass fraction of 0.23 (composition shown in Table 7.1) and a formation temperature of about 58°C.

Table 7.1. Chemical composition of the formation water given in mmol L^{-1} for 58°C with pH 5.85 (reservoir conditions, production) and for 20°C with pH 5.84 (re-injection) and in mmol kg^{-1}; the amount of total dissolved solids of the brine is 280 g L^{-1}

Constituents	Re-Injection [mmol $L^{-1}]^{20°C}$	Production [mmol $L^{-1}]^{58°C}$	[mmol kg^{-1}]
Calcium	359.1	353.0	307.0
Magnesium	95.66	94.02	81.76
Sodium	4039	3970	3452
Potassium	17.52	17.22	14.97
Chloride	4943	4859	4225
Hydrogen carbonate	0.94	0.93	0.81
Sulfate	3.92	3.85	3.35

The gas content of the brine is in the range of 180 to 250 mL L^{-1} (Table 7.2). The pH 5.85 of the water has been recalculated referring to reservoir conditions. The composition of the water reveals that it is in equilibrium concerning the mineral phases anhydrite and calcite for the formation temperature of 58°C. This is shown in Fig. 7.2 by the interdependent saturation indices of anhydrite and calcite. With decreasing temperature the solubility of both minerals increases (retrograde solubility). Despite very similar saturation indices it can be seen that the up-take capacity of the water is two orders of magnitude higher for anhydrite than for calcite. The up-take capacity is defined here as the amount of mineral additionally soluble in the formation water if temperature decreases from 58°C (initial formation temperature) to a certain temperature (x-axis).

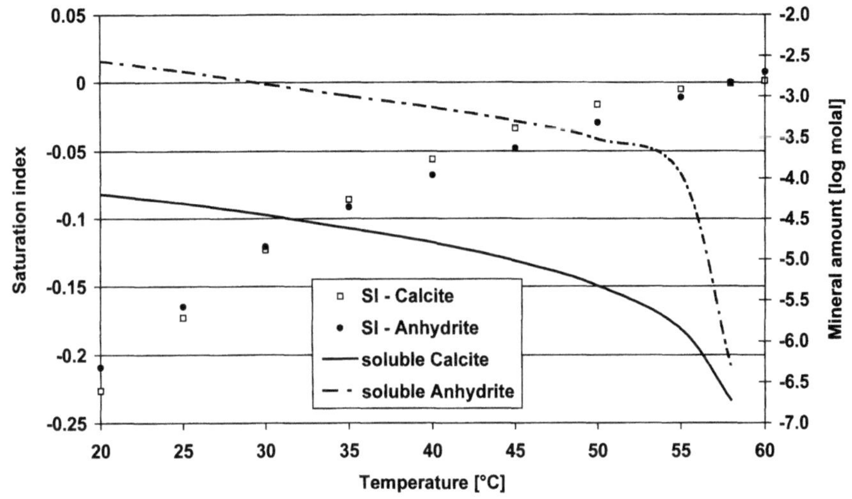

Fig. 7.2. Interdependent saturation indices of anhydrite and calcite versus temperature and the mineral up-take capacity of the reservoir water displaying the amounts of anhydrite and calcite additionally soluble in the formation water due to decreasing temperature from initial formation temperature (58°C) to a certain temperature on the x-axis

Table 7.2. Total gas content of the brine is 180 – 250 mL L^{-1}

Component	Vol %
Methane	0.8
Ethane	0.18
Nitrogen	97.0 – 98.2
Carbon dioxide	0.8 – 2.2
Helium	0.30 – 0.58

7.2.2 Conceptual Model of Injection and Production Wells

The modeled horizontal area of the Stralsund location measures 12 km x 6 km and is partly delineated by the existing impervious geological faults (Fig. 7.1). The reservoir properties applied, taken from Bartels and Iffland (2000), are summarized in Table 7.3.

A 2D simulation is carried out with a uniform thickness of the horizontal layer of 34 m (Kühn et al. 2002a). The two drillings nearest to the town are used for production and the third one for re-injection to minimize transport distances for the hot water. Production rate is 50 m^3 h^{-1} for each production well (in heating plants controlled by water-meter; reference conditions: 58°C, 1 bar, 280 g L^{-1}).

The produced water is re-injected with a temperature of 20°C. Mass conservation leads to a re-injection rate of 98 m^3 h^{-1}. For diagnostic reasons a conservative tracer is additionally injected to visualize transport of dissolved ions in the model area. The simulated period of operation of the geothermal heating plant is 80 years.

Table 7.3. Reservoir properties of the Detfurth sandstone (isotropic)

Parameter	Value	Units
Anhydrite	76.5	[mol m^{-3}]
Calcite	1170	[mol m^{-3}]
Porosity	0.239	[-]
Permeability	0.540×10^{-12}	[m^2]
Thermal capacity	2.3	[MJ m^{-3} K^{-1}]
Thermal conductivity	2.5	[W m^{-1} K^{-1}]
Temperature	58.0	[°C]
Aquifer thickness	34.0	[m]
Fractal exponents	5 / 12	according to Eq. (4.3)
Reservoir initial pressure	16.02	[MPa] 1520 m depth
Fluid salinity – mass fraction	0.23	[-]

The chemical calculations are based on the "equilibrium"-assumption that reaction rate is very fast compared to the other processes involved. This neglects reaction kinetics, due to the fact that the saturation lengths of both anhydrite and calcite are far below the extent of the smallest model cell of 10 m (Bartels et al. 2002, Schulz 1988). Resulting porosity changes ($\Delta\Phi$) are calculated from the molar volume of the minerals. Permeability changes (Δk) are derived with the help of the fractal permeability-porosity-relation [Eq. (4.3)].

$$\Delta k = f\left\{\Delta\Phi\left(\Delta c_{\text{mineral}}\right)\right\} \tag{7.1}$$

Simulations are carried out (a) under isothermal conditions (flow and re-injection with formation temperature), (b) for the non-reactive case (flow and heat transfer only), and (c) for the reactive case (flow, heat transfer, transport, and chemical reactions). This set of simulations was chosen to compare and separate the thermal and chemical effects on the injectivity of the re-injection well.

The calculations of the permeability changes by dissolution and precipitation are done with varying fractal exponents [D_f in Eq. (4.3)] according to different kinds of cementation and therefore different structural changes of the pore space. An exponent of 5, representing few, big crystals (smooth shaped grains or coatings), was determined from petrophysical laboratory measurements of the pore space structure due to cementation formed in geological time periods (Clauser et al., 2000). An exponent of 12, representing many and small crystals, was found in core flooding experiments. The experiments were set up to investigate precipitation and dissolution in a technical time scale. This assumption could be verified by REM micrographs of the pore space of core samples from the precipitation region (Bartels et al., 2002). One aim of this study is to determine the sensitivity of the reservoir injectivity to changes of the cement minerals pore space structure because this is usually not well known (applied fractal exponents listed in Table 7.3).

7.2.3 Numerical Simulation of 80 Years Heat Production

Examination and understanding of the long-term behavior of reservoir properties of the Stralsund deep aquifer requires, as far as possible, a quantitative separation of the single contribution of the various processes involved to the hydraulic changes in the aquifer which arise due to re-injection.

Due to the proximity of the geological faults to the wells (compare geological setting, Fig. 7.1), it has been considered that these reservoir boundaries may enhance the hydraulic head increase. Hence, at first an isothermal simulation of reservoir exploitation is conducted, neglecting any chemical reaction and associated effects, with re-injection of brine at reservoir temperature, to investigate the influence of the impervious faults bounding the model area. The initial hydraulic head of 1315 m increases shortly after the beginning of re-injection and remains constant at 1381 m for the entire simulation period of 80 years (Fig. 7.3, lower curve). This steady state shows that any effect of the faults on the hydraulic well head can be disregarded.

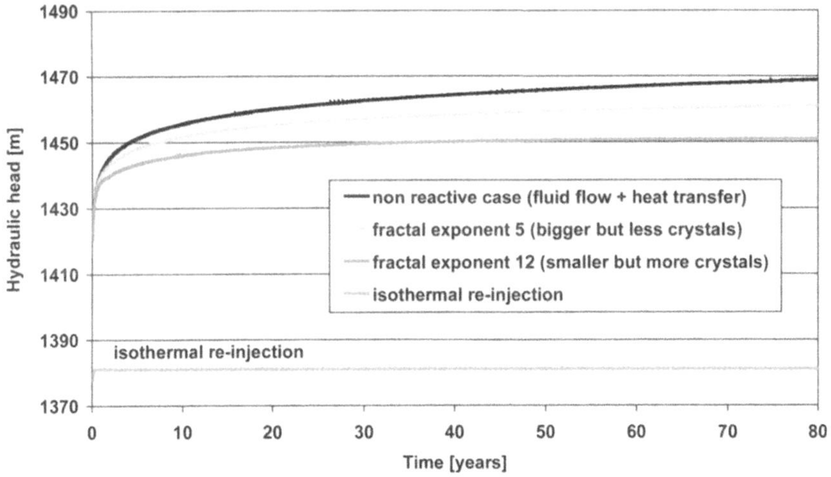

Fig. 7.3. Temporal evolution of the pressure head at the injection well shown as hydraulic head during 80 years reservoir exploitation

To quantify the effect of cooled water re-injection the next step is a simulation of the non-reactive, non-isothermal case. In Fig. 7.4 A and Fig. 7.4 B the temperature and tracer distribution is shown after 10 years cold water re-injection. On account of clarity the figure shows the influenced area between and around the wells only. It is clearly recognizable that the tracer front (marked by the 5.0 mmol L^{-1} isoline, i.e. the arithmetic mean of injected and initial homogeneous aquifer concentration) propagates about 1.7 times faster compared to the thermal front (marked by the 39°C isotherm, i.e. the arithmetic mean of the injected water and the initial formation temperature).

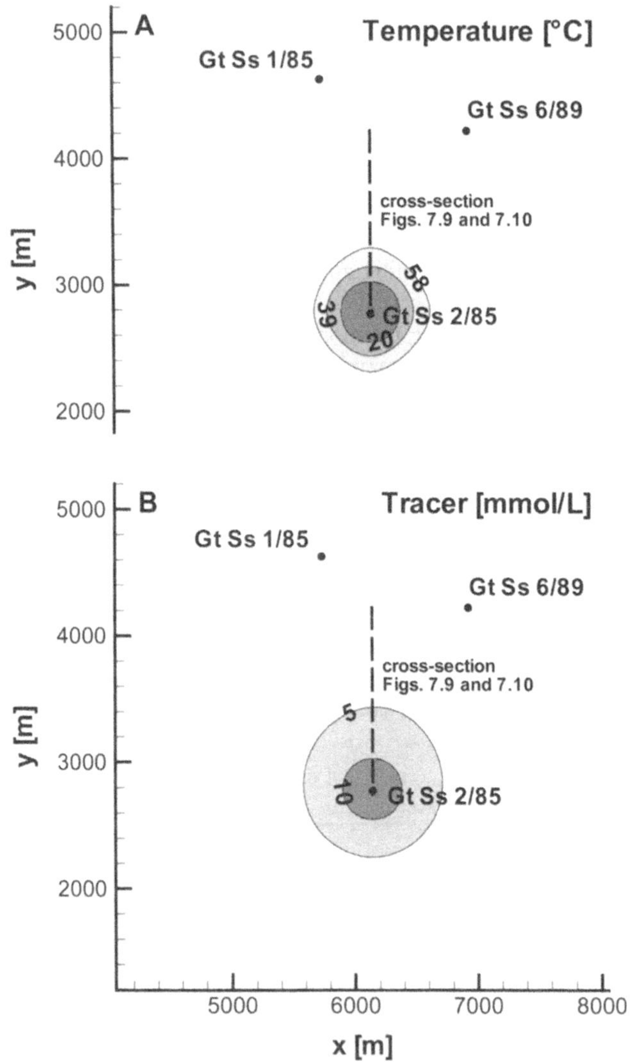

Fig. 7.4. Thermal (A) and tracer front (B) represented by the 39°C isotherm and the 5.0 mmol L^{-1} isolines (arithmetic means), respectively, propagating from the injection well Gt Ss 2/85 to the production wells Gt Ss 1/85 and Gt Ss 6/89 (black dots indicate the wells); situation after 10 years re-injection with a temperature of 20°C into the aquifer with a formation temperature of 58°C

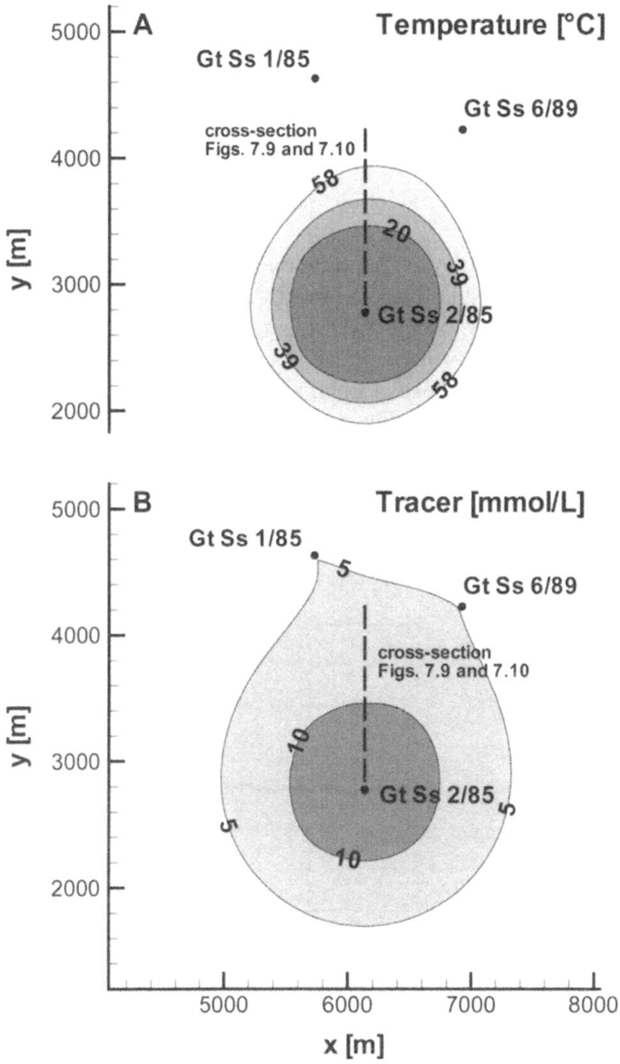

Fig. 7.5. Thermal (A) and tracer front (B) represented by the 39°C isotherm and the 5.0 mmol L^{-1} isolines (arithmetic means), respectively, propagating from the injection well Gt Ss 2/85 to the production wells Gt Ss 1/85 and Gt Ss 6/89 (black dots indicate the wells); situation after 50 years re-injection; the tracer has just reached the production wells whereas the thermal front is still away

After 50 years it can be seen that the tracer front just has reached the production wells, which means that the water firstly produced has reached the production, well again (Fig. 7.5 B). The thermal front however is still several hundred meters away from the production wells Gt Ss 6/89 and Gt Ss 1/85 (Fig. 7.5 A). It is well known that the propagation of thermal fronts in well doublets is slower than the propagation of tracer fronts (i.g. Hoopes and Harleman 1967). This is due to the fact that the re-injected water always has to cool the rock material. In this term the thermal front velocity depends on the volumetric specific heat capacity of the formation (Table 7.3) and the rate of injected brine. In general, production and re-injection wells of a geothermal plant are drilled at a distance such that the break-through time of cold brine in the production well is greater than 30 years of exploitation.

The inspection of the production temperature over the entire simulation period of 80 years yields that a detectable thermal breakthrough, a temperature decrease of 0.1°C in the production wells, occurs after about 78 years solely in well Gt Ss 6/89 whereas the temperature in well Gt Ss 1/85 is still 58°C (Fig. 7.6). But even after 80 years the temperature decrease is small and the reservoir is still above the technically exploitable temperature minimum. At this respect it has additionally to be taken into account that the 2D simulation conducted here results in a steeper production temperature decrease (worst case) compared to a simulation allowing for temperature exchange with bed and cap rock.

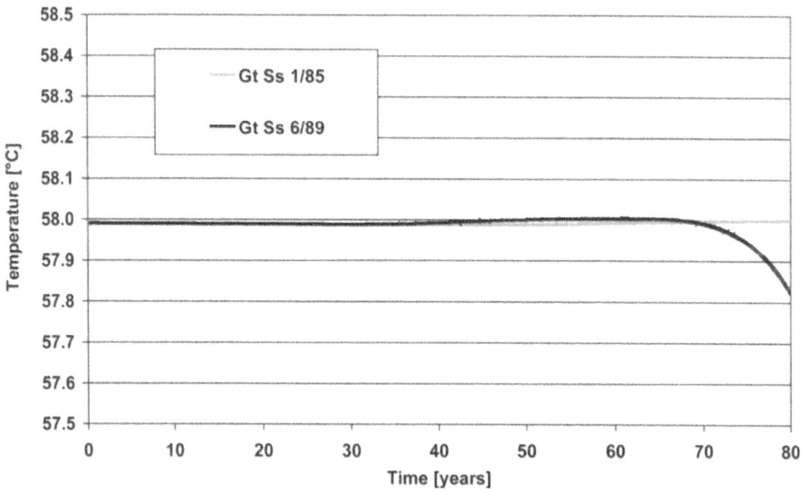

Fig. 7.6. Production temperature of the wells Gt Ss 1/85 and Gt Ss 6/89; temperature break-through occurs after about 78 years in the well Gt Ss 1/85 whereas the temperature at Gt Ss 6/89 is still 58°C after 80 years

The sandface pressure necessary for injection strongly depends on the injection temperature. With temperature decreasing from 58°C to 20°C the viscosity of the water increases from 4.8×10^{-4} kg m^{-1} s^{-1} to 1.0×10^{-3} kg m^{-1} s^{-1}. This is why the hydraulic head does not reach a steady state during the simulation period (Fig. 7.3, non reactive case). The aquifer volume filled with cold water increases with time and thus the overall pressure losses in the aquifer increase continuously.

In the following part the redistribution of the minerals anhydrite and calcite, triggered by hydraulic processes within the model area, is investigated. In Fig. 7.7 the resulting spatial distribution of the cement minerals anhydrite (A) and calcite (B) is shown after 50 years of re-injection.

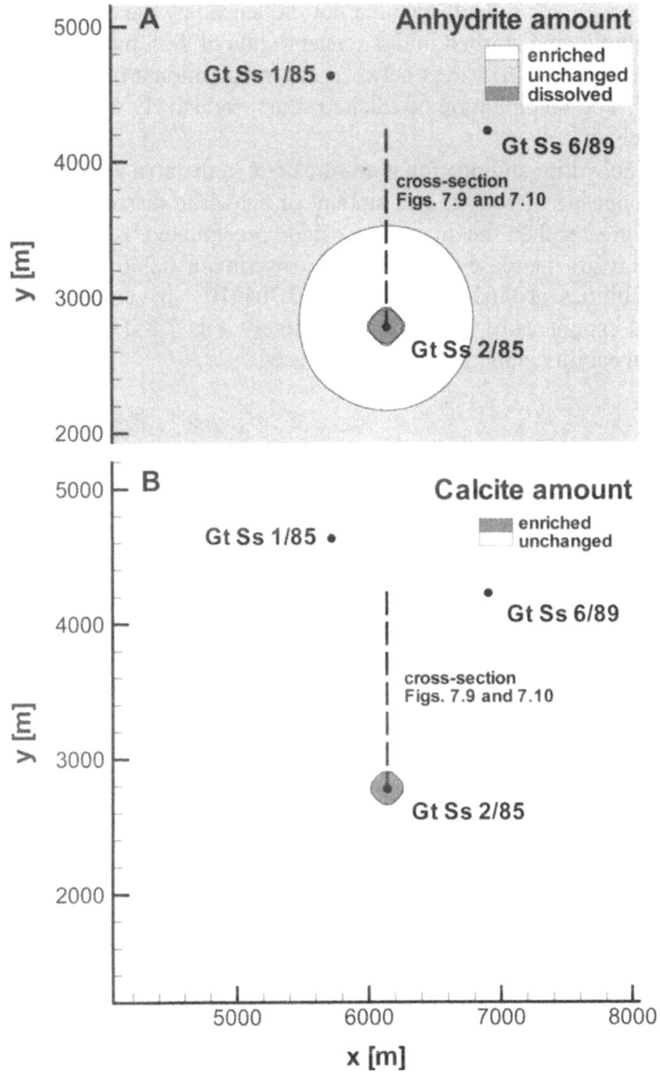

Fig. 7.7. Spatial distribution of anhydrite (A) and calcite (B) in the vicinity of the injection well Gt Ss 2/85 after 50 years brine re-injection; the region marked "dissolved" refers to an anhydrite amount of 0 mol m^{-3}; the area "enriched" refers to anhydrite between the initial concentration and the maximum amount of 77.7 mol m^{-3}; the area "unchanged" corresponds to the initial concentration of 76.5 mol m^{-3} (A); calcite varies from the initial concentration of 1170 mol m^{-3} mentioned as "unchanged" up to 1187 mol m^{-3} marked as "enriched" (B); the areas of anhydrite dissolution and calcite precipitation coincide

The distribution of anhydrite varies from 0 up to 77.7 mol m^{-3}. The dark gray patch around the injection well displays an area where anhydrite is totally dissolved. A white region where the amount of anhydrite has increased compared to the initial concentration surrounds it. The outer border of the white region coincides with the position of the thermal front after 50 years (Fig. 7.5 A, 39°C isotherm). The light gray color reflects the area not influenced by the 80 years period of exploitation with the undisturbed initial concentration of 76.5 mol m^{-3}. The distribution of calcite (Fig. 7.7 B) varies between the initial amount of 1170 mol m^{-3} and 1187 mol m^{-3}. The precipitation of calcite occurs exclusively where anhydrite has been completely dissolved.

Dissolution of anhydrite and precipitation of calcite in the area around the well, alter porosity in opposite direction. The amount of anhydrite dissolved is one order of magnitude greater than the amount of calcite precipitated. Accordingly, the net effect is a porosity increase from 0.239 to maximum 0.246 with resulting maximum permeabilities of 0.624×10^{-12} m^2 and 0.764×10^{-12} m^2 corresponding to the applied fractal exponents of 5 and 12, respectively. Fig. 7.8 delineates the areas where the permeability is increased and decreased.

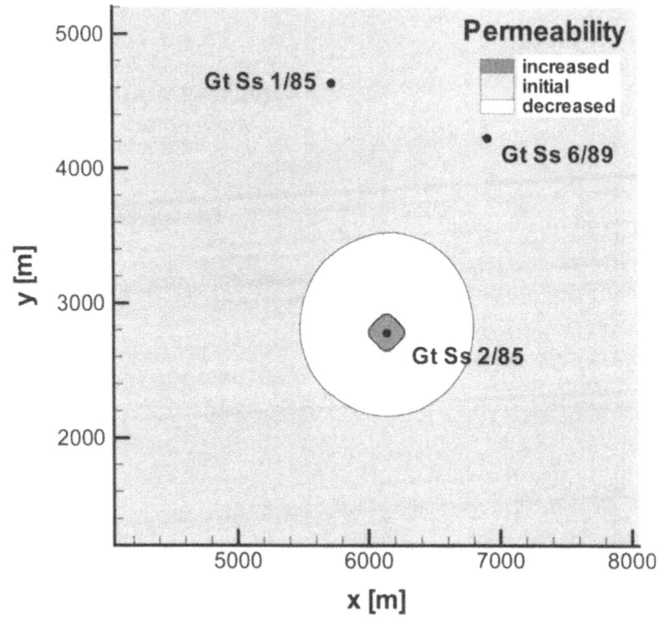

Fig. 7.8. Increased, decreased, and initial permeability in the vicinity of the injection well (Gt Ss 2/85) as result of the dissolution and precipitation of the minerals anhydrite and calcite; final permeabilities depend on the applied fractal exponents (refer to text)

As a consequence of the anhydrite precipitation in the white region (Fig. 7.8) the permeability is moderately reduced by 0.2% (for both applied fractal exponents). The region of anhydrite enrichment and permeability reduction propagates through the aquifer together with the broadening thermal front, followed by the anhydrite dissolution zone. When the cooling front propagates further and the injection temperature of 20°C is approached the redistributed anhydrite is dissolved again, accompanied by calcite precipitation.

The propagating reaction fronts are shown in more detail along the cross sections in Fig. 7.9 and Fig. 7.10 as functions of distance from the injection well Gt Ss 2/85 (x = 6140 m and y = 2775 m). Fig. 7.9 displays the distribution of pH and dissolved calcium, sulfate, and carbonate as well as the minerals anhydrite and calcite compared to the thermal and tracer front after a period of 10 years. The thermal front (20°C ⇒ 58°C) extends from 3100 m to 3400 m (y-axis) and the tracer front (0 mmol kg^{-1} ⇒ 10 mmol kg^{-1}) from 3300 m to 3600 m. Obviously the concentrations of solutes and minerals are closely related to the temperature distribution and transport processes. Near the injection well calcium has a concentration of 307.7 mmol kg^{-1} (injected conc. 306.9 mmol kg^{-1}), which increases to 309.4 mmol kg^{-1} at 2880 m. Sulfate reacts comparable to calcium with a concentration of 3.35 mmol kg^{-1} (≡ injected conc.) increasing to 5.08 mmol kg^{-1}. Within the thermal front the concentrations of calcium and sulfate decrease to 305.4 and 3.32 mmol kg^{-1}, respectively. Within the tracer front, intersecting the thermal front, calcium re-increases to 306.9 mmol kg^{-1} whereas sulfate further declines to 3.12 mmol kg^{-1}. Compared to calcium and sulfate the development of the pH and the carbonate concentration is different. pH and carbonate increase almost instantaneously after re-injection. The pH shifts from 5.92 (injected pH 5.84) to 5.98 and carbonate 0.84 mmol kg^{-1} (injected conc. 0.81 mmol kg^{-1}) to 0.86 mmol kg^{-1}.

At 2880 m the pH decreases from 5.98 to 5.97 accompanied by an increase in carbonate from 0.86 to 0.87 mmol kg^{-1}. Within both the thermal and the tracer front pH decreases to minimum 5.75 and rises again to 5.82. In contrast to pH carbonate increases firstly to 1.05 mmol kg^{-1} and decreases afterwards to 0.81 mmol kg^{-1}. As already shown (Fig. 7.7 A and B), anhydrite is totally dissolved in the near vicinity of the injection well whereas calcite increased from the initial concentration of 1170 mol m^{-3} to maximum 1187 mol m^{-3} within this area. Further downstream from the injection well anhydrite is increased to maximum 77.7 mol m^{-3} curving down to the initial concentration of 76.5 mol m^{-3} within the thermal front. Calcite remains constant with the initial concentration of 1170 mol m^{-3}.

Fig. 7.10 displays the same quantities as Fig. 7.9 but now after the period of 50 years. The thermal front (20°C ⇒ 58°C) extends from 3500 m to 3900 m. The tracer front already had passed the observed sector. Shapes of the distributions as well as the concentrations of solutes and minerals after 50 years re-injection are similar to the situation after 10 years. In difference, rise in calcium, sulfate, and carbonate as well as decrease of pH occurs further downstream (y = 2950 m) and re-increase of calcium behind the thermal front is small from 305.1 to 305.4 mmol kg^{-1}. The pH decreases from 5.98 to 5.72 within the thermal front and remains constant from there on. Carbonate concentration increases from 0.88 to 1.20 mmol kg^{-1} and smoothly curves down to 1.17 mmol kg^{-1}. The region of anhydrite dissolution, which is identical with the area of calcite increase, is enlarged as well as the region of anhydrite increase downstream up to the thermal front.

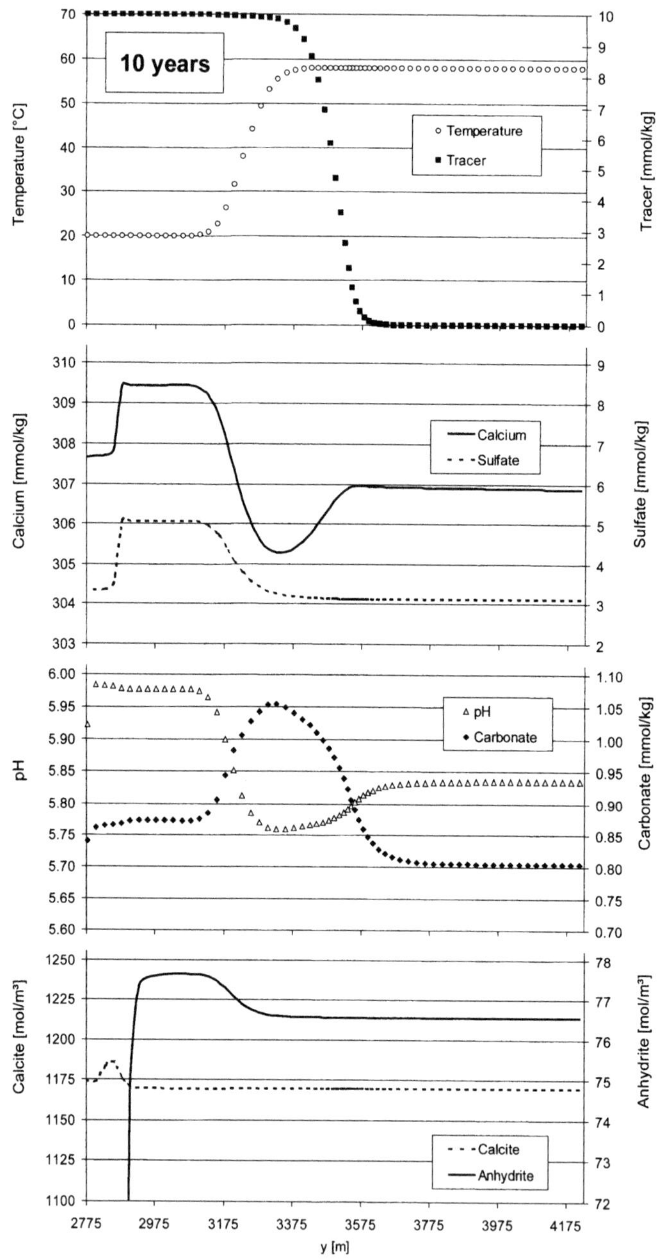

Fig. 7.9. Reaction fronts within the flow path from injection well (x = 6140 and y = 2775, compare cross-section Fig. 7.4) towards production wells; situation after 10 years brine re-injection; thermal and tracer fronts are shown compared to the pH, the dissolved amounts of calcium, sulfate, and carbonate and the mineral phases calcite and anhydrite

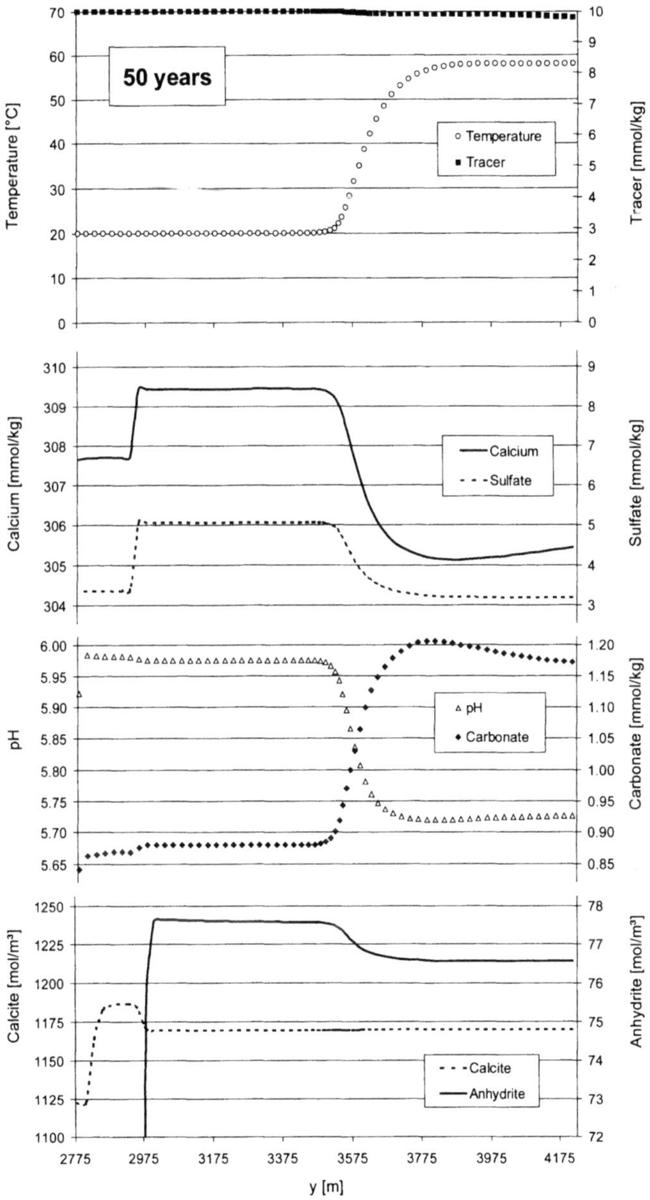

Fig. 7.10. Reaction fronts within the flow path from injection well towards production wells (x = 6140 and y = 2775, compare cross-section Fig. 7.5); situation after 50 years brine re-injection; thermal and tracer fronts are shown compared to the pH, the dissolved amounts of calcium, sulfate, and carbonate and the mineral phases calcite and anhydrite

The temperature regime during the injection process controls the chemical reactions. As anhydrite and calcite are more soluble in cold than in hot water (retrograde solubility, Fig. 7.2) injecting 20°C water into an aquifer with a formation temperature of 58°C should cause dissolution of anhydrite and calcite around the injection well. As can be seen in Fig. 7.9 and Fig. 7.10 anhydrite dissolution does occur, but simultaneously the calcite amount in the vicinity of the well is increased due to precipitation. Separate geochemical simulations with a box model revealed that with the beginning of the re-injection anhydrite and calcite start to dissolve. Due to the fact, that the uptake capacity of the geothermal brine is two orders of magnitude higher for anhydrite if temperature decreases (Fig. 7.2) and additionally the initial amount of anhydrite is one order of magnitude lower than the one of calcite (Table 7.3), the depletion of anhydrite apparently proceeds at a higher rate if unsaturated cold water passes. When anhydrite is totally dissolved the injected brine equilibrates solely with the remaining calcite and with a slightly higher amount than in the additional presence of dissolved anhydrite.

This injected water, now saturated with calcite (in the absence of anhydrite) further downstream reaches areas with both, anhydrite and calcite, in the pore space. Re-equilibration leads to dissolution of anhydrite and simultaneous precipitation of calcite. As a result the calcite amount increases slightly above the initial concentration of 1170 mol m^{-3} (Fig. 7.7 B, Fig. 7.9, and Fig. 7.10). That reaction exclusively occurs at the border between the innermost region already freed from anhydrite and the adjoining one where both minerals are still left.

Compared to the composition of the re-injected water, the mineral reactions of anhydrite and calcite described above result in transport of increased amounts of calcium, sulfate, and carbonate through the aquifer. The transport of dissolved ions is faster than the propagation of the low temperature front through the aquifer as discussed in the beginning of this section (compare Fig. 7.4, Fig. 7.5, Fig. 7.9, and Fig. 7.10). Therefore water at 20°C in equilibrium with larger amounts of calcium, sulfate, and carbonate eventually reaches the high temperature region from upstream. Due to their retrograde solubility less anhydrite and calcite should be soluble in the brine at the higher temperature. Hence, increased amounts of anhydrite are observed in the zone between the inner region where anhydrite is totally dissolved and the outer high temperature region (Fig. 7.9 and Fig. 7.10), compared to the initial concentration of 76.5 mol m^{-3}. But for calcite, contrary to the expectations raised above, dissolution is indicated by decreasing pH and increasing carbonate concentration within the thermal front. The reason for these observations is the interaction between solutes and minerals close to thermodynamic equilibrium. Within the thermal front thermodynamic equilibrium is reached by precipitation of anhydrite and simultaneous dissolution of calcite.

The pH, the concentrations of the solutes, and the temperature remain at their natural formation values downstream of the thermal front (Table 7.1, production; Fig. 7.9, after 10 years). After tracer breakthrough in the production well the distribution of the solutes has changed completely compared to their initial distribution, due to the chemical processes occurring at the injection well and the thermal front and the transport of the re-equilibrated brine through the model area (Fig. 7.10, after 50 years).

Running a Geothermal Heating Plant the most essential parameter, beside production rate and temperature, is the injection pressure. This pressure depends on permeability, thickness of the aquifer, temperature, and rate of re-injection. The trend of the equivalent hydraulic head at the injection well is shown in Fig. 7.3, which is, due to the chosen reference density (brine of 280 g L^{-1}), equivalent with the well-block pressure (the computed pressure in the grid block containing the well). The initial hydraulic head of 1315 m corresponds to the reservoir pressure of 16.02 MPa assuming a water column of constant temperature and salinity and a situation of no exploitation. For the non-reactive case (flow and heat transfer only) it can be seen that the hydraulic head increases almost up to 1470 m. Even after 80 years of heat production it is not at steady state. Taking into account chemical reactions the hydraulic head increase rate declines. A fractal exponent of 5 represents dissolution and precipitation of natural mineral cementation with structures formed by diagenesis (Clauser et al. 2000). The effect of reduced hydraulic head increase is smaller with a fractal exponent 5 compared to the fractal exponent 12. A fractal exponent of 12 represents dissolution and precipitation of many small crystals formed in a technical time scale at comparatively rapid flow (Bartels et al. 2002). It can be seen that the mineral reactions at application of the fractal exponent 12 provide an almost constant hydraulic head of 1450 meters required for re-injection at the end of the simulation period. The hydraulic head evolution at the injection well depends on the rate of permeability changes: the larger the fractal exponent, the lower the resulting hydraulic head.

The occurring chemical reactions applied with the two limiting values of the observed fractal exponents 5 and 12, cause a decrease of the hydraulic head by about 5 % and 13 %, respectively. As a result of the mineral redistribution around the injection well, the injectivity of the layer increases compared to the non-reactive case of flow. Considering the injection well head pressure this could balance moderate aging trends in well injectivity observed in operating plants but not discussed here.

In the period considered, the injectivity decrease due to the cold water viscosity is partially or even fully compensated by the dissolution of calcite and anhydrite in the formation. The absolute hydraulic head change due to the chemical reactions is smaller compared to changes caused by thermal effects, but however the hydraulic reservoir properties are improved, due to the dissolution of anhydrite around the well, whereas the precipitation of anhydrite within the thermal front is of minor importance concerning the injectivity of the re-injection well. The resulting net increase of well injectivity is sensitive to the actual pore space structure.

7.2.4 Conclusion Drawn from the Stralsund Case Study

The numerical case study of heat exploitation for district heating is carried out for the Stralsund location with its already confirmed geothermal potential. It is chosen because a complete data set of the formation parameters of the Detfurth sandstone and the high saline formation water is available. From this study focusing on injectivity changes due to exploitation of a hydrothermal reservoir it can be concluded, that:

- the evolution of the well injectivity, one of the most important technical parameter for reservoir exploitation, is influenced primarily by thermal effects, in which the re-injection of the cooled water leads to a steady reduction of the hydraulic conductivity;
- mineral dissolution and precipitation of anhydrite and calcite has no negative effect on the reservoir exploitation, but quite the reverse, well injectivity improves moderately. The resulting hydraulic head decreases with up to 13 %, compared to the non-reactive case of flow, are significant but not dominating for the process. Nevertheless, this difference can represent a large fraction of re-injection pressure requirement at the wellhead in cases where sandface pressure has to be higher than the cold water column hydrostatic pressure;
- the chemical mineral redistribution in the aquifer weakens the viscosity induced trend of increasing hydraulic head at the injection well;
- the pore space structure of both, the natural mineral cementation of the formation and the newly precipitated mineral, significantly determines the rate of permeability change due to the chemical reactions.

It can be summarized that a reservoir moderately cemented with retrograde dissolving minerals (here anhydrite and calcite) and with sufficient initial porosity and permeability will require lower pressure heads for long-term re-injection compared to a reservoir with the same initial permeability and porosity but without reactive cement minerals. Hence, from the geochemical point of view the long-term operation of a geothermal heating plant at the location Stralsund is not restricted.

With this study, adding geochemical and complex modeling aspects, previous studies have been completed, which have stated that the Stralsund location is suitable for long-term operation of a geothermal heating plant from the hydrogeological and hydrothermal point of view. Taking into account mineral reactions and resulting changes of the pore space structure and hydraulic parameters further confirms and ensures the utilizable geothermal potential at Stralsund.

References

Abe M (1993) Long term use of acidic reservoir at Onikobe geothermal power plant. In: Lee KC, Dunstall MG, Hochstein MP (eds) Proceedings of the 15th New Zealand Geothermal Workshop 1993 Geothermal Institute, University of Auckland, pp 193-197

Adams MC (1996) Chemistry of fluids from Ascension #1, a deep geothermal well on Ascension Island, South Atlantic Ocean. Geothermics 25(4/5):561-579

Adams MC, Lemieux M, Moore JN, Johnson SD (1989) Fluid chemistry and hydrogeology of the Heber geothermal system, California. Proceedings, Fourteenth Workshop on Geothermal Reservoir Engineering Stanford University, Stanford, California, January 24-26, pp 81-86

Akaku K, Kasai K, Nakatsuka K (2000) Modeling of formation of acid water discharged from high temperature geothermal reservoir. Proceedings of World Geothermal Congress 2000, Kyushu - Tohuku, Japan, May 28 - June 10, pp 895-900

Appelo CAJ, Postma D (1999) Geochemistry, groundwater and pollution. Balkema Publishers, Rotterdam

ARC (1991) Waiwera geothermal groundwater resources statement and allocation plan. Auckland Regional Council, Technical Publication No. 112

ARC (1999) Waiwera geothermal groundwater resource assessment report. Auckland Regional Council, Technical Publication No. 115

Arihara S, Arihara N (1999) Modeling silica deposition in injection wells of the Otake geothermal field, Japan. Proceedings, Twenty-Fourth Workshop on Geothermal Reservoir Engineering, Stanford University, Stanford, California, January 25-27, 2001, pp 78-85

Ármannsson H, Kristmannsdóttir H, Ólafsson M (2000) Geothermal influence on groundwater in the Lake Mývatn area, North Iceland. Proceedings of World Geothermal Congress 2000, Kyushu - Tohuku, Japan, May 28 - June 10, pp 515-520

Arnorsson S, Gunnlaugsson E, Svavarsson H (1983) The chemistry of geothermal waters in Iceland. II. Mineral equilibria and independent variables controlling water compositions. III. Chemical geothermometry in geothermal investigations. Geochimica et Cosmochimica Acta 42:547-566/567-577

ARWB (1980) Waiwera water resource survey – Preliminary water allocation/management plan. Auckland Regional Water Board, Technical Publication No. 17

ARWB (1987) Waiwera thermal groundwater allocation and management plan 1986. Auckland Regional Water Board, Technical Publication No. 39

Back W, Hanshaw BB, Pyle TE, Plummer LN, Weidie AE (1979) Geochemical significance of ground-water discharge and carbonate solution to the formation of Caleta Xel Ha, Quintana Roo, Mexico. Water Resources Research 15(6):1521-1535

Baermann A, Kröger J, Taugs R, Wüstenhagen K, Zarth M (2000a) Anhydrite Cement in Rhaetian Sandstones in Southeastern Hamburg: Morphologie and Structure. Zeitschrift für angewandte Geologie 46(3):138-144 (in German)

Baermann A, Kröger J, Zarth M (2000b) Anhydrite Cement in Rhaetian Sandstones in Hamburg: X-ray and NMR Tomographie Studies and Leaching Tests. Zeitschrift für angewandte Geologie 46(3):144-152 (in German)

Balanque MIRD (2000) Chemical reaction path modelling of vein mineralization in the Tongonan geothermal field, Leyte. Proceedings of World Geothermal Congress 2000, Kyushu - Tohuku, Japan, May 28 - June 10, pp 953-958

Baldschuhn R, Binot F, Fleig S, Kockel F(2001) Tectonic Atlas of Northwest Germany and the German North Sea Sector – Structures, Structural Development, Palaeogeography. Geologisches Jahrbuch, Reihe A, Heft 153

Baldwin SF (2002) Renewable energy: Progress and prospects. Physics Today 55(4):62-67

Balmes CP (2000) The geochemistry of the Mt. Balut Island geothermal prospect, Davao del Sur, Philippines. Proceedings of World Geothermal Congress 2000, Kyushu - Tohuku, Japan, May 28 - June 10, pp 959-964

Bartels J, Iffland J (2000) Hydraulic, thermal, and mechanical behavior of geothermal aquifers under exploitation: Analysis and assessment of petrographical / petrophysical and hydrochemical data and investigations as well as simulations. Project Final Report BMBF-0326995D, Landesamt für Umwelt, Naturschutz und Geologie Mecklenburg-Vorpommern 64 (in German)

Bartels J, Kühn M, Clauser C (2003) Numerical simulation of reactive flow using SHEMAT. In: Clauser C (ed) Numerical simulation of reactive flow in hot aquifers - SHEMAT and Processing SHEMAT, Springer Publishers, Heidelberg, pp 5-74

Bartels J, Kühn M., Schneider W., Clauser C., Pape H., Meyn V., Lajczak I. (2002) Core flooding laboratory experiment validates numerical simulation of induced permeability change in reservoir sandstone. Geophysical Research Letters 29(9):10.1029 / 2002GL014901

Bath AH, Williamson KH (1983) Isotopic and chemical evidence for water sources and mixing in the Cerro Pando geothermal area, Republic of Panama. Geothermics 12(2/3):177-184

Battistelli A, Calore C, Pruess K (1997) The simulator TOUGH2/EWASG for modelling geothermal reservoirs with brines and non-condensible gas. Geothermics 26(4):437-464

Bear J (1972) Dynamics of Fluids in Porous Media, Dover Publications, New York

Bear J (1979) Hydraulics of Groundwater. MacGraw-Hill, New York

Berner R (1980) Early diagenesis: A theoretical approach. Princeton University Press

Bethke CM (1985) A numerical model of compaction-driven groundwater flow and heat transfer and its application to the paleohydrology of intercratonic sedimentary basins. Journal of Geophysical Research 90(B8):6817-6828

Bethke CM (1996) Geochemical reaction modeling – Concepts and Applications. Oxford University Press, New York – Oxford

Bethke CM, Lee MK, Quinodoz HAM, Kreiling WN (1993) Basin modeling with BASIN2. Urbana: Univeristy of Illinois Hydrology Program

Beyene K (2000) Chemical and isotopic studies of geothermal prospects, in southern Afar Region, Ethopia. Proceedings of World Geothermal Congress 2000, Kyushu - Tohuku, Japan, May 28 - June 10, pp 977-983

Bischoff JL, Radtke AS, Rosenbauer RJ (1981) Hydrothermal alteration of greywacke by brine and seawater: roles of alteration and chlorite complexing on metal solubilities at 200°C and 350°C. Economic Geology 79:659-676

Bitzer K (1999) Two-dimensional simulation of clastic and carbonate sedimentation, consolidation, subsidence, fluid flow, heat flow and solute transport during the formation of sedimentary basins. Computer & Geosciences 25:431-447

Bjorlykke K, Mo A, Palm E (1988) Modelling of thermal convection in sedimentary basins and its relevance to diagenetic reactions. Marine and Petroleum Geology 5, 338-351

Block J, Waters OB (1968) The $CaSO_4$-Na_2SO_4-$NaCl$-H_2O System at 25°C to 100°C. Journal of Chemical Engineering 13(3):336-344

Bolt GH, Groenevelt PH (1969) Coupling phenomena as a possible cause for non-Darcian behaviour of water in soil. Bulletin International Association Hydrology 14(2), 17-26

Bortolami GC, Cravero M, Olivero GF, Ricci B, Zuppi GM (1983) Chemical and isotopic measurements of geothermal discharges in the Acqui Terme district, Piemont, Italy. Geothermics 12(2/3):185-197

Boudreau BP (1997) Diagenetic models and their implementation – Modelling transport and reactions in aquatic sediments. Springer, Berlin

Brace WF (1980) Permeability of crystalline and argillaceous rocks. International Journal of Rock Mechanics and Mining Sciences and Geomechanics Abstracts 17:241-251

Brace WF (1984) Permeability of crystalline rocks: New in situ measurements. Journal of Geophysical Research 89:4,327-4,330

Brown KL (1986) Gold deposition from geothermal disharges in New Zealand. Economic Geology 81:979-983

Brown KL, Webster JG, Christenson BW (1996) Precious metal sampling at the Ohaaki geothermal field. In: Simmons SF, Rahman MM, Watson A (eds) Proceedings of the 18th New Zealand Geothermal Workshop 1996 Geothermal Institute, University of Auckland, pp 169-174

Browne PRL (1978) Hydrothermal alteration in active geothermal fields. Ann. Rev. Earth Planet. Sci. 6:229-250

Burt DM (1981) Acidity-salinity diagrams – Application to greisen and porphyry deposits. Economic Geology 76:832-843

Butler JN (1998) Ionic Equilibrium: Solubility and pH Calculations. Wiley & Sons, New York

Campos T (1988) Geothermal resources of El Salvador – Preliminary assessment. Geothermics 17(2/3):319-332

Carman PC (1956) The Flow of Gases through Porous Media. Academic Press, New York

Cataldi R (1999) The year zero of geothermics. In: Cataldi R, Hodgson SF, Lund JW (eds) Stories from a heated Earth. Geothermal Resources Council, Davis California USA, pp 7-17

Cataldi R, Burgassi PD (1999) Flowering and decline of thermal bathing and other uses of natural heat in the Mediterranean area, from the birth of Rome to the end of the first millenium. In: Cataldi R, Hodgson SF, Lund JW (eds) Stories from a heated Earth. Geothermal Resources Council, Davis California USA, pp 147-163

Cataldi R, Hodgson SF, Lund JW (1999) Stories from a heated Earth. Geothermal Resources Council, Davis California USA

Cathles LM (1981) Fluid flow and genesis of hydrothermal ore deposits. Economic Geology 75th Anniversary Volume:424-457

Cathles LM, Smith AT (1983) Thermal constraints on the formation of Mississippi Valley-type lead-zinc deposits and their implications for episodic basin dewatering and deposit genesis. Economic Geology 78:983-1002

Chapman DS, Pollack HN (1975) Global Heat Flow: A new look. Earth and Planetary Science Letters 28:23-32

Chaturongkawanich S, Leevongchareon S (2000) The geothermal resources of Changwat Ranong Southern Thailand. Proceedings of World Geothermal Congress 2000, Kyushu - Tohuku, Japan, May 28 - June 10, pp 1049-1052

Cheng HP, Yeh GT (1998) Development and demonstrative application of a 3-D numerical model of subsurface flow, heat transfer, and reactive chemical transport: 3DHYDROGEOCHEM. Journal of Contaminant Hydrology 34:47-83

Chiodini G, Comodi P, Giaquinto S (1988) Ammonia and boric acid in steam and water. Experimental data from geothermal wells in the Phlegrean fields, Naples, Italy. Geothermics 17(5/6):711-718

Christensen S, Kirsch R, Liebsch-Dörschner T, Schenck P-F (2002) Hydrogeothermal Energy Use in Schleswig-Holstein (Germany) – Possibilities and Restrictions. In: Geothermische Vereinigung (ed) 20 Years Deep Geothermal Energy in Germany, 7. Geothermische Fachtagung, 06.11.-08.11.2002, Waren, pp 99-107 (in German)

Christenson BW, Mroczek EK, Wood CP, Arehart GB (1997) Magma-ambient production environments: PTX constraints for paleo-fluids associated with the Ngatamariki diorite intrusion. In: Simmons SF, Morgan OE, Dunstall MG (eds) Proceedings of the 19th New Zealand Geothermal Workshop 1997 Geothermal Institute, University of Auckland, pp 87-92

Clauser C (1992) Permeability of crystalline rocks. EOS, Transactions American Geophysical Union 73:233,237

Clauser C (2003) Numerical Simulation of Reactive Flow in Hot Aquifers. Springer Publishers, Heidelberg

Clauser C, Pape H, Bartels J, Kühn M (2000) Simulating coupled flow, heat transport, multi-species transport, chemical water-rock interaction and resulting changes in permeability for model geothermal reservoirs. In: Kümpel H-J (ed) Thermo-Hydro-Mechanical Coupling in Fractured Rock, 3rd Euroconference on Rock Physics and Rock, Bad Honnef - Germany, November 14. – 18., pp 53

Cox M, Browne PRL (1991) Geochemistry of ground and thermal waters in the Ngawha area and elsewhere in Northland, New Zealand. In: Freeston DH, Browne PRL, Scott GL (eds) Proceedings of the 13th New Zealand Geothermal Workshop 1991 Geothermal Institute, University of Auckland, pp 271-277

Craig H (1963) The isotopic geochemistry of water and carbon in geothermal areas. In: Tongiorgi E (ed) Nuclear geology in geothermal areas. Consiglio Nazional delle Ricerche, Laboratorio di Geologia Nuclear, Pias, pp 17-53

D'Amore F, Fancelli R, Caboi R (1987) Observations on the application of chemical geothermometers to some hydrothermal systems in Sardinia. Geothermics 16:271-282

Darcy H (1856) Les Fontaines Publiques de la Ville de Dijon. Victor Dalmont, Paris

De Gennaro M, Ferreri M, Ghiara MR, Stanzione D (1984) Geochemistry of thermal waters on the Island of Ischia (Campania, Italy). Geothermics 13(4):361-374

De Marsily G (1986) Quantitative Hydrogeology. Academic Press, San Diego

Debye P, Hückel E (1923) Theory of electrolyte solutions. Phys. Z. 24:185-206, 305-325. Cited in: Matthes G (1990) The composition of groundwater – Textbook of Hydrogeology Vol. 2, Gebrüder Borntrăger, Berlin 2. Print

Dongarra G, Hauser S, Alaimo R, Carapezz M, Tonani F (1983) Hot waters on Pantelleria Island. Geochemical features and preliminary geothermal investigations. Geothermics 12(1):49-63

Dorrington P, Brown KL (2000) Management of stibnite deposition at Ngawha. In: Simmons SF, Morgan OE, Dunstall MG (eds) Proceedings of the 22nd New Zealand Geothermal Workshop 2000 Geothermal Institute, University of Auckland, pp 163-168

Durst P (2002) Geochemical modelling of Soultz-sous-Forets Hot Dry Rock test site: Coupling fluid-rock interactions to heat and fluid transport. Institute of Geology, Faculty of Science, University of Neuchatel

Edmond JM, Damm K (1983) Hot springs on the ocean floor. Scientific American 248:70-85

Ehlers J (1993) Geological Map of Hamburg 1:25.000, Explications concerning sheet no. 2526 Allermöhe. Geologisches Landesamt Hamburg, Freie Hansestadt Hamburg (in German)

Elder JW (1967) Transient convection in a porous medium. Journal of Fluid Mechanics 27(3):609-623

Ellis AJ, Mahon WAJ (1964) Natural hydrothermal systems and experimental hot-water/rock interactions. Geochimica et Cosmochimica Acta 28:1323-1357

Ellis AJ, Mahon WAJ (1967) Natural hydrothermal systems and experimental hot-water/rock interactions (Part II). Geochimica et Cosmochimica Acta 31:519-538

Endeshaw A (1988) Current status (1987) of geothermal exploration in Ethiopia. Geothermics 17(2/3):477-488

Energy Information Administration (2001) Office of Integrated Analysis and Forecasting, International Energy Outlook 2001, DOE/EIA-0484, US Department of Energy, Washington DC (http://www.eia.doe.gov/oiaf/ieo/index.html, last access 27.03.2003)

Evans DG, Nunn JA (1989) Free thermohaline convection in sediments surrounding a salt column. Journal of Geophysical Research 94(B9):413-422

Fischedick M, Langniß O, Nitsch J (2000) After the getting off – future course of renewable energies. S. Hirzel Verlag, Stuttgart Leipzig (in German)

Fitts CR (2002) Groundwater Science. Academic Press, Amsterdam

Fjalov AN (1959) Termofizika porod. Gostoptechizdat (in Russian)

Fournier RO (1989) Geochemistry and dynamics of the Yellowstone National Park hydrothermal system. Ann. Rev. Earth Planet. Science 17:13-53

Fournier RO (1990) The interpretation of Na-K-Mg relations in geothermal waters. Geothermal Resources Council TRANSACTIONS 14(Part II):1421-1425

Fournier RO, Potter RW (1979) Magnesium correction to the Na-K-Ca chemical geothermometer. Geochimica et Cosmochimica Acta 43:1543-1550

Fournier RO, Potter RW (1982) A revised and expanded silica (quartz) geothermometer. Geothermal Resources Council Bulletin 11(10):3-9

Freeston DH (1995) Direct uses of geothermal energy 1995 – Preliminary review. Proceedings of World Geothermal Congress 1995, Florence, Italy, May 18 - 31; pp 15-25

Freeze RA, Cherry JA (1979) Groundwater. Prentice-Hall, Englewood Cliffs

Fridleifson IB (1999) Historical aspects of geothermal utilization in Iceland. In: Cataldi R, Hodgson SF, Lund JW (eds) Stories from a heated Earth. Geothermal Resources Council, Davis California USA, pp 307-319

Frisch U (1993) Pre-Pleistocene. In: Ehlers J (ed) Geological Map of Hamburg 1:25.000, Explications concerning sheet no. 2526 Allermöhe. Geologisches Landesamt Hamburg, Freie Hansestadt Hamburg, pp 14-27 (in German)

Fritz J (1989) Geothermal Project Bruchsal. Final Report – European Commission, Non Nuclear Energy (in German)

Gambolatti G, Putti M, Teatini P (1996) Land subsidence. In: Singh VP (ed) Hydrology of Disasters, Kluwer Academic Publishers, Dordrecht, pp 231-268

Gammons CH, Barnes HL (1989) The solubility of Ag_2S in near-neutral aqueous sulfide solutions at 25 to 300°C. Geochimica et Cosmochimica Acta 53:279-290

Gandino A, Piovesana, Rossi R, Zan L (1985) Preliminary evaluation of Soufrière geothermal field, St. Lucia (Lesser Antilles). Geothermics 14(4):577-590

Garven G, Freeze RA (1984) Theoretical analysis of the role of groundwater flow in the genesis of stratabound ore deposits 1. Mathematical and numerical model. American Journal of Science 284:1085-1124

Gaupp (1991) Depositional environment and diagenesis of the Main Sandstone (Middle Rhaetian), Thönse gasfield, Lower Saxony. Niedersächsische Akademie der Geowissenschaften Veröffentlichungen 6:34-48 (in German)

Gelhar LW, Axness CL (1983) Three dimensional analysis of macrodispersion in aquifers. Water Resources Research 19(1):161-180

Gelhar LW, Welty C, Rehfeldt KR (1992) A critical review of data on field-scale dispersion in aquifers. Water Resources Research 28(7):1955-1974

GEO – Geothermal Education Office (2001) Geothermal Energy Introduction. Microsoft PowerPoint presentation (http://geothermal.marin.org/index.html, last acc. 27.03.2003)

Georgieva S, Vlaskovski I (2000) Structure and geothermal potential of the Bourgas hydrothermal basin, Bulgaria. Proceedings of World Geothermal Congress 2000, Kyushu - Tohuku, Japan, May 28 - June 10, pp 1151-1156

Ghomshei MM, Croft SAS, Stauder JJ (1986) Geochemical evidence of chemical equilibria in the South Meager Creek geothermal system, British Columbia, Canada. Geothermics 15(1):49-61

Gianelli G, Grassi S (2001) Water-rock interaction in the active geothermal system of Pantelleria, Italy. Chemical Geology 181:113-130

Gianelli G, Piovesana F, Quy HH (1997) A reconnaissance geochemical study of thermal waters of south and central Vietnam. Geothermics 26(4):519-533

Gianelli G, Teklemariam M (1993) Water-rock interaction processes in the Aluto-Langano geothermal field (Ethiopia). Journal of Volcanology and Geothermal Research 56:429-445

Gibert JP, Jaudin F (1999) Using geothermal waters in France: The district heating system of Chaudes-Aigues from the Middle Ages. In: Cataldi R, Hodgson SF, Lund JW (eds) Stories from a heated Earth. Geothermal Resources Council, Davis California USA, pp 287-305

Giggenbach WF (1981) Geothermal mineral equilibria. Geochimica et Cosmochimica Acta 45:393-410

Giggenbach WF (1984) Mass transfer in hydrothermal alteration systems. Geochimica et Cosmochimica Acta 48:2693-2711

Giggenbach WF (1987) Redox processes governing the chemistry of fumarolic gas discharges from White Island, New Zealand. Applied Geochemistry 2:143-161

Giggenbach WF (1988) Geothermal solute equilibria. Derivation of Na-K-Mg-Ca indicators. Geochimica et Cosmochimica Acta 52:2749-2765

Giggenbach WF, Glover RB (1992) Tectonic regime and major processes governing the chemistry of water and gas discharges from the Rotorua geothermal field, New Zealand. Geothermics 21(1/2):121-140

Giggenbach WF, Gonfiantini R, Jangi BL, Truesdell AH (1983) Isotopic and chemical compositions of Parbati Valley geothermal discharges, North-West Himalaya, India. Geothermics 12(2/3):199-222

Giggenbach WF, LeGuern F (1976) The chemistry of magmatic gases from Erta'Ale, Ethopia. Geochimica et Cosmochimica Acta 40:25-30

Giggenbach WF, Sheppard DS, Robinson BW, Stewart MK, Lyon GL (1995) Geochemical structure and position of the Waiotapu geothermal field, New Zealand. Geothermics 23(5/6):599-644

Goff F, Grigsby CO, Trujillo PE, Counce D, Kron A (1981) Geology, water geochemistry and geothermal potential of the Jemez Springs area, Canon de San Diego, New Mexico. Journal of Volcanology and Geothermal Research 10:227-244

Goff F, Tully J (1994) Geothermal assessment of Archuleta county and the Pagosa Springs aquifer, Colorado. Geothermal Resources Council TRANSACTIONS 18:191-196

Goff SJ, Goff F, Janik CJ (1992) Tecuamburro volcano, Guatemala: Exploration geothermal gradient drilling and results. Geothermics 21(4):483-502

Goko K (2000) Structure and hydrology of the Ogiri field, West Kirishima geothermal area, Kyushu, Japan. Geothermics 29:127-149

Gonzalez CN (1986) Interpretation of downhole temperature survey at Waiwera thermal area. Geothermal Institute, University of Auckland, Report no. 86.07

Grassi S, Kolios N, Mussi M, Saradeas A (1996) Groundwater circulation in the Nea Kessani low-temperature geothermal field (NE Greece). Geothermics 25(2):231-247

Grimaud D, Huang S, Michard G, Zheng K (1985) Chemical study of geothermal waters of central Tibet (China). Geothermics 14(1):35-48
Gröber H, Erk S, Grigull U (1963) Heat transfer. Springer Publishers, Berlin (in German)
Günther A (2003a) SECTION: An AVENUE program for deriving arbitrary oriented vertical cross-sections through grid-based 3-D models. (http://arcscripts.esri.com/details.asp?dbid=12710, last access 27.03.2003)
Günther A (2003b) RESTORE: An AVENUE program for vertical restoration of geological horizons with grid based data (http://arcscripts.esri.com/details.asp?dbid=12650, last access 27.03.2003)
Haar L, Gallagher JS, Kell GS (1984) NBS/NRC Steamtables. Hemisphere Publishing Corporation, Washington, New York, London
Hanano M, Seth MS (1995) Numerical modeling of hydrothermal convection systems including super-critical fluid. Proceedings of World Geothermal Congress 1995, Florence, Italy, May 18 – 31, pp 1681-1668
Handford CR (1981) A process-sedimentary framework for characterizing recent and ancient sabkhas. Sedimentary Geology 30:255-265
Hanor JS (1987) Kilometre-scale thermohaline overturn of pore waters in the Louisiana Gulf Coast. Nature 327:501-503
Hanshaw BB, Back W, Deike RG (1971) A geochemical hypothesis for dolomitization by ground water. Economic Geology 66(3):710-724
Hardie LA (1996) Secular variation in seawater chemistry: An explanation for the coupled secular variation in the mineralogies of marine limestones and potash evaporites over the past 600 m.y. Geology 24:279-283
Hardie LA, Eugster HP (1970) The evolution of closed basin brines. Mineralogical Society of America Special Paper 3:273-290
Harrison WJ, Tempel RN (1993) Diagenetic pathways in sedimentary basins. In: Horbury AD, Robinson AG (eds) Diagenesis and Basin Development. American Association of of Petroleum Geologist's Studies in Geology 36, pp 69-86
Hazen A (1911) Discussion: Dams and sand foundations: Transactions, American Society of Civil Engineers 73:199
Hearst J, Nelson P, Paillet F (2000) Well logging for physical properties: A handbook for geophysicists, geologists and engineers, 2nd ed. Wiley, New York
Hedenquist JW (1990) The thermal and geochemical structure of the Broadlands-Ohaaki geothermal system. Geothermics 19:151-185
Hedenquist JW, Henley RW (1985) Hydrothermal eruptions in the Waiotapu geothermal system, New Zealand: Origin, breccia deposits and effect on precious metal mineralization. Economic Geology 80:1640-1668
Heiken G (1982) Geology of geothermal systems. In: Edwards LM, Chilingar GV, Rieke III HH, Fertl WH (eds) Handbook of Geothermal Energy, Gulf Publishing Company, Houston, pp 177-217
Hemley JJ, Hostetler PB, Gude AJ, Mountjoy WT (1969) Some stability relations of alunite. Economic Geology 64:599-612
Hemley JJ, Jones WR (1964) Chemical aspects of hydrothermal alteration with emphasis on hydrogen metasomatism. Economic Geology 59:538-569
Henley RW, Ellis AJ (1983) Geothermal systems, ancient and modern: A geochemical review. Earth Sci. Rev. 19:1-50
Hochstein MP (1990) Classification and assessment of geothermal resources. In: Dickson MH, Fanelli M (eds) Small Geothermal Resources: A Guide to Development and Utilization, UNITAR, New York, pp 31-57

Hochstein MP, O'Sullivan MJ, Bhadee S (1987) Modelling of San Kamphaeng geothermal reservoir. In: Geothermal Isntitute (ed) Proceedings of the 9th New Zealand Geothermal Workshop 1987 Geothermal Institute, University of Auckland, pp 159-162

Hoefner ML, Fogler HS (1988) Pore evolution and channel formation during flow and reaction in porous media. AIChE Journal 34(1):45-54

Hoopes JA, Harleman DR (1967) Wastewater recharge and dispersion in porous media. Journal of the Hydraulics Division ASCE 93 (HY5)

Huang HF, Goff F (1986) Hydrogeochemistry and reservoir model of Fuzhou geothermal field, China. Journal of Volcanology and Geothermal Research 27:203-227

Hubbert MK (1940) The theory of ground-water motion. Journal of Geology 48:785-944

Hubbert MK (1956) Darcy's law and the field equations of the flow of underground fluids. Transactions of the American Institute of Mining, Metallurgical, and Petroleum Engineers 207:222-239

Hurter S, Haenel R (2002) Atlas of Geothermal Resources in Europe. European Commission, No. EUR 17811

Huttrer GW (2000) The status of world geothermal power generation 1995-2000. Proceedings of World Geothermal Congress 2000, Kyushu - Tohuku, Japan, May 28 - June 10, pp 23-34

Huyarkon PS, Pinder GF (1983) Computational methods in subsurface flow. Academic Press, New York

Idris H (1994) A contribution to the hydrogeothermics of groundwater systems in Egypt. Hydrogeothermics 15:85-111

Ingebritsen SE, Sanford WE (1998) Groundwater in Geologic Processes. Cambridge University Press, Cambridge

Itoi R, Fukuda M, Jinno K, Shimizu S, Tomita T (1987) Numerical analysis of the injectivity of wells in the Otake geothermal field, Japan. In: Proceedings 9th New Zealand Geothermal Workshop 1987, November 4–6, Geothermal Institute, University of Auckland, Auckland, New Zealand, pp 103-108

Jaritz W (1973) Development of the salt structures of NW Germany. Geologisches Jarhbuch A 10:1-77 (in German)

Keir RS (1980) The dissolution kinetics iof biogenic calcium carbonates in seawater. Geochimica Cosmochimica Acta 44:241-252

Kharaka YK (1986) Origin and evolution of water and solutes in sedimentary basins. In: Hydrogeology of sedimentary basins: application to exploration and exploitation, Third Canadian / American Conference on Hydrogeology 1986, pp 173-195

Kissling WM (1997) Large scale hot plate models for the TVZ geothermal fields. In: Simmons SF, Morgan OE, Dunstall MG (eds) Proceedings of the 19th New Zealand Geothermal Workshop 1997 Geothermal Institute, University of Auckland, pp 51-56

Kissling WM, Brown KL, O'Sullivan MJ, White SP, Bullivant DP (1996) Modeling chloride and CO_2 chemistry in the Wairakei geothermal reservoir. Geothermics 25(3):285-305

Konikow LF, Bredehoeft JD (1992) Groundwater models can not be validated. Advances in Water Resources 15:75-83

Kralj P, Kralj P (2000) Overexploitation of geothermal wells in Murska Soboty, Northeastern Slovenia. Proceedings of World Geothermal Congress 2000, Kyushu - Tohuku, Japan, May 28 - June 10, pp 837-842

Kristmannsdóttir H (1989) Types of scaling occurring by geothermal utilization in Iceland. Geothermics 32:183-190

Krumbein WC, Graybill FA (1965) An introduction to statistical models in geology. McGraw-Hill, New York

Kühn M (1997) Successive geochemical reactions during utilization of hydrogeothermal energy. Berichte, Fachbereich Geowissenschaften, Universität Bremen, 92 (partly in German)

Kühn M (2003) Preferential Flow Path Development in an Anhydrite Cemented Sandstone. In: Clauser C (ed) Numerical simulation of reactive flow in hot aquifers - SHEMAT and Processing SHEMAT, Springer Publishers, Heidelberg, pp 231-242

Kühn M, Bartels J, Iffland J (2002a) Predicting reservoir property trends under heat exploitation: Interaction between flow, heat transfer, transport, and chemical reactions in a deep aquifer at Stralsund, Germany. Geothermics 31(6):725-749

Kühn M, Bartels J, Pape H, Schneider W, Clauser C (2002b) Modeling chemical brine-rock interaction in geothermal reservoirs. In: Stober I, Bucher K (eds) Water-Rock Interaction, Kluwer Academic Publishers, Dordrecht, pp 147-169

Kühn M, Chiang W-H (2003) Pre- and post processing with "Processing SHEMAT". In: Clauser C (ed) Numerical simulation of reactive flow in hot aquifers - SHEMAT and Processing SHEMAT, Springer Publishers, Heidelberg, pp 75-151

Kühn M, Frosch G, Kölling M, Kellner T, Althaus E, Schulz HD (1997) Laboratory experiments for the investigation of barite supersaturation of thermal brines. Grundwasser - Zeitschrift der Fachsektion Hydrogeologie 2 (3):111-117 (in German)

Kühn M, Niewöhner C, Isenbeck-Schröter M, Schulz HD (1998) Main and minor constituents in highly saline thermal waters. Water Research 32(2):265-274

Kühn M, Schneider W (2003) Injection well with reaction kinetics. In: Clauser C (ed) Numerical simulation of reactive flow in hot aquifers - SHEMAT and Processing SHEMAT, Springer Publishers, Heidelberg, pp 253-262

Kühn M, Schneider W, Bartels J, Pape H, Clauser C (1999) Numerical simulation of chemically induced permeability changes during heat mining in hot aquifers. In: Simmons SF, Morgan OE, Dunstall MG (eds) Proceedings 21st New Zealand Geothermal Workshop 1999, November 10–12, 1999, Auckland University, New Zealand, pp 223-228

Kühn M, Stöfen H (2001) Reaction front fingering in anhydrite cemented sandstone. In: Simmons S, Dunstall MG, Morgan OE (eds) Proceedings 23rd New Zealand Geothermal Workshop 2001, November 7–9, Auckland University, New Zealand, pp 207-212

Kühn M, Stöfen H, Schneider W (2001) Coupled flow, heat transfer, solute transport, and chemical reactions within the thermal water reservoir Waiwera, New Zealand. In: Seiler K-P, Wohnlich S (eds) New Approaches Characterizing Groundwater Flow, Balkema Publishers, pp 1001-1005

Kühn M, Stöfen H, Schneider W (2002c) Estimation of subsidence processes in the vicinity of geothermal plants, Grundwasser - Zeitschrift der Fachsektion Hydrogeologie 7(1):3-13 (in German)

Lahsen A (1988) Chilean geothermal resources and their possible utilization. Geothermics 17:401-408

Langmuir D, Mahoney J (1984) Chemical equilibrium and kinetics of geochemical processes in groundwater studies. In: Hitchon B, Walleck EI (eds) First Canadian-American Conference on Hydrogeology, National Water Well Association, Dublin, Ohio, pp 69-75

Langmuir D, Melchior D (1985) The geochemistry of Ca, Sr, Ba and Ra sulfates in some deep brines from the Palo Duro Basin, Texas. Geochimica et Cosmochimica Acta 49:2423-2432

Lapwood ER (1948) Convection of a fluid in a porous media. Proceedings Cambridge Philosophical Society 44, pp 508-521

Lasaga AC (1998) Kinetic Theory in the Earth Sciences. Princeton Series in Geochemistry, Princeton Universtiy Press, Princeton, New Jersey

Lawless JV, Bromley CJ, Leach TM, Licup AC, Cope DM, Recio CM (1983) Bacon-Manito geothermal field: A geoscientific exploration model. In: Geothermal Isntitute (ed) Proceedings of the 5th New Zealand Geothermal Workshop 1983 Geothermal Institute, University of Auckland, pp 97-102

Lenz G, Horn H, Kellner T (1997) Final report on geology and drilling of the bore Hamburg Allermöhe 1. Geothermie Neubrandenburg GmbH (in German)

Lerman A (1979) Geochemical processes: Water and sediment environments. Wiley

Li YH, Gregory S (1974) Diffusion of ions in sea water and in deep-sea sediments. Geochimica et Cosmochimica Acta 38:703-714

Lichtner PC (1985) Continuum model for simultaneous chemical reactions and mass transport in hydrothermal systems. Geochimica et Cosmochimica Acta 49:779-800

Lichtner PC (1996) Continuum formulation of multicomponent-multiphase reactive transport. Reviews in Mineralogy and Geochemistry 34(1):83-129

Lindal B (1973) Industrial and other applications of geothermal energy. In: Armstead HCH (ed) Geothermal Energy, UNESCO, Paris, pp 135-148

Liu Z, Sun B, Liu X (1999) Chemical dosing system at Nagqu geothermal power platn (Tibet). In: Simmons SF, Morgan OE, Dunstall MG (eds) Proceedings of the 21st New Zealand Geothermal Workshop 1999 Geothermal Institute, University of Auckland, pp 193-197

Lopez MRT, Arriaga MCS (2000) Geochemical evolution of the Los Azufres, Mexico, geothermal reservoir. Part I: Water and salts. Proceedings of World Geothermal Congress 2000, Kyushu - Tohuku, Japan, May 28 - June 10, pp 2257-2260

Lowell RP, Yao Y (2002) Anhydrite precipitation and the extent of hydrothermal recharge zones at ocean ridge crests. Journal of Geophysical Research 107(B9):2183, doi:10.1029/2001JB001289

Lu C, Reed MH, Misra KC (1992) Zinc-lead skarn mineralization at Tin Creek, Alaska: Fluid inclusions and skarn-forming reactions. Geochimica et Cosmochimica Acta 56:109-119

Lubimova EA (1968) Thermal history of the Earth. Geophys. Mon. Ser. 13:63-77

Lund JW, Freeston DH (2000) World-wide direct use of geothermal energy 2000. Proceedings of World Geothermal Congress 2000, Kyushu - Tohuku, Japan, May 28 - June 10, pp 1-21

Lunis BC, Breckenridge R (1991) Environmental considerations. In: Lienau PJ, Lunis BC (eds) Geothermal Direct Use, Engineering and Design Guidebook, Geo-Heat Center, Klamath Falls, Oregon, pp 437-445

Mahon T, Harvey C, Crosby D (2000) The chemistry of geothermal fluids in Indonesia and their relationship to water and vapour dominated systems. Proceedings of World Geothermal Congress 2000, Kyushu - Tohuku, Japan, May 28 - June 10, pp 1389-1394

Mahon WAJ (1966a) Natural hydrothermal systems and the reaction of hot water with sedimentary rocks. New Zealand Journal of Science 10:206-221

Mahon WAJ (1966b) Silica in hot water discharged from drillholes at Wairakei, New Zealand. New Zealand Journal of Science 9:135-144

Manov GG, Bates RG, Hamer WJ, Acree SF (1943) Values of the constants in the Debye-Hückel equation for activity coefficients. Journal of the American Chemical Society 65:1765-1767

Marini L, Bonaria V, Guidi M, Huziker JC, Ottonello G, Zuccolini MV (2000) Fluid geochemistry of the Acqui Terme-Visone geothermal area (Piemont, Italy). Applied Geochemistry 15:917-935

Marini L, Cioni R, Guidi M (1998) Water chemistry of San Marcos area, Guatemala. Geothermics 27(3):331-360

Marshall WI, Slusher R (1966) Thermodynamics of Calcium Sulfate Dihydrate in Aqueous Sodium Chloride Solutions, 0-110°C. Journal of Physical Chemistry 70(12):4015-4027
Martinovic M, Milivojevic M (2000) The hydrothermal model of Macva. Proceedings of World Geothermal Congress 2000, Kyushu - Tohuku, Japan, May 28 - June 10, pp 2145-2151
McCume CC, Forgler HS, Kline WE (1979) An experiment technique for obtaining permeability-porosity relationship in acidized porous media. Ind Eng Chem. Fundam 18(2):188-192
McGuinness MJ, White SP, Young RM, Ishizaki H, Ikeuchi K, Yoshida Y (1995) A model of the Kakkonda geothermal reservoir. Geothermics 24:1-48
Melchior DC, Bassett RL (1990) Chemical Modeling of Aqueous Systems II. ACS Symposium Series 416, American Chemical Society, Washington DC
Mercer JW, Thomas SD, Ross B (1982) Parameters and Variables Appearing in Repository Siting Models. GeoTrans Inc., NUREG/CR-3066
Merkel (1991) Results of hydrochemical and isotopic investigations of the brines Waren "GT/Wa 1/81" and Neubrandenburg "N 1 GT 86". WATEC Ingeieurgesellschaft, Markt Schwaben (in German, unpublished)
Meyer C, Hemley JJ (1967) Wall rock alteration. In: Barnes HL (ed) Geochemistry of hydrothermal ore deposits, Holt Rinehart & Winston, New York, pp 166-235
Moeller N, Weare JH, Greenberg JP, Duan Z (2000) Modeling technology for increasing geothermal energy productivity. Proceedings of World Geothermal Congress 2000, Kyushu - Tohuku, Japan, May 28 - June 10, pp 4059-4064
Monnin C, Ramboz C (1996) The anhydrite saturation index of the ponded brines and sediment pore waters of the Red Sea deeps. Chemical Geology 153:187-209
Moon BR, Dharam P (1988) Geothermal energy in India: Present Status and future prospects. Geothermics 17(2/3):439-449
Morse JW (1983) The kinetics of calcium carbonate dissolution and precipitation. In: Reeder RJ (ed) Reviews in Mineralogy, American Minerlogical Society 11:227-264
Mroczek EK, White SP, Graham D (2000) Deposition of silica in porous packed beds - Predicting the lifetime of reinjection aquifers. Geothermics 29:737-757
Mroczek EK, White SP, Graham D (2002) Estimating the quantity of silica deposited in the reservoir around an injection well. In: Soengkono S, Browne PRL (eds) Proceedings of the 24th New Zealand Geothermal Workshop 2002 Geothermal Institute, University of Auckland, pp 187-190
Mroczek EK, White SP, Graham DJ (1999) Comparison of amorphous silica deposition rates measured using packed pipes and a fluidised bed at FP10, Wairakei. In: Simmons SF, Morgan OE, Dunstall MG (eds) Proceedings 21st New Zealand Geothermal Workshop 1999, November 10–12, 1999, Auckland University, New Zealand, pp 67-71
Muffler LJP (1985) Geothermal energy and geothermal resources of hydrothermal convection systems. In: Englewood CNJ (ed) Energy: For ourselves and our prosperity, Prentice Hall, pp 188-207
Müller EP, Papendieck G (1975) Distribution, genesis and dynamic of deep-seated waters under special consideration of the Zechstein. Zeitschrift geologischer Wissenschaften 3(2):167-196 (in German)
Nicholson K (1993) Geothermal Fluids – Chemistry and Exploration Techniques. Springer Verlag, Berlin
Nieva D, Verma MP, Santoyo E, Portugal E, Campos A (1997) Geochemical exploration of the Chipilapa geothermal field, El Salvador. Geothermics 26(5/6):589-612
Nitzsche O, Meinrath G, Merkel B (2000) Database uncertainty as a limiting factor in reactive transport prognosis. Journal of Contaminant Hydrology 44:223-237

Noda T, Shimada K (1993) Water mixing model calculation for evaluation of deep geothermal water. Geothermics 22(3):165-180

Nordstrom DK, Ball JW (1984) Chemical models, computer programs and metal complexation in natural water. In: Kramer CJM, Duniker J (eds) International Symposium on Trace Metal Complexation in Natural Water. Martinus Nijhoff/Dr J.W. Junk Publishing Co, pp 149-169

Norton D, Knight J (1977) Transport phenomena in hydrothermal systems: Cooling plutons. American Journal of Science 277:937-981

O'Sullivan M, Pruess K, Lippmann MJ (2001) State of the art of geothermal reservoir simulation. Geothermics 30:395-429

Olivella S, Carrera J, Gens A, Alonso EE (1996) Porosity variations in saline media caused by temperature gradients coupled to multiphase flow and dissolution/precipitation. Transport in Porous Media 25:1-25

Oliver J (1986) Fluids expelled tectonically from orogenic belts: Their role in hydrocarbon migration and other geologic phenomena. Geology 14:99-102

Ormond A, Ortoleva P (2000) Numerical modeling of reaction-induced cavities in a porous rock. Journal of Geophysical Research 105(B7):16,737-16,747

Ortoleva P, Chadam J, Merino E, Sen A (1987) Geochemical self-organization II: The reactive-infiltration instability. American Journal of Science 287:1008-1040

Ortoleva PJ (1994) Geochemical Self-Organization. Oxford Monographs on Geology and Geophysics Vol. 23, Oxford University Press, New York

Özgüler ME, Kasap A (1999) The geothermal history of Anatolia, Turkey. In: Cataldi R, Hodgson SF, Lund JW (eds) Stories from a heated Earth. Geothermal Resources Council, Davis California USA, pp 51-67

Pandey OMP (1982) Terrestrial heat flow in New Zealand. Victoria University of Wellington, New Zealand, Ph.D. thesis (unpublished)

Pape H, Clauser C, Iffland J (1999) Permeability prediction for reservoir sandstones based on fractal pore space geometry. Geophysics 64(5):1447-1460

Parkhurst DL, Appelo CAJ (1999) User's guide to PHREEQC (version 2)--A computer program for speciation, batch-reaction, one-dimensional transport, and inverse geochemical calculations: U.S. Geological Survey Water-Resources Investigations Report 99-4259

Paschke SS, van der Heijde P (1996) Overview of chemical modeling in groundwater and listing of available geochemical models. International Groundwater Modeling Center, Report GWMI-96-01, Golden Co., Colorado School of Mines

Pham M, Klein C, Sanyal S, Xu T, Pruess K (2001) Reducing cost and environmental impact of geothermal power through modeling of chemical processes in the reservoir. Proceedings, Twenty-Sixth Workshop on Geothermal Reservoir Engineering, Stanford University, Stanford, California, January 29-31, 2001, SGP-TR-168, pp 216-223

Phillips OM (1991) Flow and reaction in permeable rocks. Cambridge University Press, New York

Phutela RC, Pitzer KS (1986) Thermodynamics of electrolyte mixtures. Enthalpy and the effect of temperature on the activity coefficient. Journal of Solution Chemistry 15:649

Piper AM (1944) A graphical procedure in the geochemical interpretation of water-analyses. American Geophysical Union Transactions 25:914-923

Pirajno F (1992) Hydrothermal mineral deposits – Principles and fundamental concepts for the exploration geologist. Springer Verlag, Berlin Heidelberg New York

Pitzer KS (1973) Thermodynamics of electrolytes. 1. Theoretical basis and general equations. J. Physical Chemistry 77:268-277

Pitzer KS (1975) Thermodynamics of electrolytes. V. Effects of higher-order electrostatic terms. Journal of Solution Chemistry 4(3):249-265

Pitzer KS (1991) Ion interaction approach: Theory and data correlation. In: Pitzer KS (ed) Activity Coefficients in Electrolyte Solutions, CRC Press Boca Raton, 2nd Edition, pp 76-153

Pitzer KS, Kim JJ (1974) Thermodynamics of electrolytes. IV. Activity and osmotic coefficient for mixed electrolytes. Journal of American Chemical Society 96(18):5701-5707

Pitzer KS, Mayorga G (1973) Thermodynamics of electrolytes. II. Activity and osmotic coefficient for strong electrolytes with one or both ions univalnet. Journal of Physical Chemistry 77(19):2300-2308

Pitzer KS, Mayorga G (1974) Thermodynamics of electrolytes. III. Activity and osmotic coefficients for 2-2 electrolytes. Journal of Solution Chemistry 3(7):539-546

Pitzer KS, Silvester LF (1976) Thermodynamics of electrolytes. VI. Weak electolytes including H_3PO_4. Journal of Solution Chemistry 5(4):269-277

Plumlee GS (1994) Fluid chemistry evolution and mineral deposition in the Main-Stage Creede epithermal system. Economic Geology 89:1860-1882

Plumlee GS, Goldhaber MB, Rowan EL (1995a) The potential role of magmatic gases in the genesis of Illinois-Kentucky fluorspar deposits: Implications from chemical reaction path modeling. Economic Geology 90:999-1011

Plumlee GS, Leach DL, Hofstra AH, Landis GP, Rowan EL, Viets JG (1995b) Reaction path modeling of ore deposition in Mississippi-Valley-Type Pb-Zn deposits of the Ozark region, U.S. Midcontinent. Economic Geology 89:1361-1383

Pollack HN, Hurter SJ, Johnson JR (1993) Heat flow from the Earth's interior: Analysis of the global data set. Reviews of Geophysics 31:267-280

Popper K (1959) The logic of scientific discovery. Harper and Row, New York

Pottorf RJ, Barnes HL (1983) Mineralogy, geochemistry and ore genesis of hydrothermal sediments from Atlantis II Deep, Red Sea. Economic Geology Monographs 5:198-223

Praserdvigai S (1987) Geothermal development in Thailand. Geothermics 16, 565-585

Prol-Ledesma RM, Hernandez-Lombardini SI, Lozano-Santa Cruz R (1995) Chemical variations in the rocks of La Primavera geothermal field (Mexico) related with hydrothermal alteration. In: Hochstein MP (ed) Proceedings of the 17th New Zealand Geothermal Workshop 1995 Geothermal Institute, University of Auckland, pp 47-53

Pruess K (1991) TOUGH2 - A general-purpose numerical simulator for multiphase fluid and heat flow, Rep. LBL-29400, Lawrence Berkeley Lab., Berkeley, California

Pruess K, Oldenburg C, Moridis G (1999) TOUGH2 – User's guide, Version 2.0. Rep. LBL-43134, Lawrence Berkeley Lab., Berkeley, California

Pytkowicz RM (1983) Equilibria, nonequilibria and natural waters. John Wiley & Sons, New York

Raffensberger JP, Garven G (1995a) The formation of unconformity-type uranium ore deposits 1. coupled groundwater flow and heat transport modeling. American Journal of Science 295(5):581-636

Raffensberger JP, Garven G (1995b) The formation of unconformity-type uranium ore deposits 2. Coupled hydrochemical modeling. American Journal of Science 295:639-696

Raffensberger JP, Vlassopoulos D (1999) The potential for free and mixed convection in sedimentary basins. Hydrogeology Journal 7:505-520

Ramirez AH (1988) Present status of geothermal reconnaissance studies in the Republic of Panama. Geothermics 17(2/3):355-367

Reed MH (1982) Calculation of multicomponent chemical equilibria and reaction processes in systems involving minerals, gases and aqueous phase. Geochimica et Cosmochimica Acta 46(4):513-528

Reed MH (1983) Seawater-basalt reaction and the origin of Greenstones and related ore deposits. Economic Geology 78:466-485

Reed MH, Spycher N (1984) Calculation of pH and miineral equilibria in hydrothermal waters with application to geothermometry and studies of boiling and dilution. Geochimica et Cosmochimica Acta 48:1479-1492

Reyes A, Giggenbach WF, Salera JRM, Salonga ND, Vergara MC (1993) Petrology and geochemistry of Alto Peak, a vapor-cored hydrothermal system, Leyte Province, Philippines. Geothermics 22(5/6):479-519

Reyes AG, Giggenbach WF (1999) Condensate-formation and cold water incursion in the Poihipi sector, Wairakei geothermal system. In: Simmons SF, Morgan OE, Dunstall MG (eds) Proceedings of the 21st New Zealand Geothermal Workshop 1999 Geothermal Institute, University of Auckland, pp 29-36

Rickard D, Sjöberg EL (1983) Mixed kinetic control of calcite dissolution rates. American Journal of Science 283:815-830

Rockel I (1986) Taking the waters. Early spas in New Zealand. Government Printing Office Publishing

Rose WI, Chuan RL, Giggenbach WF, Kyle PR, Symonds RB (1986) Rates of sulfur dioxide emissions from White Island volcano, New Zealand, and an estimate of the total flux of major gaseous species. Bulletin of Volcanology 48:181-188

Rougerie F, Andrié C, Jean-Baptiste P (1991) Helium-3 inside atoll barrier reef interstitial water : A clue for geothermal endoupwelling. Geophys. Res. Letters 18:109-112

Sanford WE, Konikow LF (1989a) Simulation of calcite dissolution and porosity changes in salt water mixing zones in coastal aquifers. Water Resources Research 25:655-667

Sanford WE, Konikow LF (1989b) Porosity development in coastal carbonate aquifers. Geology 17:249-252

Saripalli KP, Meyer PD, Bacon DH, Freedman VL (2001) Changes in hydrologic properties of aquifer media due to chemical reactions: A Review. Critical Reviews in Environmental Science and Technology 31(4):311-349

Sasada M, Doi N, Muffler LJP, Hedenquist JW (1998) Special Issue: Deep geothermal systems Japanese National project at Kakkonda. Geothermics 27(5/6):505-721

Satman A, Ugur Z, Onur M (1999) The effect of calcite deposition on geothermal well inflow performance. Geothermics 28:425-444

Saunders JA, Schoenly PA (1995) Boiling, colloid nucleation and aggregation, and the genesis of bonanza Au-Ag ores of the Sleeper deposit, Nevada. Mineralium Deposita 30:199-210

Saxena VK, Gupta ML (1985) Aquifer chemistry of thermal waters of the Godavari Valley, India. Journal of Volcanology and Geothermal Research 25:181-191

Saxena VK, Gupta ML (1987) Geochemistry of the thermal waters of Salbardi and Tatapani, India. Geothermics 16:705-716

Schack A (1953) The industrial heat transfer. Steel and Iron, Düsseldorf (in German)

Schechter RS, Gidley JL (1969) The change in pore size distribution from surface reaction in porous media. AIChE Journal 15(3):339-350

Scheidegger AE (1961) General theory of dispersion in porous media. Journal of Geophysical Research 66:3,273-3,278

Schellschmidt R, Clauser C, Sanner B (2000) Geothermal energy use in Germany at the turn of the millennium. Proceedings of World Geothermal Congress 2000, Kyushu - Tohuku, Japan, May 28 - June 10, pp 427-432

Schenberger DM, Barnes HL (1989) The solubility of gold in aqueous sulfide solution from 150 to 300°C. Geochimica et Cosmochimica Acta 53:269-278

Schildknecht F, Schneider W (1987) About the validity of Darcy's law under low hydraulic gradients in sediments characterized by cohesion. Geologisches Jahrbuch, Reihe C, Heft 48 (in German)

Schmitz J, Flixeder F (1993) Structure of a classic chalk oil field and production enhancement by horizontal drilling, Reitbrook, NW Germany. In: Spencer AM (ed) Generation, Accumulation and Production of Europe's Hydrocarbons III, Special Publication of the European Association of Petroleum Geoscientists No. 3, pp 141-154

Schroeder JA, Purser BA (1986) Reef Diagenesis. Springer Verlag, New York

Schulz (1988) Laboratory measurments of the saturation length as a measure of the solubility kinetic of carbonates in groundwater. Geochimica et Cosmochimica Acta 52:2651-2657 (in German)

Sekioka M (1999) Japanese geothermal waters throughout history. In: Cataldi R, Hodgson SF, Lund JW (eds) Stories from a heated Earth. Geothermal Resources Council, Davis California USA, pp 393-405

Severne CM (1998) Recent geochemical investigations of the Tokaanu-Waihi geothermal field. In: Simmons SF, Morgan OE, Dunstall MG (eds) Proceedings of the 20th New Zealand Geothermal Workshop 1998 Geothermal Institute, University of Auckland, pp 251-257

Severne CM (1999) Traditional use of geothermal resources by New Zealand Maori. In: Cataldi R, Hodgson SF, Lund JW (eds) Stories from a heated Earth. Geothermal Resources Council, Davis California USA, pp 435-449

Seyfried WE, Bischoff JL (1981) Experimental seawater-basalt interaction at 300°C, 500 bars; chemical change, secondary mineral formation and implications for the transport of heavy metals. Geochimica et Cosmochimica Acta 45:135-147

Shanker R, Absar A, Srivasta GC, Bajpai IP (2000) Tapoban hot water system, NW Himalaya, India – A unique example of heating under normal gradient conditions. In: Simmons SF, Morgan OE, Dunstall MG (eds) Proceedings of the 22nd New Zealand Geothermal Workshop 2000 Geothermal Institute, University of Auckland, pp 9-12

Sheppard DS, Giggenbach WF (1980) Chemistry of the well discharges at Ngawha. In: Geothermal Institute (ed) Proceedings of the 2nd New Zealand Geothermal Workshop 1980 Geothermal Institute, University of Auckland, pp 91-95

Sheppard DS, Mroczek EK (2002) CO2 fluxes from geothermal systems: Assessing the effects of exploitation, and the carbon tax implications. In: Soengkono S, Browne PRL (eds) Proceedings of the 24th New Zealand Geothermal Workshop 2002 Geothermal Institute, University of Auckland, pp 17-21

Sibson RH, Moore JM, Rankin AH (1975) Seismic pumping – a hydrothermal fluid transport mechanism. Journal of the Geological Society London 131:653-659

Simmons SF, Browne PRL (2000) Hydrothermal minerals and precious metals in the Broadlands-Ohaaki geothermal system: Implications for understanding low-sulfidation epithermal environments. Economic Geology 95:971-999

Simsek S (1985) Geothermal model of Denizli, Sarayköy – Buldan area. Geothermics 14:393-417

Sjöberg EL (1978) Kinetics and mechanism of calcite dissolution in aqueous solutions at low temperatures. Stockholm Contribution Geology XXXII, 1, pp 1-92

Soler JM, Lasaga AC (1998) An advection-dispersion-reaction model for bauxite formation. Journal of Hydrology 209:311-330

Sorey ML (1978) Numerical modeling of liquid geothermal systems. U.S. Geological Survey Professional Paper 1044-D

Sorey ML, Covard EM (1997) Hydrologic investigations in the Mammoth Corridor, Yellowstone National Park and vicinity, U.S.A. Geothermics 26(2):221-249

Sorey ML, Suemnicht GA, Sturchio NC, Nordquist GA (1991) New evidence on the hydrothermal system Long Valley caldera, California, from wells, fluid sampling, electrical geophysics, and age determinations of hot-spring deposits. Journal of Volcanology and Geothermal Research 48:229-263

Spencer RJ, Eugster HP, Jones BF (1985b) Geochemistry of Great Salt Lake, Utah, II: Pleistocene-holocene evolution. Geochimica et Cosmochimica Acta 49:739-747

Spencer RJ, Eugster HP, Jones BF, Rettig SL (1985a) Geochemistry of Great Salt Lake, Utah, I: Hydrochemistry since 1850. Geochimica et Cosmochimica Acta 49:727-737

Spycher N, Reed MH (1992) Users guide for CHILLER: A program for computing water-rock interactions, boiling, mixing and other reaction processes in aqueous-mineral-gas systems. University of Oregon, Eugene, Oregon

Spycher NF, Reed MH (1989) Evolution of a Broadlands-type epithermal ore fluid along alternative P-T paths: Implications for the transport and deposition of base, precious and volatile metals. Economic Geology 84:328-359

Steefel CI, Lasaga AC (1990) Evolution of dissolution patterns – Permeability change due to coupled flow and reaction. In: Melchior DC, Bassett RL (eds) ACS Symposium Series, Chemical Modeling of Aqueous Systems II, No. 416, Chap. 2, American Chemical Society, Washington DC, pp 213-225

Stöfen H (2000) Development of a heat transfer and solute transport model as a tool for sustainable management of the thermal water resource Waiwera, New Zealand. Master thesis, Technische Universität Hamburg-Harburg (unpublished)

Stöfen H, Kühn M (2003) Waiwera Coastal Geothermal System. In: Clauser C (ed) Numerical simulation of reactive flow in hot aquifers - SHEMAT and Processing SHEMAT, Springer Publishers, Heidelberg, pp 297-316

Stöfen H, Kühn M, Schneider W (2000) A heat transfer and solute transport model as a tool for sustainable management of the geothermal reservoir Waiwera, New Zealand. In: Simmons SF, Morgan OE, Dunstall MG (eds) Proceedings of the 22nd New Zealand Geothermal Workshop 2000 Geothermal Institute, University of Auckland, pp 273-278

Stumm W, Morgan JJ (1996) Aquatic Chemistry – Chemical Equilibria and Rates in Natural Waters. John Wiley & Sons, New York

Sunaryo D, Hantono S, Ganda, Nugroho (1993) Exploration results of the Ulubelu geothermal prospect, South Sumatra, Indonesia. In: Lee KC, Dunstall MG, Hochstein MP (eds) Proceedings of the 15th New Zealand Geothermal Workshop 1993 Geothermal Institute, University of Auckland, pp 103-106

Sundhoro H, Nasution A, Simanjuntak (2000) Sembalun Bumbung geothermal area, Lombok Island, West Nusatenggara, Indonesia: An integrated exploration. Proceedings of World Geothermal Congress 2000, Kyushu - Tohuku, Japan, May 28 - June 10, pp 1785-1790

Svalova VB (1999) Geothermal legends through history in Russia and the former USSR: Abridge to the past. In: Cataldi R, Hodgson SF, Lund JW (eds) Stories from a heated Earth. Geothermal Resources Council, Davis California USA, pp 337-355

Svanbjörnsson A, Matthíasson J, Frímannsson H, Arnórsson S, Björnsson S, Stefánsson V, Saemundsson K (1983) Overview of geothermal development at Olkaria in Kenya. Proceedings Ninth Workshop Geothermal Reservoir Engineering Stanford University, Stanford, California, December 1983, SGP-TR-74, pp 65-71

Swartzendruber D (1962) Non-Darcy flow behaviour in liquid-saturated porous media. Journal Geophysical Research 67:5205-5213

Szita G, Kocsis E (2000) How to do geothermal projects in Hungary? Dreams and realities. Proceedings of World Geothermal Congress 2000, Kyushu - Tohuku, Japan, May 28 - June 10, pp 3573-3578

Taylor HP (1983) Oxygen and hydrogen isotope studies of hydrothermal interactions at submarine and subaerial spreading centers. In: Rona PA, Bostrom K, Laubier L, Smith KL (eds) Hydrothermal processes at seafloor spreading centers. Plenum, New York, pp 83-140

Taylor SR (1964) Abundance of chemical elements in the continental crust, a new table. Geochimica et Cosmochimica Acta 28:1273-1285

Thomas DM (1986) Geothermal resources assessment in Hawaii. Geothermics 15(4):435-514

Titley SR (1990) Evolution and style of fracture permeability in intrusion-centered hydrothermal systems. In: Bredehoeft JD, Norton DL (eds) The role of Fluids in Crustal Processes. National Academy Press, Washington DC, pp 50-63

Tole MP (1988) Low enthalpy geothermal system in Kenya. Geothermics 17(5/6):777-783

Topinka (1997) Active Volcanoes and Plate Tectonics, "Hot Spots" and the "Ring of Fire". USGS/CVO (http://vulcan.wr.usgs.gov/Imgs/Gif/PlateTectonics/Maps/, last access 27.03.2003)

Traganos G, Jux U, Steuber T (1995) Isotopic characteristics of geothermal waters and fossil spring deposits in Mygdonia basin, Northern Greece. Geothermics 24(1):61-80

Truesdell AH (1976) Summary of section III geochemical techniques in exploration. In: Proceedings Second United Nations Symposium on the Development and Use of Geothermal Resources, San Francisco, 1975, 1, pp 1iii-1xxxix

Van der Lee J, De Windt L (2001) Present state and future directions of modeling of geochemistry in hydrogeological systems. Journal of Contaminant Hydrology 47:265-282

Walter AL, Frind EO, Blower DW, Ptacek CJ, Molson JW (1994) Modeling of multicomponent reactive transport in groundwater. 1. Model development and evaluation. Water Resources Research 30:3137-3148

Wang HF, Anderson MP (1982) Introduction to groundwater modeling: Finite difference and finite element methods. W.H. Freeman Company, San Francisco

Ward RF, Kendall CGStC, Harris PM (1986) Late permian (Guadalupian) facies and their association with hydrocarbons. American Association of Petroleum Geologists Bulletin 70:239-262

Warren JK (1989) Evaporite Sedimentology. Prentice Hall, Englewood Cliffs, New Jersey

Warren JK, Kendall GCStC (1985) Comparison of marine sabkhas (subaerial) and salina (subaqueous) evaporites: Modern and ancient. American Association of Petroleum Geologists Bulletin 69:1013-1023

Weeber JH, Brown LJ, White PA, Russell WJ, Thorpe HR (2001) A history of groundwater development in New Zealand. In: Rosen MR, White PA (eds) Groundwaters of New Zealand, New Zealand Hydrological Society, The Caxton Press, Christchurch, New Zealand, pp 5-44

Wei C, Ortoleva P (1990) Reaction front fingering in carbonate-cemented sandstones. Earth-Science Rev 29:183-198

Weir GJ, White SP (1996) Surface deposition from fluid flow in a porous medium. Transport in Porous Media 25:79-96

White A (1986) Chemical and isotopic characteristics of fluids within the Baca geothermal reservoir, Valles Caldera, New Mexico. Journal of Geophysical Research 91:1855-1866

White AF, Peterson ML (1991) Chemical equilibrium and mass balance relationships associated with the Long Valley hydrothermal system, California, U.S.A. Journal of Volcanology and Geothermal Research 48:283-302

White DE (1981) Active geothermal systems and hydrothermal ore deposits. Economic Geology 75:392-423

White DE (1992) The Beowawe Geysers, Nevada, before geothermal development. US Geological Survey Bulletin 1998

White SP (1995) Multiphase non-isothermal transport of systems of reacting chemicals. Water Resources Research 31(7):1761-1772

White SP, Christenson BW (2000) Modelling the alteration halo of a diorite intrusion. Proceedings of World Geothermal Congress 2000, Kyushu - Tohuku, Japan, May 28 - June 10, pp 3969-3974

White SP, Christenson BW (2002) Modelling two-phase flow above a diorite intrusion. In: Soengkono S, Browne PRL (eds) Proceedings of the 24th New Zealand Geothermal Workshop 2002 Geothermal Institute, University of Auckland, pp 97-102

White SP, Mroczek EK (1998) Permeability changes during the evolution of a geothermal field due to dissolution and precipitation of quartz. Transport in Porous Media 33:81-101

White WB (1988) Geomorphology and Hydrology of Karst Terrain. Oxford University Press, New York

Wolery TJ, Daveler SA (1992) EQ6, a computer program for reaction path modeling of aqueous geochemical systems: theoretical manual, user's guide, and related documentation. Lawrence Livermore National Laboratory

Wolery TJ, Jackson KJ (1990) Activity coefficients in aqueous salt solutions. In: Melchior DC, Bassett RL (eds) ACS Symposium Series, Chemical Modeling of Aqueous Systems II, No. 416, Chap. 2, American Chemical Society, Washington DC, pp 16-29

Wood CP, Mroczek EK, Carey BS (1997) The boundary of Wairakei geothermal field: Geology and chemistry. In: Simmons SF, Morgan OE, Dunstall MG (eds) Proceedings of the 19th New Zealand Geothermal Workshop 1997 Geothermal Institute, University of Auckland, pp 25-30

Wood JR, Hewett TA (1982) Fluid convection and mass transfer in porous snadstones – a theoretical model. Geochimica et Cosmochimica Acta 46:1707-1713

Wurster P (1965) Crustal movement, sea level fluctuation, and climatic change in the German Triassic times. Geologische Rundschau 54:224-240 (in German)

Xu M, Eckstein Y (1997) Statistical analysis of the relationships between dispersivity and other physical properties of porous media. Hydrogeology Journal 5(4):4-20

Xu T, Pruess K (2001) On fluid flow and mineral alteration in fractured caprock of magmatic hydrothermal systems. Journal Geophysical Research 106 (B2):2121-2138

Yusa Y, Ohsawa S (2000) Age of the Beppu hydrothermal system, Japan. Proceedings of World Geothermal Congress 2000, Kyushu - Tohuku, Japan, May 28 - June 10, pp 3005-3008

Zarrouk SJ, O'Sullivan MJ (2001) The effect of chemical reactions on the transport properties of porous media. In: Simmons S., Dunstall M.G., Morgan O.E. (eds) Proceedings 23rd New Zealand Geothermal Workshop 2001, November 7–9, Auckland University, New Zealand, pp 231-236

Zhonghe P (2000) Isotope geochemistry of geothermal waters in Northern North China basin: Implications on deep fluid migration. Proceedings of World Geothermal Congress 2000, Kyushu - Tohuku, Japan, May 28 - June 10, pp 1559-1564

Zhu C, Anderson G (2002) Environmental Applications of Geochemical Modeling. Cambridge University Press, Cambridge

Zongyu C (2000) Analyzing the long-term exploitation stability of Beijing geothermal field from geothermal water chemistry. Proceedings of World Geothermal Congress 2000, Kyushu - Tohuku, Japan, May 28 - June 10, pp 1053-1058

List of Symbols

Listed are common symbols in alphabetical order. Other less frequently used symbols are defined where they appear in the text.

Symbol	Quantity	Unit used here
A	Area	m^2
a_i	Ion activity	-
C	Molecular or ion concentration	mol L^{-1}
$C_{p(f)}$	Fluid specific heat capacity	J kg^{-1} K^{-1}
$C_{p(r)}$	Rock specific heat capacity	J kg^{-1} K^{-1}
C^{Φ}	Pitzer ion interaction parameter	-
D	Dispersion Coefficient	m^2 s^{-1}
D_m	Molecular diffusion coeff. (porous medium)	m^2 s^{-1}
D_W	Molecular diffusion coefficent (open water)	m^2 s^{-1}
d_m	Mean grain diameter	m
Da	Damköhler number	-
G	Gibbs free energy	kJ mol^{-1}
g	Gravitational acceleration (= 9.81)	m s^{-2}
H	Enthalpy	kJ mol^{-1}
h	Hydraulic potential, head	m
I	Ionic strength	mol L^{-1}
IAP	Ion activity product	-
$\underline{K}$, K	Hydraulic conductivity	m s^{-1}
$\underline{k}$, k	Permeability	m^2, mD (milli Darcy)
K_{eq}	Chemical equilibrium constant	-
k_{reac}	Reaction rate	s^{-1}
L, l	Characteristic length	m
m_i	Solute concentration	mol kg^{-1} Water
n	Porosity (Chap. 3)	-
n_e	Effective porosity (Chap. 3)	-
n_i	Number of moles, porosity	mol
P	Pressure	kg m^{-1} s^{-2}
Pe	Peclet number	-
Q_s	Source or sink of solutes	mol L^{-1} s^{-1}
q, q_x, q_y, q_z	Volumetric flow rate per unit area, Darcy velocity	m s^{-1}, m a^{-1}
q_d	Diffusive flux	m s^{-1}
q_{free}	Flow rate due to free thermal convection	m yr^{-1}
q_h	Heat flux	W m^{-2}
R	Ideal gas constant (= 8.31441)	J K^{-1} mol^{-1}
Ra	Rayleigh number	-

S	Entropy	$J\ mol^{-1}\ K^{-1}$
s_0	Solid surface	m^2
SI	Saturation index	-
T	Temperature	°C or K
t	Time	s
v, v_x, v_y, v_z	Average linear fluid velocity	$m\ s^{-1}$
z	Elevation head	m
α_L	Longitudinal dispersion length	m
α_T	Transverse dispersion length	m
$\beta^{(0)}, \beta^{(1)}, \beta^{(2)}$,	Pitzer ion interaction parameter	-
γ_i	Ion activity coefficient	-
Θ_{ij}	Pitzer ion interaction parameter	-
λ_e	Effective thermal conductivity	$W\ m^{-1}\ K^{-1}$
λ_f	Fluid thermal conductivity	$W\ m^{-1}\ K^{-1}$
λ_r	Rock thermal conductivity	$W\ m^{-1}\ K^{-1}$
μ	Dynamic viscosity	Pa s
μ_i	Chemical potential	$kJ\ mol^{-1}$
ρ	Density	$kg\ m^{-3}$
ρ_f	Fluid density	$kg\ m^{-3}$
ρ_W	Density of water	$kg\ m^{-3}$
τ	Tortuosity	-
ϕ_{ij}	Pitzer ion interaction parameter (Chap. 3)	-
ϕ, ϕ_0, ϕ_c	Porosity, initial porosity, critical porosity (Chap. 4-7)	-
Ψ	Pitzer ion interaction parameter	-

List of Minerals

Listed are the minerals mentioned within the manuscript with chemical formula in alphabetical order.

Mineral	Formula
acanthite	Ag_2S
adularia	$KAlSi_3O_8$
albite (Na-feldspar)	$NaAlSi_3O_8$
alunite	$KAl_3(SO_4)_2(OH)_6$
amphiboles	$(Ca,Na)_2(Mg,Fe,Al)_5Si_8O_{22}(OH)_2$
andesine	$Na_2CaAl_4Si_8O_{24}$
anhydrite	$CaSO_4$
anorthite	$CaAl_2Si_2O_8$
apatite	$Ca_5(PO_4)_3(OH,F,Cl)$
barite	$BaSO_4$
bauxite	$Al_2O_3 \cdot 2H_2O$
biotite	$K(Mg,Fe)_3AlSi_3O_{10}(OH,F)_2$
bischofite	$MgCl_2 \cdot 6H_2O$
bloedite	$Na_2Mg(SO_4)_2 \cdot 4H_2O$
bornite	Cu_5FeS_4
calcite	$CaCO_3$
carnallite	$KMgCl_3 \cdot 6H_2O$
chalcedony	SiO_2
chalcocite	Cu_2S
chalcopyrite	$CuFeS_2$
chlorite	$(Mg,Fe,Al)_6(Al,Si)_4O_{10}(OH)_8$
clinopyroxenes	$Ca(Mg,Fe)Si_2O_6 - Na(Al,Fe)Si_2O_6$
clinozoisite	$Ca_2Al_3Si_3O_{12}(OH)$
covellite	CuS
cristobalite	SiO_2
diopside	$(CaMg)Si_2O_6$
dolomite	$CaMg(CO_3)_2$
epidote	$Ca_2FeAl_2Si_3 O_{12}(OH)$
fayalite	Fe_2SiO_4
fluorite	CaF_2
galena	PbS
gibbsite	$Al(OH)_3$
glauberite	$Na_2Ca(SO_4)_2$
goethite	$FeOOH$
gypsum	$CaSO_4 \cdot 2H_2O$
halite	$NaCl$

hexahydrite	$MgSO_4 \cdot 6H_2O$
illite	$(K,H_3O)(Al,Mg,Fe)_2(Si,Al)_4O_{10}[(OH)_2,(H2O)]$
kainite	$KMgClSO_4 \cdot 3H_2O$
kaolinite	$Al_2Si_2O_5(OH)_4$
kieserite	$MgSO_4 \cdot H_2O$
laumontite	$CaAl_2Si_4O_{12} \cdot 4H_2O$
magnetite	Fe_3O_4
microcline (K-feldspar)	$KAlSi_3O_8$
mirabilite	$Na_2SO_4 \cdot 10H_2O$
molybdenite	MoS_2
montmorillinite (smectite)	$(\frac{1}{2}Ca,Na)(Al,Mg,Fe)_4(Si,Al)_8O_{20}(OH)_4 \cdot nH_2O$
muscovite	$KAl_3Si_3O_{10}(OH)_2$
olivine	Mg_2SiO_4
orthopyroxene	$Mg,Fe)_2Si_2O_6$
paragonite	$NaAl_3Si_3O_{10}(OH)_2$
plagioclase	$(Na,Ca)(Si,Al)_4O_8$
polyhalite	$K_2MgCa_2(SO_4)_4$
pyrite	FeS_2
pyrophyllite	$Al_2Si_4O_{10}(OH)_2$
pyrrhotite	$Fe_{(1-x)}S$ ($x = 0 - 0.17$)
quartz	SiO_2
rutile	TiO_2
saponite	$Ca_{0.25}(Mg,Fe)_3(Si,Al)_4O_{10}(OH)_2 \cdot nH_2O$
sericite (K-mica)	$KAl_3Si_3O_{10}(OH)_2$ or fine-grained muscovite, illite, paragonite
serpentine	$Mg_3Si_2O_5(OH)_4$
sphalerite	ZnS
sylvite	KCl
tachyhydrite	$Mg_2CaCl_6 \cdot 12H_2O$
talc	$Mg_3Si_4O_{10}(OH)_2$
thenardite	Na_2SO_4
topaz	$Al_2SiO_4(F,OH)_2$
tourmaline	$(Na,Ca)(Mg,Li,Al,Fe^{II},Fe^{III})_3(Al,Mg,Cr)_6B_3Si_6(OH,O,F)_4$
tremolite	$Ca_2(Mg,Fe)_4AlSi_7AlO_{22}(OH)_2$
trona	$Na_3H(CO_3)_2 \cdot 2H_2O$
wairakite	$CaAl_2Si_4O_{12} \cdot H_2O$
wollastonite	$CaSiO_3$

List of Numerical Codes

The numerical codes mentioned in the text are cited here in alphabetical order.

Program	Reference
3DHYDROGEOCHEM	Cheng and Yeh 1998
BASIN2	Bethke et al. 1993
CHEM-TOUGH	White 1995
CHILLER	Reed 1982, Spycher and Reed 1992
EQ3/6	Wolery and Daveler 1992
FRACCHEM	Durst 2002
PHREEQC	Parkhurst and Appelo 1999
Processing SHEMAT	Kühn and Chiang 2003
RST2D	Raffensberger and Garven 1995a, 1995b
SHEMAT	Bartels et al. 2003, Clauser 2003
SOLVEQ	Reed 1982
TOUGH / EWASG	Battistelli et al. 1997
TOUGH2	Pruess 1991
TOUGHREACT	Xu and Pruess 2001

Appendix

In the following section the compilation of geothermal waters from the literature study (Chap. 2) is listed. These analyses meet the accuracy requirement of an ionic strength in the range of ± 5 %. The compilation is based on the constituents Na, K, Ca, Mg, Cl, SO_4, and HCO_3 / CO_3 / CO_2. Data for pH, temperature, and SiO_2 are listed if available.

The table is built up in a double paged listing. All samples are shown with their reference, location, site, temperature, pH, sodium, potassium, calcium, magnesium, hydrogencarbonate, carbonate, carbon dioxide, and silica data. The samples are alphabetically ordered by location.

- Reference: first author and year of the publication to be found in the references,
- Location: geothermal field location abbreviated with "continent / country / district" following the ISO 3166 (International Standardisation Organisation, Table A.1),
- Site: particular naming and kind of the drawing (e.g. well or spring),
- T: temperature of the sample as published,
- pH: measured pH of geothermal water if mentioned in the paper,
- unit: Unit of concentration of the particular water sample
- NA / K / CA / MG / CL / SO4 / HCO3 / CO3 / CO2 / SIO2: concentration of sodium, potassium, calcium, magnesium, chloride, sulfate, hydrogencarbonate, carbonate, carbon dioxide, and silica.

Table A.1. Used country codes from ISO 3166 (alphabetical by shorts)

Short	Country	Short	Country	Short	Country
BG	Bulgaria	CA	Canada	CL	Chile
CN	China	CO	Colombia	EG	Egypt
SV	El Salvador	ET	Ethopia	DE	Germany
GR	Greece	GT	Guatemala	HU	Hungary
IS	Iceland	IN	India	ID	Indonesia
IT	Italy	JP	Japan	MX	Mexico
PA	Panama	PH	Philippines	LC	Saint Lucia
SI	Slovenia	TH	Thailand	TR	Turkey
GB	United Kingdom	US	United States	VN	Vietnam
YU	Yugoslavia				

Reference	Location	Site	T	pH	unit
Idris (1994)	AFR/EG/DakhlaOasis	Balat 2, well	36	6.6	ppm
Idris (1994)	AFR/EG/DakhlaOasis	Maasara 2, well	35	6.9	ppm
Idris (1994)	AFR/EG/DakhlaOasis	Asmant 2, well	32	6.5	ppm
Idris (1994)	AFR/EG/DakhlaOasis	Mutt 2, well	37	6.4	ppm
Idris (1994)	AFR/EG/DakhlaOasis	ElRashda 3, well	34	6.6	ppm
Idris (1994)	AFR/EG/DakhlaOasis	ElKalamoun 2, well	36	6.4	ppm
Idris (1994)	AFR/EG/DakhlaOasis	Budkhulu 2, well	34	6.3	ppm
Idris (1994)	AFR/EG/DakhlaOasis	ElMawhoob 2, well	35	6.7	ppm
Idris (1994)	AFR/EG/DakhlaOasis	Ons ElAin, Mutt, well	30	6.7	ppm
Idris (1994)	AFR/EG/DakhlaOasis	ElAbeed, ElMawhoub, well	37	6.7	ppm
Idris (1994)	AFR/EG/KhargaOasis	Mahariq, well	31		ppm
Idris (1994)	AFR/EG/KhargaOasis	Kharga Gomhoria R, well	39		ppm
Idris (1994)	AFR/EG/KhargaOasis	Naser 1, well	39	7.3	ppm
Idris (1994)	AFR/EG/KhargaOasis	Bulaq 4, well	38	7.2	ppm
Idris (1994)	AFR/EG/KhargaOasis	Baris, well	36	7.2	ppm
Idris (1994)	AFR/EG/KhargaOasis	Bulaq 4A, well	32	7.1	ppm
Endeshaw (1988)	AFR/ET/Aluto-Langano	La-1, well	88	9.6	ppm
Endeshaw (1988)	AFR/ET/Aluto-Langano	La-3, well	315	9.3	ppm
Endeshaw (1988)	AFR/ET/Aluto-Langano	La-4, well	231	9.5	ppm
Endeshaw (1988)	AFR/ET/Aluto-Langano	La-5, well	208	9.0	ppm
Endeshaw (1988)	AFR/ET/Aluto-Langano	La-6, well	335	9.0	ppm
Endeshaw (1988)	AFR/ET/Aluto-Langano	La-8, well	271	9.1	ppm
Gianelli (1993)	AFR/ET/Aluto-Langano	La-2, well	110	9.4	ppm
Gianelli (1993)	AFR/ET/Aluto-Langano	La-4, well	233	8.4	ppm
Gianelli (1993)	AFR/ET/Aluto-Langano	La-6, well	335	6.9	ppm
Gianelli (1993)	AFR/ET/Aluto-Langano	La-7, well	226	8.2	ppm
Beyene (2000)	AFR/ET/SouthernAfar/Dofan	Dofan H. sp1, sp		8.2	ppm
Beyene (2000)	AFR/ET/SouthernAfar/Dofan	Dofan H. sp2, sp		8.4	ppm
Beyene (2000)	AFR/ET/SouthernAfar/Dofan	Dofan H. sp3, sp		8.3	ppm
Beyene (2000)	AFR/ET/SouthernAfar/Fanatale	Fantale II. sp1, sp		8.2	ppm
Beyene (2000)	AFR/ET/SouthernAfar/Fanatale	Fantale II. sp2, sp		8.5	ppm
Beyene (2000)	AFR/ET/SouthernAfar/Meteka	Meteka H.sp2, sp		8.4	ppm
Beyene (2000)	AFR/ET/SouthernAfar/Meteka	Meteka H.sp3, sp		8.2	ppm
Beyene (2000)	AFR/ET/SouthernAfar/Wonji	Hippo Pool-1,sp		8.1	ppm
Beyene (2000)	AFR/ET/SouthernAfar/Wonji	Hippo Pool-2,sp		8.0	ppm
Beyene (2000)	AFR/ET/SouthernAfar/Wonji	Hippo Pool-3,sp		8.0	ppm
Beyene (2000)	AFR/ET/SouthernAfar/Wonji	Hippo Pool-4,sp		8.4	ppm
Beyene (2000)	AFR/ET/SouthernAfar/Wonji	Wonji G.D.W, well		7.3	ppm
Tole (1988)	AFR/KE/Narosura	034/002, sp	31	7.0	ppm
Svanbjörnsson (1983)	AFR/KE/Olkaria	OW-10, well		8.6	ppm
Svanbjörnsson (1983)	AFR/KE/Olkaria	OW-12, well		9.1	ppm
Ghomshei (1986)	AME/CA/BC/South Meager Creek	Angel Creek	4	9.6	ppm
Ghomshei (1986)	AME/CA/BC/South Meager Creek	No good Creek	4	8.2	ppm
Ghomshei (1986)	AME/CA/BC/South Meager Creek	Well M2-75D	10	7.7	ppm
Ghomshei (1986)	AME/CA/BC/South Meager Creek	Well M6-79 D	28	6.1	ppm
Ghomshei (1986)	AME/CA/BC/South Meager Creek	Well M12-80 D	11	7.1	ppm
Ghomshei (1986)	AME/CA/BC/South Meager Creek	South Fork SF 44	6	7.6	ppm
Ghomshei (1986)	AME/CA/BC/South Meager Creek	Upper South Fork Swamp	6	6.8	ppm
Ghomshei (1986)	AME/CA/BC/South Meager Creek	West Meager WM-35	8	7.0	ppm
Ghomshei (1986)	AME/CA/BC/South Meager Creek	Meager Creek Hot Spring N	53	7.1	ppm
Ghomshei (1986)	AME/CA/BC/South Meager Creek	Meager Creek Hot Spring N	53	7.1	ppm
Ghomshei (1986)	AME/CA/BC/South Meager Creek	EMR 301-2	30	6.7	ppm
Ghomshei (1986)	AME/CA/BC/South Meager Creek	M1-74 D	56	7.3	ppm
Ghomshei (1986)	AME/CA/BC/South Meager Creek	Placid Springs	45	6.9	ppm
Ghomshei (1986)	AME/CA/BC/South Meager Creek	No Good Spring No.1 (S,19	35	6.4	ppm
Ghomshei (1986)	AME/CA/BC/South Meager Creek	No Good Spring No.12 (C,	30	6.8	ppm
Ghomshei (1986)	AME/CA/BC/South Meager Creek	Weirbox MC-1 23/10/1982	100	8.3	ppm
Ghomshei (1986)	AME/CA/BC/South Meager Creek	Nitrogen Lift MC-3 11/11/	100	9.0	ppm
Lahsen (1988)	AME/CL/ElTatio	Ju-7, sp	66	7.6	mg/l

Na	K	Mg	Ca	Cl	SO4	HCO3	CO3	CO2	SIO2
19.00	22.00	7.80	8.80	34.00	33.00	47.60			
27.00	11.00	13.60	24.00	62.60	62.50	41.00			
29.00	7.50	11.00	28.80	70.00	57.50	39.00			
29.00	7.50	11.00	27.00	68.00	58.50	36.60			
25.00	10.00	9.70	20.00	54.00	49.50	36.60			
33.50	11.00	10.20	27.00	76.00	60.00	36.60			
23.50	10.00	8.30	20.00	48.00	46.80	39.00			
21.50	10.00	10.00	23.00	46.00	57.50	43.90			
40.00	7.70	17.00	49.60	116.00	100.00	42.50			
38.00	7.70	6.30	16.80	52.00	50.00	75.50			
98.00	22.00	21.00	16.00	68.00	11.00	324.00			
21.00	24.00	21.40	14.40	38.00	29.00	130.00			
31.00	31.00	13.00	15.00	59.00	13.00	114.00			
39.00	26.00	14.00	21.00	80.00	13.00	108.00			
64.00	29.00	12.00	26.00	103.00	53.00	98.00			
92.00	36.00	38.00	58.00	220.00	132.00	88.00			
563.00	39.00	0.10	1.00	230.00	19.00	744.00	162.00		84.00
675.00	157.00	0.10	1.00	310.00	282.00	830.00	163.00	2052.00	556.00
758.00	230.00	0.50	5.00	479.00	473.00	375.00	199.00	5432.00	558.00
1060.00	148.00	0.50	6.00	720.00	168.00	1300.00	157.00		317.00
934.00	150.00	0.20	6.00	454.00	204.00	1442.00	175.00	2500.00	418.00
670.00	53.00	0.40	6.00	550.00	73.00	536.00	107.00	2376.00	186.00
89.00	20.00	0.20	1.00	21.00	6.00	208.00			40.00
1015.00	138.00	0.40	3.80	671.00	131.00	1647.00			339.00
688.00	223.00	0.30	0.80	459.00	372.00	854.00			659.00
854.00	47.00	0.60	2.60	302.00	31.00	1769.00			150.00
357.20	16.70	0.10	3.20	172.60	168.10	427.10			124.50
347.60	17.50	1.00	3.80	171.80	164.80	420.60	5.60		122.60
395.00	15.60	0.70	8.80	204.60	216.40	401.10			112.60
456.20	21.80	1.20	1.50	178.60	89.30	806.80			64.40
422.70	20.90	2.20	2.00	146.10	84.00	734.20	14.40		64.10
274.00	11.40	1.90	3.70	123.20	94.30	398.40	4.40		61.20
300.70	11.60	1.70	3.50	144.00	105.60	428.50			59.90
229.30	14.50	0.70	2.90	23.50	30.80	493.00			104.00
223.70	14.10	0.70	2.80	24.90	25.20	496.00			103.10
221.60	14.00	0.70	2.30	25.70	23.10	496.60			102.80
238.60	16.00	0.80	3.40	27.50	26.80	515.60	8.60		110.60
24.70	9.20	1.70	17.40	5.60	13.40	112.60			85.60
16.00	7.50	28.00	12.00	0.40	182.00			53.00	22.00
734.00	147.40	0.03	0.92	1140.20	30.10			34.40	734.00
476.00	72.20	0.02	0.62	629.90	44.10			71.10	880.00
2.00	1.46	11.20	26.10	3.20	41.00	86.40			4.90
2.90	2.12	16.60	70.50	1.30	85.00	182.00			10.40
23.00	6.90	18.00	33.00	0.60	23.00	233.00			15.50
10.00	6.80	41.00	209.00	0.90	16.00	883.00			24.00
3600.00	136.00	240.00	490.00	4230.00	1820.00	3534.00			16.40
3000.00	150.00	88.00	210.00	3290.00	1280.00	2093.00			51.00
2270.00	113.00	67.00	237.00	2350.00	840.00	2647.00			108.00
700.00	69.00	65.00	300.00	870.00	400.00	1344.00			140.00
370.00	38.00	22.00	69.00	550.00	120.00	415.00			230.00
370.00	34.00	18.00	82.00	520.00	120.00	347.00			203.00
820.00	48.00	97.00	410.00	1060.00	900.00	1372.00			87.00
2230.00	87.00	93.00	390.00	2420.00	1980.00	1469.00			174.00
560.00	55.00	34.00	130.00	760.00	180.00	740.00			173.00
320.00	32.00	16.00	88.00	470.00	110.00	310.00			120.00
175.00	22.00	14.00	76.00	196.00	69.00	382.00			101.00
1260.00	97.00	0.80	40.00	1990.00	120.00	72.00			370.00
1010.00	71.00	1.30	35.00	1370.00	410.00	98.00			240.00
300.00	14.00	1.00	304.00	243.00	1100.00	58.00			53.00

Reference	Location	Site	T	pH	unit
Lahsen (1988)	AME/CL/ElTatio	Su-2, sp	83	8.1	mg/l
Lahsen (1988)	AME/CL/ElTatio	Chi-1, sp	30	7.4	mg/l
Lahsen (1988)	AME/CL/ElTatio	Pu-98, sp	86	6.9	mg/l
Lahsen (1988)	AME/CL/ElTatio	Li-1, sp	69	7.9	mg/l
Lahsen (1988)	AME/CL/ElTatio	Ta-226, sp	83	7.0	mg/l
Lahsen (1988)	AME/CL/ElTatio	7, sp	77	7.4	mg/l
Lahsen (1988)	AME/CL/ElTatio	SL-4, sp	46	8.9	mg/l
Lahsen (1988)	AME/CL/ElTatio	Ls-, sp	95	2.1	mg/l
Lahsen (1988)	AME/CL/ElTatio	Pe-2, sp	68	6.9	mg/l
Marini (1998)	AME/GT/SanMarco	3, sp	20	8.4	ppm
Marini (1998)	AME/GT/SanMarco	4, sp	20	8.4	ppm
Marini (1998)	AME/GT/SanMarco	5, sp	87	7.6	ppm
Marini (1998)	AME/GT/SanMarco	6, sp	24	7.2	ppm
Marini (1998)	AME/GT/SanMarco	7, sp	23	7.2	ppm
Marini (1998)	AME/GT/SanMarco	8, sp	47	6.6	ppm
Marini (1998)	AME/GT/SanMarco	9, sp	50	6.3	ppm
Marini (1998)	AME/GT/SanMarco	10, sp	42	5.9	ppm
Marini (1998)	AME/GT/SanMarco	11, sp	45	8.0	ppm
Marini (1998)	AME/GT/SanMarco	12, sp	50	7.6	ppm
Marini (1998)	AME/GT/SanMarco	13, sp	58	8.0	ppm
Marini (1998)	AME/GT/SanMarco	14, sp	37	8.0	ppm
Marini (1998)	AME/GT/SanMarco	15, sp	27	8.1	ppm
Marini (1998)	AME/GT/SanMarco	16, sp	50	6.9	ppm
Marini (1998)	AME/GT/SanMarco	17, sp	56	6.7	ppm
Marini (1998)	AME/GT/SanMarco	18, sp	30	6.5	ppm
Marini (1998)	AME/GT/SanMarco	19, sp	15	6.5	ppm
Marini (1998)	AME/GT/SanMarco	23, sp	24	6.4	ppm
Marini (1998)	AME/GT/SanMarco	28, sp	50	7.0	ppm
Marini (1998)	AME/GT/SanMarco	29, sp	33	6.7	ppm
Marini (1998)	AME/GT/SanMarco	34, sp	17	6.9	ppm
Marini (1998)	AME/GT/SanMarco	35, sp	14	6.8	ppm
Marini (1998)	AME/GT/SanMarco	36, sp	18	6.8	ppm
Marini (1998)	AME/GT/SanMarco	37, sp	14	7.0	ppm
Marini (1998)	AME/GT/SanMarco	38, sp	16	6.8	ppm
Marini (1998)	AME/GT/SanMarco	40, sp	69	7.1	ppm
Marini (1998)	AME/GT/SanMarco	41, sp	70	7.1	ppm
Marini (1998)	AME/GT/SanMarco	42, sp	63	7.4	ppm
Marini (1998)	AME/GT/SanMarco	43, sp	94	7.9	ppm
Marini (1998)	AME/GT/SanMarco	44, sp	94	8.2	ppm
Marini (1998)	AME/GT/SanMarco	45, sp	94	8.2	ppm
Marini (1998)	AME/GT/SanMarco	46, sp	93	8.2	ppm
Marini (1998)	AME/GT/SanMarco	47, sp	44	7.8	ppm
Marini (1998)	AME/GT/SanMarco	48, sp	45	7.0	ppm
Marini (1998)	AME/GT/SanMarco	49, sp	48	6.9	ppm
Marini (1998)	AME/GT/SanMarco	50, sp	43	6.7	ppm
Marini (1998)	AME/GT/SanMarco	51, sp	55	7.5	ppm
Marini (1998)	AME/GT/SanMarco	52, sp	41	7.5	ppm
Marini (1998)	AME/GT/SanMarco	53, sp	18	6.3	ppm
Marini (1998)	AME/GT/SanMarco	54, sp	18	7.2	ppm
Marini (1998)	AME/GT/SanMarco	55, sp	18	6.9	ppm
Marini (1998)	AME/GT/SanMarco	56, sp	20	7.4	ppm
Marini (1998)	AME/GT/SanMarco	58, sp	76	6.7	ppm
Marini (1998)	AME/GT/SanMarco	59, sp	60	7.4	ppm
Marini (1998)	AME/GT/SanMarco	60, sp	64	7.8	ppm
Marini (1998)	AME/GT/SanMarco	61, sp	63	6.6	ppm
Marini (1998)	AME/GT/SanMarco	62, sp	36	7.1	ppm
Marini (1998)	AME/GT/SanMarcos	1, sp	92	8.3	ppm
Goff (1992)	AME/GT/Tecumburro	TCB1-90-2	33	7.4	ppm
Goff (1992)	AME/GT/Tecumburro	TCB1-90-3	61	7.3	ppm

Na	K	Mg	Ca	Cl	SO4	HCO3	CO3	CO2	SIO2
1115.00	187.00	8.00	98.00	1780.00	215.00	238.00			157.00
116.00	24.00	14.00	192.00	115.00	572.00	96.00			67.00
1569.00	115.00	1.00	79.00	2744.00	89.00	73.00			258.00
300.00	37.00	20.50	44.00	297.00	290.00	137.00			198.00
4540.00	530.00	0.30	162.00	8233.00	44.00	29.00			260.00
800.00	92.00	2.60	75.00	1300.00	136.00	40.00			114.00
58.00	1.00	0.10	5.30	12.00	77.00	54.00			52.00
22.00	7.40	24.00	50.00	19.00	2555.00	10.00			373.00
734.00	29.00	3.00	313.00	1564.00	217.00	24.00			6.80
24.20	4.52	4.09	8.29	16.00	13.90	69.60			90.20
20.20	4.11	4.13	8.14	9.50	12.40	68.30			88.50
475.00	24.40	1.12	17.10	549.00	132.00	227.00			196.00
8.92	2.80	3.07	6.21	0.35	2.01	55.50			89.10
9.15	2.78	3.09	6.21	0.41	2.45	56.10			91.60
27.00	6.05	2.44	4.03	1.03	9.04	90.90			173.00
27.30	10.70	4.22	8.78	0.91	8.53	126.00			169.00
28.30	7.50	7.55	17.60	1.11	9.30	165.00			160.00
15.70	6.77	2.90	9.78	1.39	7.86	82.40			158.00
28.50	6.39	1.95	4.92	0.84	15.00	89.10			169.00
177.00	16.50	1.35	11.00	169.00	48.00	160.00			206.00
126.00	12.60	2.94	11.80	117.00	42.50	141.00			181.00
18.50	6.20	4.73	10.50	0.87	12.00	93.40			134.00
186.00	12.00	4.46	18.20	191.00	97.10	157.00			130.00
170.00	12.10	4.49	11.70	164.00	48.90	176.00			147.00
18.00	3.20	4.97	10.70	0.65	8.61	102.00			124.00
5.50	2.62	2.87	6.36	0.81	0.75	47.00			59.60
10.70	4.73	6.32	10.60	0.54	1.05	90.30			97.90
31.00	6.80	2.46	4.20	1.05	2.58	105.00			176.00
19.20	3.05	1.95	7.20	0.77	9.72	72.00			139.00
6.19	3.28	4.08	6.79	1.46	1.54	52.50			78.50
6.09	2.27	2.88	5.78	2.86	1.82	40.90			48.60
7.04	3.38	4.20	8.54	0.55	1.75	61.60			71.80
4.11	1.55	1.81	3.74	2.10	5.15	25.00			58.00
6.39	3.59	5.70	11.40	0.25	1.21	80.50			61.10
116.00	5.43	0.99	7.90	87.70	44.80	147.00			166.00
97.70	7.86	1.12	5.68	61.00	37.20	147.00			153.00
79.40	4.95	0.80	5.09	33.00	29.60	148.00			133.00
233.00	16.30	0.40	15.00	256.00	136.00	69.00			145.00
546.00	71.00	0.01	3.14	746.00	166.00	1.50			545.00
533.00	57.10	0.01	6.64	736.00	163.00	1.40			462.00
538.00	57.90	0.01	6.50	740.00	165.00	1.00			461.00
75.10	7.45	5.93	24.80	80.00	49.10	111.00			111.00
291.00	14.60	4.63	60.90	281.00	71.70	434.00			116.00
94.90	10.40	4.77	13.30	86.00	42.50	132.00			146.00
87.40	11.60	12.80	29.50	81.00	18.90	239.00			117.00
428.00	20.20	6.54	33.20	415.00	83.70	492.00			134.00
284.00	14.90	8.50	55.90	264.00	59.50	437.00			76.00
5.63	3.55	3.08	9.93	1.68	3.09	56.80			87.30
8.81	2.74	4.87	7.65	0.36	1.56	70.80			78.50
9.31	3.03	4.62	7.37	0.42	3.30	67.70			81.30
9.91	2.72	4.87	7.13	0.31	2.59	69.60			82.80
365.00	24.60	1.58	52.20	460.00	127.00	235.00			214.00
344.00	32.70	3.18	73.30	420.00	141.00	331.00			198.00
322.00	11.80	0.45	38.00	418.00	104.00	159.00			160.00
398.00	36.00	5.03	86.90	508.00	103.00	422.00			160.00
19.00	8.41	8.87	13.90	1.69	17.40	123.00			114.00
446.00	24.60	0.09	8.79	549.00	143.00	61.10			232.00
100.00	12.60	0.23	12.90	67.80	55.10	131.00			46.00
238.00	27.30	0.25	12.40	183.00	115.00	256.00			268.00

Reference	Location	Site	T	pH	unit
Goff (1992)	AME/GT/Tecumburro	TCB1-90-4a	96	7.0	ppm
Goff (1992)	AME/GT/Tecumburro	TCB1-90-4c	105	7.4	ppm
Goff (1992)	AME/GT/Tecumburro	TCB1-90-6	240	7.9	ppm
Goff (1992)	AME/GT/Tecumburro	TCB1-90-9	240	7.0	ppm
Gandino (1985)	AME/LC/Qualibou	124, sp	32	3.5	eq/l
Gandino (1985)	AME/LC/Qualibou	132, sp	25	4.5	eq/l
Gandino (1985)	AME/LC/Qualibou	135, sp	20	7.0	eq/l
Prol-Ledesma (1995)	AME/MX/LaPrimavera	PR-9, well		8.3	ppm
Lopez (2000)	AME/MX/LosAzufres	Az-02, well			ppm
Lopez (2000)	AME/MX/LosAzufres	Az-02, well			ppm
Lopez (2000)	AME/MX/LosAzufres	Az-05, well			ppm
Lopez (2000)	AME/MX/LosAzufres	Az-05, well			ppm
Lopez (2000)	AME/MX/LosAzufres	Az-09, well			ppm
Lopez (2000)	AME/MX/LosAzufres	Az-09, well			ppm
Lopez (2000)	AME/MX/LosAzufres	Az-13, well			ppm
Lopez (2000)	AME/MX/LosAzufres	Az-16, well			ppm
Lopez (2000)	AME/MX/LosAzufres	Az-18, well			ppm
Lopez (2000)	AME/MX/LosAzufres	Az-22, well			ppm
Lopez (2000)	AME/MX/LosAzufres	Az-22, well			ppm
Lopez (2000)	AME/MX/LosAzufres	Az-33, well			ppm
Lopez (2000)	AME/MX/LosAzufres	Az-33, well			ppm
Lopez (2000)	AME/MX/LosAzufres	Az-46, well			ppm
Lopez (2000)	AME/MX/LosAzufres	Az-46, well			ppm
Lopez (2000)	AME/MX/LosAzufres	Az-16D, well			ppm
Lopez (2000)	AME/MX/LosAzufres	Az-16D, well			ppm
Lopez (2000)	AME/MX/LosAzufres	Az-16D, well			ppm
Ramirez (1988)	AME/PA/Calobre	PA-75, sp	25	7.6	mg/l
Bath (1983)	AME/PA/Catalina	6, bh	45	6.6	mg/l
Bath (1983)	AME/PA/Catalina	7, sp	33	5.9	mg/l
Bath (1983)	AME/PA/Catalina	1, sp	32	6.1	mg/l
Bath (1983)	AME/PA/Catalina	4, sp	27	6.0	mg/l
Ramirez (1988)	AME/PA/ChiriqueAbajo	PA-113, sp	25	7.4	mg/l
Bath (1983)	AME/PA/Cotito	43, sp	67	6.7	mg/l
Bath (1983)	AME/PA/Cotito	44, sp	67	6.7	mg/l
Bath (1983)	AME/PA/Cotito	6, sp	53	6.5	mg/l
Ramirez (1988)	AME/PA/ElValle	PA-109, sp	31	8.4	mg/l
Ramirez (1988)	AME/PA/ElValle	PA-111, sp	29	8.4	mg/l
Ramirez (1988)	AME/PA/ElValle	PA-117, sp	38	8.2	mg/l
Ramirez (1988)	AME/PA/HoacasdelQuije	PA-123, sp	25	7.8	mg/l
Ramirez (1988)	AME/PA/LaYegvada	PA-99, sp	30	8.2	mg/l
Ramirez (1988)	AME/PA/LaYegvada	PA-101, sp	28	8.3	mg/l
Ramirez (1988)	AME/PA/LosPozos	PA-79, sp	53	7.6	mg/l
Ramirez (1988)	AME/PA/LosPozos	PA-81, sp	23	7.9	mg/l
Ramirez (1988)	AME/PA/LosPozos	PA-83, sp	23	7.9	mg/l
Ramirez (1988)	AME/PA/LosPozos	PA-85, sp	44	7.7	mg/l
Ramirez (1988)	AME/PA/LosPozos	PA-87, sp	51	7.8	mg/l
Bath (1983)	AME/PA/LosPozos	3, bh	66	6.7	mg/l
Ramirez (1988)	AME/PA/PuebloNuevo	PA-103, sp	39	7.9	mg/l
Ramirez (1988)	AME/PA/SanFrancisco	PA-69, sp	25	7.8	mg/l
Ramirez (1988)	AME/PA/SanJuan	PA-65, sp	25	7.2	mg/l
Ramirez (1988)	AME/PA/Tambo	PA-107, sp	24	7.5	mg/l
Ramirez (1988)	AME/PA/unnamed	PA-91, sp	25	7.8	mg/l
Ramirez (1988)	AME/PA/unnamed	PA-93, sp	26	8.2	mg/l
Ramirez (1988)	AME/PA/VientoAbajo	PA-89, sp	27	7.7	mg/l
Ramirez (1988)	AME/PA/VientoAbajo	PA-95, sp	26	7.7	mg/l
Campos (1988)	AME/SV/AguaCaliente	Agua Caliente, sp	38	8.1	ppm
Campos (1988)	AME/SV/AguaCaliente	Agua Caliente, sp	64	8.4	ppm
Campos (1988)	AME/SV/Caluco	Caluco, sp	38	7.6	ppm
Campos (1988)	AME/SV/Carolina	Carolina, sp	100	8.4	ppm

Na	K	Mg	Ca	Cl	SO4	HCO3	CO3	CO2	SIO2
98.70	11.40	0.32	4.60	62.80	54.20	115.00			112.00
9.90	3.30	0.57	28.80	3.80	6.00	140.00			23.00
34.20	3.80	0.14	9.70	18.20	17.80	83.40			86.00
79.00	6.70	0.29	20.80	37.00	35.90	176.00			154.00
0.00	0.00	0.01	0.01	0.00	0.03	0.00			0.00
0.00	0.00	0.00	0.00	0.00	0.00	0.00			0.00
0.00	0.00	0.00	0.00	0.00	0.00	0.01			0.00
650.00	145.00	0.01	1.80	1030.00	54.00	200.00			1200.00
1376.00	316.00	0.40	9.60	2449.50	28.00	162.20			689.00
3722.40	510.10	0.05	302.90	6284.60	23.62	2.51			986.30
1664.00	427.40	0.20	12.80	2964.80	43.90	46.40			1089.90
1631.60	353.80	0.00	7.15	2995.10	26.78	1.39			1521.80
2740.00	900.00	10.60	152.00	5427.10	320.00	63.80			118.20
1954.40	456.30	0.03	14.48	3607.70	17.80	1.59			1434.30
1456.30	350.00	26.10	26.40	2481.40	65.30	80.50			1245.60
1340.00	214.00	0.55	17.20	2339.90	59.00	61.90			1096.00
1697.10	379.80	0.05	64.61	3121.20	55.82	6.78			642.50
1500.00	418.80	0.03	18.60	2580.60	52.00	74.30			1007.70
1897.20	492.20	0.00	22.50	3641.60	9.25	18.01			1546.90
2157.10	421.90	0.20	30.10	3742.70	58.60	126.20			993.40
4254.90	747.50	0.05	376.80	7424.20	27.84	4.96			814.30
1824.00	358.60	0.02	23.70	3180.10	35.50	113.20			737.10
3734.40	614.40	0.04	287.16	7011.10	27.46	7.64			941.00
2270.30	445.80	0.80	60.90	3904.00	86.50	83.30			394.00
4538.00	816.00	0.12	275.20	8595.80	25.00	65.03			800.00
3106.00	557.00	0.04	97.10	5596.90	32.00	11.90			614.20
6.20	1.70	2.00	9.20	1.90	0.50	54.90			29.00
2750.00	241.00	210.00	480.00	4500.00	234.00	1674.00			175.00
330.00	35.00	22.00	60.00	500.00	34.00	275.00			126.00
600.00	68.00	52.00	120.00	1060.00	63.00	567.00			119.00
610.00	88.00	65.00	140.00	900.00	57.00	692.00			119.00
77.00	4.50	2.70	9.80	71.80	47.00	79.20			23.00
2080.00	194.00	58.00	440.00	3350.00	847.00	1019.00			139.00
2080.00	194.00	55.00	440.00	3325.00	777.00	1013.00			136.00
1610.00	146.00	45.00	380.00	2550.00	648.00	857.00			121.00
2630.00	111.00	71.00	264.00	2594.00	1750.00	1911.00			35.00
2670.00	111.00	68.50	262.00	2591.00	1770.00	1911.00			30.00
529.00	28.00	62.50	102.00	863.00	23.70	529.00			134.00
245.00	12.10	18.80	86.80	483.00	53.40	109.00			51.00
83.00	5.80	21.80	135.00	3.30	10.60	707.00			90.00
78.00	5.50	20.00	140.00	2.90	11.00	668.00	8.40		89.00
472.00	2.70	2.50	260.00	225.00	1220.00	48.80			85.00
10.40	0.60	6.70	41.80	0.40	53.30	119.60			38.00
14.80	0.90	6.60	43.20	3.00	61.20	117.10			35.00
455.00	2.40	2.60	244.00	231.00	1150.00	81.00			85.00
438.00	2.60	2.50	241.00	223.00	1100.00	56.10			84.00
2870.00	222.00	83.00	336.00	4473.00	460.00	1367.00			208.00
317.00	14.90	77.90	603.00	232.00	1070.00	1365.00			135.00
3.60	0.60	3.80	11.00	0.70	2.40	63.40			26.00
4.60	1.20	1.40	5.20	3.30	4.40	29.30			19.00
8.30	2.00	3.20	8.80	7.80	3.30	46.40			38.00
5.10	1.60	2.70	13.50	0.80	4.20	61.00			26.00
1490.00	143.00	70.10	331.00	2293.00	135.00	1196.00			40.00
38.00	4.90	2.90	9.30	24.80	10.60	103.90			78.00
6.30	5.10	4.10	22.20	0.90	22.50	85.40			59.00
196.00	27.80	27.60	39.60	231.00	58.00	286.00			118.00
160.00	2.00	0.18	9.20	19.40	272.00	65.00			68.00
109.00	29.00	86.30	57.60	58.00	125.00	644.00			105.00
156.00	6.60	0.39	12.00	18.90	285.00	70.00			135.00

Reference	Location	Site	T	pH	unit
Campos (1988)	AME/SV/Chilanguera	CH-1, well	99	7.8	ppm
Nieva (1997)	AME/SV/Chipilapa	F724, sp or well	30	8.5	mg/l
Nieva (1997)	AME/SV/Chipilapa	F438, sp or well	32	8.5	mg/l
Nieva (1997)	AME/SV/Chipilapa	P448, sp or well	30	8.6	mg/l
Nieva (1997)	AME/SV/Chipilapa	P457, sp or well	30	8.2	mg/l
Nieva (1997)	AME/SV/Chipilapa	M2, sp or well	30	8.5	mg/l
Nieva (1997)	AME/SV/Chipilapa	F717, sp or well	62	8.3	mg/l
Nieva (1997)	AME/SV/Chipilapa	F730, sp or well	58	8.7	mg/l
Nieva (1997)	AME/SV/Chipilapa	F733, sp or well	51	8.6	mg/l
Nieva (1997)	AME/SV/Chipilapa	F733E, sp or well	51	8.5	mg/l
Nieva (1997)	AME/SV/Chipilapa	F734, sp or well	58	8.7	mg/l
Nieva (1997)	AME/SV/Chipilapa	F735, sp or well	59	8.7	mg/l
Nieva (1997)	AME/SV/Chipilapa	F757, sp or well	40	8.6	mg/l
Nieva (1997)	AME/SV/Chipilapa	F760, sp or well	38	8.4	mg/l
Nieva (1997)	AME/SV/Chipilapa	P468, sp or well	44	7.9	mg/l
Nieva (1997)	AME/SV/Chipilapa	F719, sp or well	42	8.3	mg/l
Nieva (1997)	AME/SV/Chipilapa	F740, sp or well	37	8.0	mg/l
Nieva (1997)	AME/SV/Chipilapa	F741, sp or well	29	7.8	mg/l
Nieva (1997)	AME/SV/Chipilapa	P412, sp or well	34	8.6	mg/l
Nieva (1997)	AME/SV/Chipilapa	P413, sp or well	36	8.6	mg/l
Nieva (1997)	AME/SV/Chipilapa	M1, sp or well	44	8.4	mg/l
Nieva (1997)	AME/SV/Chipilapa	F726, sp or well		8.6	mg/l
Nieva (1997)	AME/SV/Chipilapa	F727, sp or well	33	8.5	mg/l
Nieva (1997)	AME/SV/Chipilapa	F728, sp or well	24	7.0	mg/l
Nieva (1997)	AME/SV/Chipilapa	F729, sp or well	28	8.4	mg/l
Nieva (1997)	AME/SV/Chipilapa	F747, sp or well	39	8.5	mg/l
Nieva (1997)	AME/SV/Chipilapa	P469, sp or well	28	8.1	mg/l
Nieva (1997)	AME/SV/Chipilapa	P526, sp or well	25	8.5	mg/l
Nieva (1997)	AME/SV/Chipilapa	M28, sp or well	81	7.5	mg/l
Nieva (1997)	AME/SV/Chipilapa	F722, sp or well	46	8.1	mg/l
Nieva (1997)	AME/SV/Chipilapa	F754, sp or well	60	8.7	mg/l
Nieva (1997)	AME/SV/Chipilapa	P410, sp or well	29	8.5	mg/l
Nieva (1997)	AME/SV/Chipilapa	P414, sp or well	41	8.7	mg/l
Nieva (1997)	AME/SV/Chipilapa	P414, sp or well	42	7.6	mg/l
Nieva (1997)	AME/SV/Chipilapa	F744, sp or well	22	8.2	mg/l
Nieva (1997)	AME/SV/Chipilapa	P523, sp or well	26	8.4	mg/l
Campos (1988)	AME/SV/Coatepeque	Coatepeque, sp	70	7.6	ppm
Campos (1988)	AME/SV/Conchagua	Conchagua, sp	62	8.1	ppm
Campos (1988)	AME/SV/Durazneno	Durazneno, sp	98	8.0	ppm
Campos (1988)	AME/SV/ElSalitre	El Salitre, sp	42	8.6	ppm
Campos (1988)	AME/SV/LaCeiba	La Ceiba, sp	48	8.2	ppm
Campos (1988)	AME/SV/LasBurras	TR2, well	297	7.0	ppm
Campos (1988)	AME/SV/LosToles	TE-1, well	110	8.2	ppm
Campos (1988)	AME/SV/Metapán	Metapán, sp	79	7.8	ppm
Campos (1988)	AME/SV/NombredeJesús	Nombre de Jesús, sp	88	7.6	ppm
Campos (1988)	AME/SV/Obrajuelo	Obrajuelo, sp	98	7.9	ppm
Campos (1988)	AME/SV/ParrasLempa	Parras Lempa, sp	78	8.2	ppm
Campos (1988)	AME/SV/Playón	CH-1, well	220	6.2	ppm
Campos (1988)	AME/SV/SanLorenzo	San Lorenzo, sp	43	7.3	ppm
Campos (1988)	AME/SV/SanSimón	San Simón, sp	47	8.2	ppm
Campos (1988)	AME/SV/SantaRosa	Santa Rosa, sp	85	8.3	ppm
Campos (1988)	AME/SV/SanVicente	SV1, well	230	7.5	ppm
Adams (1989)	AME/US/CA/Heber	5, well			ppm
Adams (1989)	AME/US/CA/Heber	6, well			ppm
Adams (1989)	AME/US/CA/Heber	9, well			ppm
Adams (1989)	AME/US/CA/Heber	10, well			ppm
Adams (1989)	AME/US/CA/Heber	11, well			ppm
Adams (1989)	AME/US/CA/Heber	12, well			ppm
Adams (1989)	AME/US/CA/Heber	13, well			ppm

Na	K	Mg	Ca	Cl	SO4	HCO3	CO3	CO2	SIO2
470.00	22.00	0.32	130.00	579.00	477.00	40.30			120.00
22.60	5.60	11.70	25.00	6.50	12.50	156.00	2.50		90.00
37.00	6.90	11.00	22.00	23.20	14.00	150.00	5.40		94.00
37.20	9.60	13.10	28.80	26.90	12.00	183.00	7.20		85.00
38.20	7.40	13.60	29.00	27.70	12.00	200.00			89.00
35.40	6.20	10.00	24.60	23.90	13.30	156.00	3.60		96.00
201.00	23.00	3.60	13.90	75.00	29.90	461.00			143.00
187.00	23.60	3.50	15.10	81.80	27.90	394.00	22.80		156.00
128.00	20.80	7.00	34.00	80.70	30.40	337.00	10.80		137.00
132.00	21.00	7.20	34.00	78.20	29.70	358.00	4.80		152.00
164.00	23.00	4.70	23.40	80.20	29.30	387.00	10.80		150.00
175.00	22.80	4.20	20.40	86.90	28.70	376.00	15.60		143.00
109.00	17.50	4.60	22.40	46.70	26.00	265.00	13.80		128.00
100.00	15.20	4.60	18.70	37.20	22.20	256.00	6.00		128.00
174.00	20.00	6.00	23.80	61.80	25.40	447.00			132.00
154.00	25.20	12.10	29.80	185.00	17.80	254.00	2.40		128.00
229.00	39.00	10.30	28.00	342.00	19.90	244.00			133.00
70.00	15.70	12.80	26.80	104.00	10.70	162.00			96.00
139.00	25.20	12.60	36.80	183.00	16.40	220.00	12.00		133.00
154.00	24.40	10.00	30.40	170.00	9.50	232.00	12.00		118.00
148.00	23.60	10.60	28.40	187.00	17.10	265.00	4.80		136.00
29.90	9.50	11.20	53.00	0.40	95.40	158.00	6.10		81.00
29.30	9.60	10.80	49.00	0.30	98.30	153.00	3.60		76.00
12.00	2.70	8.30	36.40	3.50	33.40	133.00			64.00
11.50	2.20	8.40	39.50	4.00	40.60	132.00	1.80		64.00
23.40	6.20	14.20	64.00	2.30	138.00	168.00	1.20		121.00
15.10	4.50	8.20	28.20	2.00	48.40	101.00			87.00
12.50	4.10	10.80	22.60	2.70	2.60	135.00	3.00		102.00
49.80	17.80	5.90	24.60	2.60	126.00	85.00			219.00
165.00	23.40	6.50	21.40	136.00	21.40	301.00			146.00
167.00	22.20	7.00	15.00	118.00	22.00	331.00			146.00
71.50	16.60	16.00	36.40	140.00	12.70	139.00	15.00		101.00
166.00	23.60	7.00	22.40	149.00	20.90	248.00	4.20		144.00
164.00	24.30	7.80	22.80	152.00	20.50	297.00			139.00
10.20	3.30	3.90	21.20	1.90	27.50	76.00			54.00
25.00	24.40	31.50	85.00	83.10	30.90	308.00	1.80		96.00
286.00	38.00	55.50	47.40	476.00	130.00	303.00			107.00
130.00	8.50	27.50	48.00	210.00	52.00	201.00			86.70
875.00	23.40	0.17	218.00	1539.00	186.00	21.40			114.00
106.00	1.00	0.25	2.40	13.40	144.00	63.70			60.00
197.00	28.40	9.90	24.40	242.00	23.50	289.00			153.00
3750.00	835.00	0.43	67.00	6777.00	13.00	22.00			1001.00
560.00	15.80	0.07	116.00	709.00	482.00	38.80			
272.00	6.10	0.23	34.00	183.00	396.00	90.70			77.00
175.00	9.40	8.00	44.90	22.70	456.00	88.00			156.00
650.00	19.50	0.10	150.00	1090.00	240.00	52.00			124.00
420.00	12.80	0.10	148.00	572.00	470.00	34.00			110.00
2410.00	186.00	48.00	137.00	4237.00	51.80	122.20			352.00
216.00	40.50	12.60	29.50	320.00	18.00	225.00			141.00
80.00	24.00	36.00	51.00	98.00	61.00	330.00			157.00
161.00	5.30	3.90	13.90	24.10	313.00	95.00			100.00
1620.00	240.00	0.40	82.00	2827.00	59.00	63.20			419.00
4019.00	333.00	1.76	750.00	7758.00	65.90	30.60		186.00	237.00
4034.00	340.00	2.12	747.00	7738.00	63.20	31.60		197.00	224.00
4060.00	332.00	2.05	717.00	7815.00	64.00	30.20		217.00	243.00
4012.00	342.00	1.96	686.00	7746.00	73.70	30.40		207.00	230.00
4242.00	354.00	2.81	655.00	7835.00	62.90	30.10		251.00	221.00
4261.00	349.00	2.42	736.00	7866.00	64.40	30.40		212.00	232.00
4140.00	324.00	2.31	565.00	7625.00	59.70	30.00		189.00	275.00

Reference	Location	Site	T	pH	unit
Adams (1989)	AME/US/CA/Heber	14, well			ppm
Adams (1989)	AME/US/CA/Heber	16, well			ppm
Adams (1989)	AME/US/CA/Heber	101, well			ppm
Adams (1989)	AME/US/CA/Heber	102, well			ppm
Adams (1989)	AME/US/CA/Heber	103, well			ppm
Adams (1989)	AME/US/CA/Heber	104, well			ppm
Adams (1989)	AME/US/CA/Heber	105, well			ppm
Adams (1989)	AME/US/CA/Heber	106, well			ppm
Adams (1989)	AME/US/CA/Heber	107, well			ppm
Goff (1994)	AME/US/CO/Archuleta	CH91-16, well	71	7.8	ppm
Goff (1994)	AME/US/CO/Archuleta	CH91-21, sp	27	7.7	ppm
Goff (1994)	AME/US/CO/Archuleta	CH91-22, well	47	7.7	ppm
Goff (1994)	AME/US/CO/Archuleta	SB91-9, well	39	7.7	ppm
Goff (1994)	AME/US/CO/Archuleta	DM91-085, sp	54	8.1	ppm
Goff (1994)	AME/US/CO/Archuleta	LC91-07W, well	45	8.0	ppm
Goff (1994)	AME/US/CO/Archuleta	LC91-08W, well	28	7.9	ppm
Goff (1994)	AME/US/CO/Archuleta	ED91-15W, well	63	9.6	ppm
Goff (1994)	AME/US/CO/Archuleta	PS91-1A, well	65	7.5	ppm
Goff (1994)	AME/US/CO/Archuleta	PS91-3A, well	58	7.6	ppm
Goff (1994)	AME/US/CO/Archuleta	PS91-4A, well	49	7.6	ppm
Goff (1994)	AME/US/CO/Archuleta	PS91-5A, sp	54	7.7	ppm
Goff (1994)	AME/US/CO/Archuleta	PS91-6A, well	50	7.7	ppm
Goff (1994)	AME/US/CO/Archuleta	PS91-8A, well	50	7.7	ppm
Goff (1994)	AME/US/CO/Archuleta	PS91-10A, well	33	7.7	ppm
Goff (1994)	AME/US/CO/Archuleta	PS91-13A, well	54	7.3	ppm
Goff (1994)	AME/US/CO/Archuleta	PS91-14A, well	57	7.7	ppm
Goff (1994)	AME/US/CO/Archuleta	PS91-17A, well	55	7.7	ppm
Goff (1994)	AME/US/CO/Archuleta	PS91-18, sp	41	7.0	ppm
Goff (1994)	AME/US/CO/Archuleta	PS91-19W, well	22	6.7	ppm
Sorey (1991)	AME/US/LongValley	MBP-1, well	170	6.2	mg/l
Sorey (1991)	AME/US/LongValley	MBP-3, well	175	6.1	mg/l
Sorey (1991)	AME/US/LongValley	RDO-8, well	202	5.9	mg/l
White (1991)	AME/US/LongValley	MBP-1, well		6.6	ppm
White (1991)	AME/US/LongValley	MBP-3, well		6.6	ppm
White (1991)	AME/US/LongValley	RDO-8, well		5.9	ppm
White (1991)	AME/US/LongValley	CW-2, well		6.3	ppm
Kharaka (1986)	AME/USA/CA/Lafayette	St. Un. A#9, Weeks Island	117	6.2	ppm
Kharaka (1986)	AME/USA/CA/Lafayette	Edna delcombre #1,Tiger I	114	6.3	ppm
Kharaka (1986)	AME/USA/CA/Sacramento	19-1,Malton-Black Butte		7.6	ppm
Kharaka (1986)	AME/USA/CA/San Joaquin	21-28, Wheeler Ridge		6.9	ppm
Thomas (1986)	AME/USA/HI/OahuMaili	2508-02, well	29	7.8	ppm
Thomas (1986)	AME/USA/HI/OahuMaili	2508-02, well	29	7.7	ppm
Thomas (1986)	AME/USA/HI/OahuMaili	2607-01, well	24	7.5	ppm
Thomas (1986)	AME/USA/HI/OahuMaili	2712-01, well		6.9	ppm
Thomas (1986)	AME/USA/HI/OahuMaili	2712-01, well		7.1	ppm
Thomas (1986)	AME/USA/HI/OahuMaili	2712-01, well			ppm
Thomas (1986)	AME/USA/HI/OahuMaili	2808-01, well	27	7.8	ppm
Thomas (1986)	AME/USA/HI/OahuMaili	2550-01, well		6.2	ppm
Thomas (1986)	AME/USA/HI/OahuMaili	2043-01, well	30	7.3	ppm
Thomas (1986)	AME/USA/HI/OahuMaili	2042-13, well	25	7.4	ppm
Thomas (1986)	AME/USA/HI/OahuMaili	4837-01, well	24	7.4	ppm
Thomas (1986)	AME/USA/HI/OahuMaili	4937-01, well	26	7.1	ppm
Thomas (1986)	AME/USA/HI/OahuMaili	2487-01, well	24	7.3	ppm
Thomas (1986)	AME/USA/HI/OahuMaili	2881-01, well	38	7.3	ppm
Thomas (1986)	AME/USA/HI/OahuMaili	2982-01, well	93	6.8	ppm
Kharaka (1986)	AME/USA/MS/Salt Dome Basin	Geiger-Cupp Unit 9-13 No.	88	5.4	ppm
Kharaka (1986)	AME/USA/MS/Salt Dome Basin	W.M. Geiger No.2-1, Reedy	102	5.1	ppm
Kharaka (1986)	AME/USA/MS/Salt Dome Basin	W.L. West 6-11 No.1, West	118	5.5	ppm
Goff (1981)	AME/USA/NM/SanDiego	VA-7, sp	70	6.3	mg/l

Na	K	Mg	Ca	Cl	SO4	HCO3	CO3	CO2	SIO2
4091.00	327.00	2.72	687.00	7801.00	59.90	31.30		238.00	210.00
4147.00	317.00	3.42	617.00	7758.00	55.00	33.90		304.00	207.00
3967.00	366.00	2.72	836.00	8105.00	77.80	28.20		230.00	207.00
3925.00	347.00	1.83	816.00	7667.00	74.00	31.60		163.00	208.00
3766.00	303.00	1.77	843.00	7429.00	83.40	29.90		127.00	220.00
3840.00	338.00	1.79	810.00	7571.00	74.50	29.60		179.00	206.00
3683.00	333.00	1.88	830.00	7598.00	71.70	29.60		163.00	212.00
3907.00	359.00	1.60	794.00	7673.00	66.20	28.10		167.00	212.00
3746.00	281.00	1.54	985.00	7318.00	100.00	20.90		145.00	233.00
14.20	4.40	15.50	151.00	6.37	354.00	110.00			39.00
19.70	11.40	27.00	213.00	6.67	493.00	165.00			24.00
21.60	9.62	24.40	309.00	7.53	737.00	125.00			33.00
165.00	20.90	31.00	614.00	17.90	1757.00	144.00			47.00
571.00	40.00	5.97	76.00	116.00	934.00	328.00			98.00
153.00	9.98	11.80	64.70	2.23	308.00	254.00			21.00
81.00	6.21	21.00	98.40	1.21	293.00	223.00			16.00
575.00	10.60	0.32	1.40	235.00	2.00	1120.00			7.00
766.00	73.00	25.00	225.00	167.00	1480.00	680.00			59.00
802.00	79.00	25.60	228.00	157.00	1550.00	629.00			58.00
825.00	75.00	25.40	239.00	152.00	1560.00	751.00			55.00
732.00	69.00	27.50	248.00	152.00	1545.00	676.00			57.00
786.00	77.00	22.80	216.00	139.00	1590.00	635.00			52.00
784.00	72.00	22.80	217.00	141.00	1575.00	639.00			49.00
786.00	73.00	27.60	214.00	105.00	1690.00	570.00			32.00
783.00	76.00	27.00	247.00	168.00	1620.00	792.00			56.00
794.00	74.00	26.40	249.00	164.00	1505.00	775.00			57.00
782.00	72.00	23.90	261.00	152.00	1495.00	758.00			53.00
750.00	70.00	26.60	232.00	161.00	1555.00	740.00			54.00
730.00	82.00	37.90	275.00	72.80	1700.00	596.00			27.00
370.00	34.00	0.60	2.60	260.00	130.00	359.00			250.00
360.00	33.00	0.60	2.50	250.00	120.00	358.00			270.00
369.00	43.00	0.20	7.40	280.00	159.00	376.00			250.00
392.00	36.00	0.14	1.60	244.00	119.00	425.00			210.00
378.00	38.00	0.18	2.00	238.00	116.00	423.00			245.00
380.00	48.00	0.35	18.00	263.00	179.00	483.00			277.00
290.00	20.00	0.10	1.40	210.00	88.00	290.00			130.00
78000.00	1065.00	1140.00	10250.00	143000.00	6.40	450.00			48.00
40000.00	265.00	270.00	1860.00	67900.00	220.00	1050.00			57.00
7510.00	28.40	148.00	331.00	12700.00	0.90	417.00			18.00
7450.00	135.00	27.00	5550.00	21450.00	50.00	2210.00			46.00
92.00	7.90	102.00	36.00	292.00	22.00	338.00			92.00
126.00	9.60	108.00	41.00	382.00	25.00	313.00			89.00
38.00	2.80	12.00	13.00	46.00	8.50	113.00			65.00
48.00	4.30	30.00	17.00	82.00	14.00	176.00			80.00
55.00	4.10	34.00	19.00	97.00	16.00	183.00			75.00
50.00	3.70	30.00	17.00	83.00	14.00	171.00			74.00
120.00	3.20	28.00	66.00	160.00	222.00	97.00			63.00
13.00	0.60	4.30	6.60	14.00	8.20	46.00			2.00
28.00	1.10	2.80	14.00	25.00	5.40	84.00			22.00
920.00	36.00	110.00	150.00	1700.00	220.00	224.00			26.00
263.70	7.60	34.70	40.70	339.30	63.70	408.00			46.90
255.00	11.90	78.00	112.00	669.00	76.00	141.00			73.50
64.00	5.00	3.80	12.40	105.60	22.00	42.00			
1188.00	68.00	102.00	84.00	2042.00	69.00	132.00			24.00
2757.00	300.00	137.00	283.00	5257.00	335.00	30.00			970.00
57200.00	1000.00	2310.00	31700.00	158000.00	68.00	146.00			30.00
61700.00	990.00	3050.00	48600.00	198000.00	64.00	206.00			28.00
54800.00	6500.00	3350.00	33900.00	170000.00	161.00	197.00			34.00
614.00	75.20	4.56	182.00	829.00	36.10	723.00			93.00

Reference	Location	Site	T	pH	unit
Goff (1981)	AME/USA/NM/SanDiego	VA-8, sp	49	6.4	mg/l
Goff (1981)	AME/USA/NM/SanDiego	VA-9, sp	48	6.4	mg/l
Goff (1981)	AME/USA/NM/SanDiego	VA-10, sp	55	7.0	mg/l
Goff (1981)	AME/USA/NM/SanDiego	VA-12, sp	49	6.4	mg/l
Goff (1981)	AME/USA/NM/SanDiego	VA-15, well	61	6.7	mg/l
Goff (1981)	AME/USA/NM/SanDiego	VA-16, sp	50	6.6	mg/l
Goff (1981)	AME/USA/NM/SanDiego	VA-17, sp	72	6.7	mg/l
Goff (1981)	AME/USA/NM/SanDiego	VA-18, sp	36	7.5	mg/l
Goff (1981)	AME/USA/NM/SanDiego	VA-19, well	68	6.6	mg/l
Kharaka (1986)	AME/USA/TX/Corpus Christi	Portland A-3, Portland	123	6.8	ppm
Kharaka (1986)	AME/USA/TX/Corpus Christi	Taylor E-2, East Midway	128	6.4	ppm
Kharaka (1986)	AME/USA/TX/Houston-Galveston	Houston "FF" #1, Halls Ba	150	6.8	ppm
Kharaka (1986)	AME/USA/TX/Mc Allen-Pharr	Kelly A-1, Pharr	127	6.8	ppm
Kharaka (1986)	AME/USA/TX/Mc Allen-Pharr	La Blanca #12, La Blanca	148	7.3	ppm
Kharaka (1986)	AME/USA/TX/Offshore	A-11A, High Island		7.5	ppm
Kharaka (1986)	AME/USA/TX/Offshore	A-11B, High Island		7.4	ppm
Kharaka (1986)	AME/USA/TX/Offshore	A-10A, High Island		7.3	ppm
Kharaka (1986)	AME/USA/TX/Offshore	A-9A, High Island		7.3	ppm
Kharaka (1986)	AME/USA/TX/Offshore	C-8A, High Island		7.6	ppm
Kharaka (1986)	AME/USA/TX/Offshore	C-14A, High Island		7.5	ppm
Kharaka (1986)	AME/USA/TX/Offshore	B-14A, High Island		7.1	ppm
Kharaka (1986)	AME/USA/TX/Offshore	B-2A, High Island		7.1	ppm
Kharaka (1986)	AME/USA/TX/Offshore	B-12A, High Island		7.0	ppm
Kharaka (1986)	AME/USA/TX/Offshore	A-45-1, High Island		6.8	ppm
Sorey (1997)	AME/USA/WY/YellowstoneN.P.	Growler Spring, sp	93	6.8	mg/l
Sorey (1997)	AME/USA/WY/YellowstoneN.P.	3C, sp	93	6.7	mg/l
Sorey (1997)	AME/USA/WY/YellowstoneN.P.	Sheepeater, sp	73	6.5	mg/l
Sorey (1997)	AME/USA/WY/YellowstoneN.P.	Y-10, well	73	6.2	mg/l
Sorey (1997)	AME/USA/WY/YellowstoneN.P.	BC-1, sp	32	5.8	mg/l
Sorey (1997)	AME/USA/WY/YellowstoneN.P.	CUT, sp	53	6.7	mg/l
Sorey (1997)	AME/USA/WY/YellowstoneN.P.	La Duke Hot Spring, sp	68	6.5	mg/l
Fournier (1989)	AME/USA/WY/YellowstoneN.P.	Y-10, well	70	7.5	ppm
Fournier (1989)	AME/USA/WY/YellowstoneN.P.	Ear Spring, sp	94	8.5	ppm
Grimaud (1985)	ASI/CN/Capu/Tibet	6AH29, sp	82	8.1	mmol/kg
Huang (1986)	ASI/CN/Fuzhou/FujianPr.	0-2, well	25	7.3	mg/l
Huang (1986)	ASI/CN/Fuzhou/FujianPr.	0-3, well	20	7.7	mg/l
Huang (1986)	ASI/CN/Fuzhou/FujianPr.	I-3, well	29	7.3	mg/l
Huang (1986)	ASI/CN/Fuzhou/FujianPr.	II-2, well	66	8.6	mg/l
Huang (1986)	ASI/CN/Fuzhou/FujianPr.	II-3, well	72	7.5	mg/l
Huang (1986)	ASI/CN/Fuzhou/FujianPr.	II-4, well	30	7.1	mg/l
Huang (1986)	ASI/CN/Fuzhou/FujianPr.	III+1, well	42	7.1	mg/l
Huang (1986)	ASI/CN/Fuzhou/FujianPr.	G15, well	68	7.1	mg/l
Grimaud (1985)	ASI/CN/Gulu/Tibet	2AH4, sp	85	8.1	mmol/kg
Grimaud (1985)	ASI/CN/Gulu/Tibet	2AH5, sp	84	8.7	mmol/kg
Grimaud (1985)	ASI/CN/Gulu/Tibet	2AH11, sp	83	8.5	mmol/kg
Grimaud (1985)	ASI/CN/Gulu/Tibet	2AH12, sp	68	7.4	mmol/kg
Zhonghe (2000)	ASI/CN/Jidong/NorthChinaB.	Qy-1, well		7.9	mg/l
Zhonghe (2000)	ASI/CN/Jidong/NorthChinaB.	Tn-2, well		7.8	mg/l
Zhonghe (2000)	ASI/CN/Jidong/NorthChinaB.	Xg-3, well		7.3	mg/l
Zhonghe (2000)	ASI/CN/Jidong/NorthChinaB.	L21x10, well		8.3	mg/l
Zhonghe (2000)	ASI/CN/Jidong/NorthChinaB.	L25x13, well		8.3	mg/l
Zhonghe (2000)	ASI/CN/Jidong/NorthChinaB.	N34x1, well		7.4	mg/l
Zhonghe (2000)	ASI/CN/Jidong/NorthChinaB.	L20x2, well		7.3	mg/l
Zhonghe (2000)	ASI/CN/Jidong/NorthChinaB.	L90x2, well		8.2	mg/l
Zhonghe (2000)	ASI/CN/Jidong/NorthChinaB.	Hc-1, well		7.1	mg/l
Grimaud (1985)	ASI/CN/Jugu/Tibet	13AH39, sp	50	8.3	mmol/kg
Grimaud (1985)	ASI/CN/Longma/Tibet	5AH27, well	37	7.5	mmol/kg
Liu (1999)	ASI/CN/Nagqu/Tibet	ZK1102, well	115	5.8	ppm
Liu (1999)	ASI/CN/Nagqu/Tibet	ZK1202, well	117	5.8	ppm

Na	K	Mg	Ca	Cl	SO4	HCO3	CO3	CO2	SIO2
458.00	53.00	9.57	154.00	653.00	37.60	697.00			81.00
938.00	183.00	24.40	340.00	1503.00	38.40	1514.00			50.00
656.00	74.20	5.40	152.00	904.00	40.90	711.00			93.00
609.00	70.00	7.82	129.00	903.00	41.80	738.00			100.00
185.00	29.90	9.31	120.00	243.00	38.00	479.00			24.00
494.00	57.80	7.50	128.00	653.00	40.60	708.00			72.00
612.00	70.30	4.48	114.00	936.00	43.20	714.00			83.00
690.00	74.00	4.52	115.00	968.00	45.40	699.00			85.00
546.00	61.60	5.76	122.00	705.00	45.00	642.00			70.00
6500.00	68.00	15.00	89.00	9270.00	110.00	1600.00			93.00
13250.00	72.00	48.00	330.00	21000.00	42.00	1180.00			132.00
20500.00	180.00	170.00	800.00	34500.00	16.00	409.00			110.00
9420.00	240.00	18.00	4225.00	22000.00	7.00	114.00			90.00
2680.00	46.00	3.30	150.00	3950.00	57.00	400.00			86.00
26300.00	160.00	788.00	1320.00	44400.00	12.40	494.00			40.10
28400.00	172.00	869.00	1370.00	48400.00	9.90	546.00			38.80
19800.00	139.00	1030.00	1030.00	33400.00	16.00	568.00			45.40
53700.00	244.00	1270.00	3810.00	99000.00	7.40	308.00			30.00
15400.00	110.00	499.00	679.00	26200.00	9.90	480.00			40.80
13300.00	103.00	349.00	850.00	23000.00	9.90	443.00			60.00
47700.00	250.00	1040.00	2840.00	84600.00	0.50	293.00			31.60
31800.00	167.00	1010.00	2250.00	56200.00	0.40	350.00			32.90
30400.00	183.00	873.00	2010.00	53800.00	15.30	402.00	6.97		41.20
30900.00	0.39	482.00	2690.00	53300.00	11.50	2140.00			65.80
394.00	103.00	0.01	2.40	695.00	41.20	32.00			602.00
278.00	64.00	0.12	13.60	467.00	64.80	33.00			304.00
93.50	36.40	12.90	141.00	97.00	84.50	570.00			83.60
156.00	64.50	76.10	492.00	174.00	808.00	1040.00			94.60
113.00	48.80	91.40	507.00	42.80	869.00	1240.00			34.90
210.00	25.70	73.20	334.00	37.50	1228.00	250.00			76.10
240.00	24.50	64.10	332.00	44.80	1249.00	295.00			56.70
161.00	69.00	80.00	450.00	171.00	800.00	997.00			88.00
319.00	27.00	0.01	0.82	415.00	19.00	146.00			371.00
17.50	1.33	0.00	0.04	8.40	0.90	9.20			4.83
37.60	4.20	3.89	12.80	28.00	24.50	92.10			48.00
65.20	2.20	4.13	32.40	59.50	64.80	100.60			30.00
61.60	5.00	4.98	49.10	71.60	123.90	84.80			26.00
112.00	4.50	0.73	13.00	74.80	105.10	74.40			28.00
143.50	2.70	0.85	11.00	82.20	159.90	61.00			80.00
87.20	7.40	5.35	19.40	82.60	97.00	73.80			28.00
107.60	4.20	3.80	18.40	91.80	121.00	65.90			36.00
143.50	1.85	0.61	11.00	85.70	152.70	55.50			70.00
41.00	2.98	0.00	0.03	23.80	0.27	22.70			5.13
45.40	3.22	0.00	0.02	25.40	0.30	25.30			5.73
42.00	2.98	0.00	0.02	23.30	0.18	22.90			5.83
39.20	2.98	0.19	0.48	22.00	0.18	23.10			4.70
237.59	0.42	1.24	7.12	81.54	120.05	364.17	11.23		16.92
144.90	0.25	0.62	6.10	28.83	85.44	239.10	8.98		17.46
109.42	0.25	0.25	5.70	19.60	66.90	204.96			15.21
338.28	3.98	1.61	3.05	196.01	6.69	510.60	45.00		54.01
366.47	4.73	0.99	3.05	163.73	4.83	630.13	56.16		51.70
252.22	3.74	2.10	20.95	92.24	336.23	183.00			29.49
515.59	17.93	20.02	71.19	547.68	171.38	548.16			113.81
382.05	7.72	3.09	7.12	149.89	114.84	628.10	33.70		78.32
245.55	112.07	19.16	220.69	140.07	809.73	298.90			83.05
5.08	0.11	0.01	0.04	0.13	0.12	4.81			0.93
9.55	0.17	0.14	0.37	0.17	0.59	9.60			1.10
1004.00	66.00	6.00	22.50	248.00	200.00	2000.00		5240.00	75.00
1004.00	66.00	6.00	22.50	248.00	200.00	2000.00		5240.00	75.00

Reference	Location	Site	T	pH	unit
Grimaud (1985)	ASI/CN/Nagqu/Tibet	1AH1, sp	43	7.2	mmol/kg
Grimaud (1985)	ASI/CN/Qiaga/Tibet	7AH31, sp	47	9.1	mmol/kg
Grimaud (1985)	ASI/CN/Quxiang/Tibet	10AH36, sp	50	6.9	mmol/kg
Grimaud (1985)	ASI/CN/Quzai/Tibet	3AH14, sp	86	8.1	mmol/kg
Zongyu (2000)	ASI/CN/Xiaotangshan/Beijing	1, well	52	6.9	ppm
Zongyu (2000)	ASI/CN/Xiaotangshan/Beijing	4, well	49	6.9	ppm
Zongyu (2000)	ASI/CN/Xiaotangshan/Beijing	7, well	55	7.0	ppm
Zongyu (2000)	ASI/CN/Xiaotangshan/Beijing	8, well	49	7.0	ppm
Zongyu (2000)	ASI/CN/Xiaotangshan/Beijing	9, well	44	6.9	ppm
Zongyu (2000)	ASI/CN/Xiaotangshan/Beijing	11, well	59	6.9	ppm
Zongyu (2000)	ASI/CN/Xiaotangshan/Beijing	12, well	54	6.9	ppm
Grimaud (1985)	ASI/CN/Yangbajing/Tibet	4AH16, well	70	7.7	mmol/kg
Grimaud (1985)	ASI/CN/Yangbajing/Tibet	4AH18, well	160	8.3	mmol/kg
Grimaud (1985)	ASI/CN/Yangbajing/Tibet	4AH19, well	140	7.5	mmol/kg
Grimaud (1985)	ASI/CN/Yangbajing/Tibet	4AH20, well	165	8.2	mmol/kg
Grimaud (1985)	ASI/CN/Yangbajing/Tibet	4AH21, well	145	8.1	mmol/kg
Grimaud (1985)	ASI/CN/Yangbajing/Tibet	4AH23, well	155	8.5	mmol/kg
Grimaud (1985)	ASI/CN/Yangbajing/Tibet	4AH24, well	150	8.8	mmol/kg
Grimaud (1985)	ASI/CN/Yangbajing/Tibet	4AH25, well	140	8.2	mmol/kg
Mahon (2000)	ASI/ID/Darajat	S3, sp	77	3.0	ppm
Mahon (2000)	ASI/ID/Darajat	S4, sp	40	4.5	ppm
Mahon (2000)	ASI/ID/Kamojang	S1, sp	96	2.9	ppm
Mahon (2000)	ASI/ID/Salak	W12, well	100	6.7	ppm
Sundhoro (2000)	ASI/ID/SembalunBumbung/Lombo	Aik Kalak, sp		8.2	mg/l
Sundhoro (2000)	ASI/ID/SembalunBumbung/Lombo	Aik Sebau, sp		7.6	mg/l
Mahon (2000)	ASI/ID/Wayang	W13, well	100	6.3	ppm
Saxena (1985)	ASI/IN/Agnigundala/Godavari	hot spring, sp	62	7.5	ppm
Giggenbach (1983)	ASI/IN/Balargah/ParbatinVal.	TS-1, sp	45	7.5	ppm
Saxena (1985)	ASI/IN/Bhimdole/Godavari	Bhimdole, bh	52	7.4	ppm
Saxena (1985)	ASI/IN/Bhuttayagudem/Godavar	Bhuttayagudem, bh	64	7.4	ppm
Saxena (1985)	ASI/IN/Buga/Godavari	main hot spring, sp	45	7.8	ppm
Saxena (1985)	ASI/IN/Buga/Godavari	warm spring, sp	40	7.7	ppm
Moon (1988)	ASI/IN/Chumathang/NWHimalaya	17, sp	66	6.3	ppm
Moon (1988)	ASI/IN/Chumathang/NWHimalaya	72, sp	45	6.5	ppm
Moon (1988)	ASI/IN/Chumathang/NWHimalaya	101, sp	83	7.6	ppm
Moon (1988)	ASI/IN/Chumathang/NWHimalaya	CGW-1, well	85	7.9	ppm
Moon (1988)	ASI/IN/Chumathang/NWHimalaya	CGW-2, well		6.9	ppm
Moon (1988)	ASI/IN/Chumathang/NWHimalaya	39, sp	87	8.0	ppm
Moon (1988)	ASI/IN/Chumathang/NWHimalaya	40, sp	49	7.7	ppm
Giggenbach (1983)	ASI/IN/Jan/ParbatinVal.	TS-1, sp	33	6.8	ppm
Moon (1988)	ASI/IN/Jan/ParbatiVal.	Jan,sp	34	5.8	ppm
Giggenbach (1983)	ASI/IN/Jeori/SutlejVal.	Jeori, sp	59	7.7	ppm
Giggenbach (1983)	ASI/IN/Kasol/ParbatinVal.	GW-2, well	72	7.6	ppm
Moon (1988)	ASI/IN/Kasol/ParbatiVal.	2, sp	75	7.4	ppm
Moon (1988)	ASI/IN/Kasol/ParbatiVal.	3, sp	76	7.9	ppm
Giggenbach (1983)	ASI/IN/Khirganga/ParbatinVal	Khirganga, sp	49	8.3	ppm
Saxena (1985)	ASI/IN/Manguru/Godavari	TW-1, well	38	7.8	ppm
Saxena (1985)	ASI/IN/Manguru/Godavari	TW-2, well	37	7.8	ppm
Giggenbach (1983)	ASI/IN/Manikaran/ParbatiVal.	TS-7, sp	95	8.4	ppm
Giggenbach (1983)	ASI/IN/Manikaran/ParbatiVal.	TS-7, sp	94	7.4	ppm
Giggenbach (1983)	ASI/IN/Manikaran/ParbatiVal.	TS-12, sp	84	8.4	ppm
Giggenbach (1983)	ASI/IN/Manikaran/ParbatiVal.	TS-21, sp	46	7.9	ppm
Giggenbach (1983)	ASI/IN/Manikaran/ParbatiVal.	GW-2, well	95	8.3	ppm
Moon (1988)	ASI/IN/Manikaran/ParbatVal.	1, sp	87	7.5	ppm
Moon (1988)	ASI/IN/Manikaran/ParbatVal.	4, sp	88	7.6	ppm
Moon (1988)	ASI/IN/Manikaran/ParbatVal.	7, sp	96	6.1	ppm
Moon (1988)	ASI/IN/Manikaran/ParbatVal.	10, sp	96	7.5	ppm
Moon (1988)	ASI/IN/Manikaran/ParbatVal.	11, sp	82	7.7	ppm
Moon (1988)	ASI/IN/Manikaran/ParbatVal.	20, sp	37	7.5	ppm

Na	K	Mg	Ca	Cl	SO4	HCO3	CO3	CO2	SIO2
45.80	1.46	0.20	0.50	6.80	1.08	40.30			1.17
4.64	0.04	0.00	0.06	0.95	0.78	1.80			0.98
25.60	2.52	0.42	0.12	3.20	0.20	26.60			2.38
25.50	1.38	0.25	0.68	13.70	2.41	9.90			1.90
78.28	14.17	13.74	45.09	26.59	70.88	274.60			32.80
87.59	17.50	13.37	51.10	33.68	83.82	292.89			38.00
89.66	20.56	14.59	47.09	39.00	84.41	305.09			40.00
93.79	16.94	15.80	49.70	35.45	100.94	298.98			40.00
71.66	14.94	14.59	49.10	23.04	95.59	271.53			30.00
80.76	15.56	12.16	43.09	31.97	83.82	247.10			40.00
79.72	15.94	15.20	47.90	33.68	82.35	283.70			32.00
18.50	1.23	0.02	0.38	13.80	0.28	7.20			2.71
20.10	1.64	0.01	0.04	15.70	0.34	6.70			4.13
20.70	1.11	0.00	0.09	14.60	0.28	7.10			2.90
19.70	1.52	0.01	0.09	14.80	0.34	6.40			4.10
20.20	1.32	0.01	0.18	14.60	0.37	7.40			2.93
20.70	1.70	0.00	0.06	15.70	0.37	6.40			4.03
24.60	1.58	0.01	0.06	17.90	0.45	8.40			3.93
20.80	1.45	0.02	0.28	15.30	0.38	7.60			3.33
13.00	5.70	10.00	17.50	3.00	1150.00				200.00
6.30	2.60	6.20	1.60	14.00	430.00			100.00	150.00
20.00	5.00	14.60	30.30	17.00	335.00				212.00
3675.00	876.00	1.00	268.00	6810.00	20.00			37.00	495.00
316.66	5.00	8.46	332.94	211.55	1287.50	89.80			41.00
208.33	4.38	11.10	169.11	549.19	62.50	42.15			33.00
11250.00	3060.00	0.60	885.00	22160.00	75.00				355.00
321.00	23.60	0.90	32.50	430.00	147.00	38.00			143.00
16.00	6.00	17.00	55.00	7.00	41.00	258.00			45.00
44.00	4.40	4.70	41.20	100.00	38.00	64.00			43.00
142.00	20.40	8.80	97.50	217.00	150.00	143.00			63.00
95.00	7.40	7.90	50.00	50.00	147.00	171.00			67.00
100.00	8.00	18.20	25.00	50.00	134.00	171.00			43.00
620.00	60.00	3.00	14.00	401.00	133.00	860.00			200.00
660.00	100.00	3.00	26.00	427.00	140.00	934.00			140.00
600.00	100.00	9.00	9.00	396.00	128.00	1159.00			140.00
360.00	24.00	2.00	22.00	84.00	240.00	522.00			175.00
400.00	30.00	4.00	40.00	81.00	230.00	751.00			120.00
340.00	20.00	1.00	4.00	77.00	233.00	456.00			140.00
370.00	24.00	2.00	22.00	84.00	248.00	534.00			160.00
203.00	26.00	42.00	70.00	395.00	56.00	299.00		964.00	18.00
210.00	28.00	45.00	70.00	418.00	67.00	284.00			24.00
885.00	36.00	23.20	109.00	1180.00	167.00	367.00			80.00
32.00	10.00	9.00	55.00	48.00	56.00	168.00			54.00
37.00	10.00	22.00	54.00	46.00	50.00	246.00			55.00
40.00	10.00	23.00	56.00	46.00	50.00	246.00			60.00
329.00	13.00	2.30	28.00	308.00	39.00	396.00			52.00
200.00	8.00	1.20	20.00	84.00	185.00	214.00			60.00
196.00	9.60	3.00	25.00	84.00	171.00	214.00			60.00
96.00	19.00	3.30	52.00	138.00	41.00	210.00		981.00	83.00
93.00	21.00	3.30	51.00	130.00	35.00	187.00		977.00	75.00
97.00	19.00	3.50	51.00	139.00	41.00	209.00			79.00
22.00	8.00	9.20	70.00	30.00	39.00	232.00			35.00
94.00	18.00	3.80	53.00	135.00	39.00	205.00		864.00	76.00
90.00	18.00	4.00	47.00	74.00	31.00	235.00			90.00
94.00	18.00	6.00	47.00	84.00	30.00	235.00			100.00
95.00	19.00	10.00	51.00	90.00	38.00	246.00			90.00
96.00	19.00	9.00	56.00	90.00	34.00	287.00			90.00
100.00	18.00	5.00	53.00	105.00	34.00	258.00			90.00
64.00	14.00	4.00	47.00	67.00	39.00	205.00			100.00

Reference	Location	Site	T	pH	unit
Moon (1988)	ASI/IN/Manikaran/ParbatVal.	29, sp	45	7.5	ppm
Moon (1988)	ASI/IN/Manikaran/ParbatVal.	MG-1,bh	77	7.9	ppm
Giggenbach (1983)	ASI/IN/Nathpa/SutlejVal.	TS 1, well	57	8.0	ppm
Saxena (1985)	ASI/IN/Pagdaru/Godavari	drill-hole1, bh	39	7.7	ppm
Saxena (1985)	ASI/IN/Pagdaru/Godavari	drill-hole2, bh	33	7.7	ppm
Giggenbach (1983)	ASI/IN/Pali	Pali, sp	43	6.7	ppm
Moon (1988)	ASI/IN/Puga/NWHimalayan	GW-7, well	135	6.9	ppm
Moon (1988)	ASI/IN/Puga/NWHimalayan	GW-8, well	85	7.9	ppm
Giggenbach (1983)	ASI/IN/Puga/NWHimalayan	GW-2, well	80	8.9	ppm
Giggenbach (1983)	ASI/IN/Puga/NWHimalayan	GW-25, well	120	8.9	ppm
Giggenbach (1983)	ASI/IN/Pulga/ParbatinVal.	TS-1, sp	44	7.0	ppm
Saxena (1987)	ASI/IN/Salbardi	Salbardi1, sp	47	7.5	ppm
Saxena (1987)	ASI/IN/Salbardi	Salbardi2, sp	45	7.4	ppm
Saxena (1987)	ASI/IN/Salbardi	Salbardi3, sp	44	7.4	ppm
Saxena (1987)	ASI/IN/Salbardi	Salbardi4, sp	45	7.4	ppm
Shanker (2000)	ASI/IN/Tapoban/NWHimalaya	AGW-3, bh	91		mg/l
Shanker (2000)	ASI/IN/Tapoban/NWHimalaya	AGW-6, bh	66		mg/l
Saxena (1987)	ASI/IN/Tatapani	Tatapani1, sp	86	7.8	ppm
Saxena (1987)	ASI/IN/Tatapani	Tatapani2, sp	65	7.6	ppm
Saxena (1987)	ASI/IN/Tatapani	Tatapani3, sp	49	7.6	ppm
Saxena (1987)	ASI/IN/Tatapani	Tatapani4, sp	78	7.7	ppm
Moon (1988)	ASI/IN/VajrabhaigThana/WestC	5, sp	60		ppm
Giggenbach (1983)	ASI/IN/Vashist/ParbatinVal.	Vashist, sp	45	8.0	ppm
Yusa (2000)	ASI/JP/Beppu/Kyushu	BGRL (50m), well	21	7.3	mg/l
Yusa (2000)	ASI/JP/Beppu/Kyushu	BGRL (200m), well	126	7.2	mg/l
Yusa (2000)	ASI/JP/Beppu/Kyushu	BGRL (250m), well	148	8.8	mg/l
Yusa (2000)	ASI/JP/Beppu/Kyushu	BGRL (300m), well	102	8.4	mg/l
Noda (1993)	ASI/JP/Hohi/Kyushu	DB-7, well	13	7.7	mg/l
Noda (1993)	ASI/JP/Hohi/Kyushu	DB-9, well		7.6	mg/l
Noda (1993)	ASI/JP/Hohi/Kyushu	DY-1, well	117	8.8	mg/l
Noda (1993)	ASI/JP/Hohi/Kyushu	DY-5, well		8.4	mg/l
Noda (1993)	ASI/JP/Hohi/Kyushu	DY-6, well		8.0	mg/l
Abe (1993)	ASI/JP/Onikobe	103, well		7.3	mg/l
Abe (1993)	ASI/JP/Onikobe	111, well		8.0	mg/l
Abe (1993)	ASI/JP/Onikobe	127, well		5.6	mg/l
Abe (1993)	ASI/JP/Onikobe	129, well		8.1	mg/l
Abe (1993)	ASI/JP/Onikobe	130, well		3.4	mg/l
Abe (1993)	ASI/JP/Onikobe	131, well		3.4	mg/l
Abe (1993)	ASI/JP/Onikobe	133, well		3.0	mg/l
Abe (1993)	ASI/JP/Onikobe	105, well		3.3	mg/l
Abe (1993)	ASI/JP/Onikobe	GO-10, well		3.3	mg/l
Abe (1993)	ASI/JP/Onikobe	GO-11, well		2.6	mg/l
Goko (2000)	ASI/JP/WestKirishima/Kyushu	N56-K T-5, well		2.9	mg/l
Goko (2000)	ASI/JP/WestKirishima/Kyushu	KE1-2, well		9.2	mg/l
Goko (2000)	ASI/JP/WestKirishima/Kyushu	KE1-3, well		8.5	mg/l
Goko (2000)	ASI/JP/WestKirishima/Kyushu	KE1-4, well		8.8	mg/l
Goko (2000)	ASI/JP/WestKirishima/Kyushu	KE1-6, well		8.3	mg/l
Goko (2000)	ASI/JP/WestKirishima/Kyushu	KE1-9, well		2.8	mg/l
Goko (2000)	ASI/JP/WestKirishima/Kyushu	KE1-11, well		2.4	mg/l
Goko (2000)	ASI/JP/WestKirishima/Kyushu	KE1-19S, well		8.7	mg/l
Goko (2000)	ASI/JP/WestKirishima/Kyushu	KE1-21, well		7.9	mg/l
Goko (2000)	ASI/JP/WestKirishima/Kyushu	KE1-23, well		8.3	mg/l
Goko (2000)	ASI/JP/WestKirishima/Kyushu	NT-A1, well		8.7	mg/l
Goko (2000)	ASI/JP/WestKirishima/Kyushu	NT-A2, well		8.8	mg/l
Goko (2000)	ASI/JP/WestKirishima/Kyushu	NT-A3, well		8.8	mg/l
Goko (2000)	ASI/JP/WestKirishima/Kyushu	NT-A4, well		8.7	mg/l
Goko (2000)	ASI/JP/WestKirishima/Kyushu	NT-B1, well		8.8	mg/l
Goko (2000)	ASI/JP/WestKirishima/Kyushu	NT-B2, well		8.8	mg/l
Goko (2000)	ASI/JP/WestKirishima/Kyushu	NT-B3, well		8.7	mg/l

Na	K	Mg	Ca	Cl	SO4	HCO3	CO3	CO2	SIO2
16.00	6.00	17.00	55.00	7.00	41.00	258.00			45.00
40.00	12.00	6.00	36.00	53.00	28.00	129.00			60.00
201.00	13.00	0.70	19.00	206.00	24.00	303.00			98.00
111.00	8.90	3.00	25.00	50.00	73.00	200.00			48.00
95.00	7.40	7.90	50.00	50.00	121.00	171.00			35.00
413.00	7.00	0.90	490.00	1490.00	136.00	7.00			41.00
580.00	80.00	1.00	5.00	464.00	127.00	884.00			160.00
580.00	80.00	1.30	11.00	411.00	100.00	903.00			175.00
618.00	81.00	0.20	2.00	447.00	156.00	704.00			207.00
600.00	90.00	1.50	5.00	468.00	172.00	799.00			237.00
1044.00	68.00	7.20	21.00	1135.00	43.00	1020.00			68.00
100.00	10.00	6.00	30.00	60.00	35.00	220.00			114.00
90.00	8.00	7.00	32.00	52.00	30.00	218.00			110.00
92.00	7.00	10.00	40.00	52.00	57.00	232.00			105.00
94.00	8.00	9.00	38.00	58.00	52.00	225.00			102.00
12.00	6.00	26.00	90.00	15.00	27.00	363.00			91.00
15.00	9.00	25.00	95.00	15.00	30.00	380.00			84.00
135.00	7.50	0.60	5.00	67.00	85.00	132.00			142.00
130.00	6.20	0.40	7.50	50.00	105.00	150.00			131.00
140.00	8.00	1.20	10.00	67.00	105.00	150.00			128.00
138.00	8.00	0.80	6.00	71.00	80.00	142.00			140.00
285.00	32.00	1.60	152.00	704.00	155.00				55.00
174.00	9.00	0.50	8.00	123.00	73.00	207.00			69.00
32.30	3.60	10.50	20.50	17.50	34.00	127.00			99.00
154.00	16.80	48.20	65.40	65.70	84.00	714.00			191.00
380.00	23.00	7.10	1.40	282.00	166.00	390.00			353.00
229.00	17.90	6.10	5.80	64.60	85.00	487.00			246.00
5.40	2.20	3.00	11.70	3.00	1.50	56.90			50.20
170.00	48.10	5.73	45.90	0.30	181.00	460.00			150.00
625.00	47.20	0.05	40.70	998.00	67.50	19.00			283.00
1060.00	134.00	0.01	8.80	1630.00	58.60	149.00			663.00
995.00	85.80	0.34	39.30	1600.00	230.00	27.00			597.00
1400.00	250.00	2.70	290.00	2780.00	75.00	25.00			360.00
1300.00	210.00	4.60	320.00	2780.00	77.00	49.00			380.00
1700.00	340.00	2.60	340.00	3630.00	13.00				720.00
1900.00	310.00	0.81	310.00	3490.00	17.00	61.00			650.00
1900.00	440.00	33.00	450.00	4910.00	19.00				690.00
1400.00	360.00	34.00	470.00	4480.00	19.00				660.00
2000.00	430.00	63.00	650.00	4600.00	33.00				610.00
1950.00	439.00	48.40	408.00	5630.00	31.00	28.00			515.00
1787.50	503.50	137.10	408.40	5286.00	55.20				510.00
3120.00	810.00	270.00	1460.00	9570.00	201.00				144.00
772.00	161.00	2.40	3.90	1190.00	404.00				1008.00
200.00	17.00	0.01	0.30	32.00	194.00	244.00		191.40	422.00
420.00	55.60	0.01	8.30	660.00	145.00	31.50		23.30	567.00
417.00	61.00	0.01	9.70	571.00	156.00	4.00		3.00	574.00
653.00	78.30	0.01	19.50	1075.00	80.00	2.50		1.90	508.00
842.00	196.00	2.10	10.80	1205.00	375.00			2.60	1023.00
765.00	167.00	4.80	17.40	980.00	697.00			1.80	1036.00
456.00	76.30	0.01	16.00	594.00	190.00	7.00		5.00	609.00
702.00	77.90	0.01	40.70	1198.00	188.00	7.00		6.00	602.00
439.00	53.40	0.01	19.90	582.00	228.00	14.00		10.00	544.00
441.00	53.40	0.01	11.00	596.00	198.00	16.00		10.00	628.00
452.00	63.00	0.01	10.00	640.00	194.00	12.00		9.00	603.00
478.00	64.60	0.01	9.70	633.00	195.00	16.00		11.00	614.00
478.00	61.30	0.01	11.50	635.00	201.00	9.00		7.00	571.00
466.00	64.80	0.01	9.70	654.00	194.00	13.00		10.00	615.00
483.00	64.90	0.01	10.10	633.00	193.00	12.00		9.00	611.00
475.00	64.30	0.01	9.90	626.00	194.00	12.00		9.00	602.00

Reference	Location	Site	T	pH	unit
Goko (2000)	ASI/JP/WestKirishima/Kyushu	NT-B4, well		8.6	mg/l
Goko (2000)	ASI/JP/WestKirishima/Kyushu	NT-C1, well		8.5	mg/l
Goko (2000)	ASI/JP/WestKirishima/Kyushu	NT-C2, well		8.7	mg/l
Reyes (1993)	ASI/PH/AltoPeak/Leytepr.	Well AP-1D, well		6.6	ppm
Reyes (1993)	ASI/PH/AltoPeak/Leytepr.	Well AP-1D, well		6.9	ppm
Reyes (1993)	ASI/PH/AltoPeak/Leytepr.	Well AP-1D, well		6.8	ppm
Reyes (1993)	ASI/PH/AltoPeak/Leytepr.	Well AP-1D, well		7.0	ppm
Reyes (1993)	ASI/PH/AltoPeak/Leytepr.	Well AP-1D, well		7.0	ppm
Reyes (1993)	ASI/PH/AltoPeak/Leytepr.	Well AP-1D, well		7.6	ppm
Reyes (1993)	ASI/PH/AltoPeak/Leytepr.	Well AP-1D, well		7.5	ppm
Reyes (1993)	ASI/PH/AltoPeak/Leytepr.	Well AP-1D, well		7.3	ppm
Reyes (1993)	ASI/PH/AltoPeak/Leytepr.	Well AP-1D, well		7.3	ppm
Reyes (1993)	ASI/PH/AltoPeak/Leytepr.	Well AP-2D, well		6.6	ppm
Reyes (1993)	ASI/PH/AltoPeak/Leytepr.	Well AP-2D, well		6.9	ppm
Reyes (1993)	ASI/PH/AltoPeak/Leytepr.	Well AP-2D, well		7.0	ppm
Reyes (1993)	ASI/PH/AltoPeak/Leytepr.	Well AP-2D, well		6.6	ppm
Reyes (1993)	ASI/PH/AltoPeak/Leytepr.	Well AP-2D, well		6.5	ppm
Reyes (1993)	ASI/PH/AltoPeak/Leytepr.	Well AP-2D, well		6.6	ppm
Reyes (1993)	ASI/PH/AltoPeak/Leytepr.	Well AP-2D, well		7.2	ppm
Reyes (1993)	ASI/PH/AltoPeak/Leytepr.	Well AP-2D, well		6.5	ppm
Reyes (1993)	ASI/PH/AltoPeak/Leytepr.	Well AP-2D, well		7.0	ppm
Reyes (1993)	ASI/PH/AltoPeak/Leytepr.	Well AP-4D, well		7.1	ppm
Reyes (1993)	ASI/PH/AltoPeak/Leytepr.	Well AP-4D, well			ppm
Reyes (1993)	ASI/PH/AltoPeak/Leytepr.	Well AP-4D, well		7.4	ppm
Reyes (1993)	ASI/PH/AltoPeak/Leytepr.	Well AP-4D, well		7.0	ppm
Reyes (1993)	ASI/PH/AltoPeak/Leytepr.	Well AP-4D, well		7.1	ppm
Reyes (1993)	ASI/PH/AltoPeak/Leytepr.	Well AP-4D, well		7.6	ppm
Reyes (1993)	ASI/PH/AltoPeak/Leytepr.	Well AP-5D, well		6.9	ppm
Reyes (1993)	ASI/PH/AltoPeak/Leytepr.	Well AP-5D, well		7.0	ppm
Reyes (1993)	ASI/PH/AltoPeak/Leytepr.	Well AP-5D, well		7.8	ppm
Lawless (1983)	ASI/PH/Bacon-Manito/Luzon	Pawa, sp	66	6.6	ppm
Lawless (1983)	ASI/PH/Bacon-Manito/Luzon	CN-1, well		7.7	ppm
Lawless (1983)	ASI/PH/Bacon-Manito/Luzon	PAL-2D, well		7.0	ppm
Lawless (1983)	ASI/PH/Bacon-Manito/Luzon	MO-1, well		3.1	ppm
Lawless (1983)	ASI/PH/Bacon-Manito/Luzon	CN-2D, well		3.6	ppm
Balmes (2000)	ASI/PH/Mt.BalutIs./DavaodelS	G97-DVS-05w, well	75	2.7	ppm
Balmes (2000)	ASI/PH/Mt.BalutIs./DavaodelS	G97-DVS-06w, well	38	6.9	ppm
Balmes (2000)	ASI/PH/Mt.BalutIs./DavaodelS	G97-DVS-08w, well	74	2.8	ppm
Balmes (2000)	ASI/PH/Mt.BalutIs./DavaodelS	G97-DVS-09w, well	29	7.3	ppm
Balmes (2000)	ASI/PH/Mt.BalutIs./DavaodelS	G97-DVS-10w, well	39	6.7	ppm
Chaturongkawa. (2000)	ASI/TH/BanPornRang/Changw.	RN3, sp, well	55	8.4	mg/l
Chaturongkawa. (2000)	ASI/TH/BanThungYo/Changw.	RN2, sp, well	40	8.3	mg/l
Praserdvigai (1987)	ASI/TH/Fang	F-1, well	99	9.0	ppm
Praserdvigai (1987)	ASI/TH/Fang	FGTE-3, well	98		ppm
Praserdvigai (1987)	ASI/TH/Fang	FGTE-5, well	105		ppm
Praserdvigai (1987)	ASI/TH/SanKampaeng	SK-1, well	98	8.9	ppm
Praserdvigai (1987)	ASI/TH/SanKampaeng	SK-2, well	100	8.9	ppm
Hochstein (1987)	ASI/TH/SanKampaeng	SP5, sp, well	97	8.1	ppm
Hochstein (1987)	ASI/TH/SanKampaeng	GTE, well	120	8.5	ppm
Gianelli (1997)	ASI/VN/Bagoi	Bagoi, well	42	8.1	mg/l
Gianelli (1997)	ASI/VN/BinhChau	Binh Chau, well	81	6.8	mg/l
Gianelli (1997)	ASI/VN/BinhChau	Binh Chau, well	66	6.6	mg/l
Gianelli (1997)	ASI/VN/FuocLong	Fuoc Long, sp	54	8.6	mg/l
Gianelli (1997)	ASI/VN/NghiaTang	Nghia Tang, sp	67	7.5	mg/l
Gianelli (1997)	ASI/VN/NinhHoa	Ninh Hoa, sp	67	8.9	mg/l
Gianelli (1997)	ASI/VN/PhuNinh	Phu Ninh, sp	65	6.5	mg/l
Gianelli (1997)	ASI/VN/ThachBich	Thach Bich, sp	65	7.8	mg/l
Gianelli (1997)	ASI/VN/TriemDuc	Triem Duc, sp	74	7.9	mg/l
Gianelli (1997)	ASI/VN/TuBong	Tu Bong, well	66	7.1	mg/l

Na	K	Mg	Ca	Cl	SO4	HCO3	CO3	CO2	SIO2
466.00	64.00	0.01	9.40	633.00	194.00	10.00		8.00	611.00
478.00	69.40	0.01	8.30	660.00	194.00	2.10		1.50	647.00
448.00	69.60	0.01	15.50	583.00	188.00	9.00		7.00	612.00
2071.00	381.00	5.00	94.00	3734.00	29.00	234.00		949.00	555.00
2108.00	419.00	4.00	88.00	3741.00	20.00	260.00		875.00	581.00
2245.00	456.00	3.30	98.00	4226.00	20.00	264.00		858.00	594.00
2857.00	610.00	2.00	168.00	5307.00	13.00	234.00		828.00	650.00
1734.00	298.00	2.90	100.00	3652.00	22.00			932.00	489.00
2282.00	428.00	3.30	194.00	4592.00	33.00	156.00		943.00	468.00
4304.00	807.00	4.90	420.00	8564.00	34.00	206.00		912.00	642.00
6335.00	1216.00	5.00	685.00	12515.00	36.00	137.00		928.00	663.00
4466.00	867.00	2.90	452.00	8761.00	49.00	176.00		978.00	647.00
1049.00	298.00	1.73	18.00	1958.00	22.00	7.00		933.00	782.00
1371.00	312.00	0.24	9.00	2472.00	31.00	143.00		972.00	913.00
1421.00	305.00	0.08	11.00	2507.00	53.00	120.00		945.00	969.00
1480.00	340.00	0.88	17.00	2625.00	11.00	108.00		946.00	1004.00
1450.00	338.00	0.14	17.00	2723.00	5.00	109.00		950.00	973.00
1560.00	375.00	0.24	21.00	2813.00	3.00	102.00		959.00	1018.00
1540.00	340.00	0.04	10.00	2787.00	11.00	118.00		939.00	990.00
1461.00	345.00	0.14	15.00	2756.00	8.00	118.00		969.00	987.00
1490.00	356.00	0.07	16.00	2752.00	7.00	134.00			1024.00
3056.00	373.00	0.82	238.00	5745.00	41.00			950.00	546.00
3104.00	325.00	0.88	238.00	5000.00	38.00	66.00		972.00	537.00
3795.00	396.00	1.05	288.00	6525.00	42.00	47.00		945.00	619.00
3444.00	364.00	0.96	261.00	6028.00	36.00	70.00		963.00	564.00
3519.00	367.00	0.90	269.00	6214.00	35.00	48.00		958.00	580.00
3492.00	364.00	0.90	162.00	6124.00	35.00	57.00		961.00	560.00
3117.00	318.00	0.85	82.00	5106.00	183.00	174.00		931.00	467.00
3203.00	330.00	0.74	67.00	5207.00	149.00	156.00		934.00	472.00
3268.00	318.00	0.68	67.00	5150.00	153.00	129.00		933.00	471.00
814.00	147.00	8.90	68.00	1418.00	37.00	118.00			182.00
4486.00	914.00	0.07	172.00	8668.00	22.50	2.29		43.00	797.00
4975.00	990.00	5.20	85.00	8727.00	124.00	171.00		2583.00	912.00
4422.00	638.00	6.40	251.00	8125.00	226.00			23.00	466.00
3096.00	242.00	12.70	8.00	4261.00	1435.00			3568.00	620.00
164.00	52.00	23.50	69.40	279.00	471.00				230.00
118.00	11.90	28.60	41.00	123.00	26.00	362.00			116.00
124.00	33.60	20.90	101.00	224.00	564.00				224.00
10500.00	344.00	1270.00	379.00	18600.00	2620.00	122.00			0.80
159.00	19.20	54.10	111.00	216.00	66.90	606.00			134.00
46.90	3.00	0.01	44.30	10.00	44.90	190.00			72.00
46.40	3.20	0.03	44.10	11.00	44.90	189.00			75.50
122.00	7.90	0.04	0.72	23.00	44.00	185.00		75.03	191.00
100.00	6.00	0.10	3.40	21.90	29.00	146.00	36.00		
110.00	7.20	0.10	3.20	27.30	39.00	159.00	24.00	24.20	
151.00	13.50	0.18	2.24	26.00	35.00	312.00			148.00
151.00	14.50	0.12	0.70	20.00	40.00	278.00			150.00
164.00	13.30	0.10	1.20	40.00	12.00	328.00			148.00
166.00	14.60	0.03	1.40	27.00	8.00	387.00			180.00
175.00	7.00	0.15	5.30	153.00	33.00	151.30			67.50
965.00	38.50	1.40	208.00	1606.00	375.00	100.70			111.00
998.00	39.00	1.43	200.00	1535.00	357.00	88.50			125.00
112.00	3.50	0.02	0.80	44.50	29.30	149.50			87.50
161.00	6.40	0.03	2.90	167.00	27.50	140.00			90.00
72.00	2.50	0.08	0.70	14.30	28.30	96.40			84.00
520.00	14.50	4.40	216.00	1035.00	115.00	67.00			68.00
117.00	5.50	0.06	2.45	88.90	15.60	169.60			95.00
138.00	3.00	0.02	1.30	35.40	24.00	241.00			68.00
188.00	7.30	0.06	5.20	225.00	38.20	131.20			93.30

Reference	Location	Site	T	pH	unit
Gianelli (1997)	ASI/VN/VinhHao	Vinh Hao, well	36	6.9	mg/l
Gianelli (1997)	ASI/VN/VinhHao	Vinh Hao, well	33	6.6	mg/l
Gianelli (1997)	ASI/VN/VinhThinh	Vinh Thinh, sp	70	8.0	mg/l
Georgieva (2000)	EUR/BG/Medovo/BourgasB.	B-22, well		7.4	ppm
Georgieva (2000)	EUR/BG/Medovo/BourgasB.	B-25, well		7.0	ppm
Georgieva (2000)	EUR/BG/SunnyBeach/BourgasB.	B-1, well		7.5	ppm
Georgieva (2000)	EUR/BG/SunnyBeach/BourgasB.	B-76, well		7.4	ppm
Georgieva (2000)	EUR/BG/SunnyBeach/BourgasB.	B-77, well		7.4	ppm
Georgieva (2000)	EUR/BG/SunnyBeach/BourgasB.	B-79, well		7.4	ppm
Lenz (1997)	EUR/DE/Hamburg	Allermöhe, bh	123	6.4	mg/l
Fritz (1989)	EUR/DE/Hamburg	GB2,bh	102	6.0	mg/l
Bartels (2000)	EUR/DE/Karlshagen	GtKhn1/88, bh		5.2	mg/l
Bartels (2000)	EUR/DE/Karlshagen	GtKhn2/87, bh		5.2	mg/l
Kühn (1997)	EUR/DE/Neubrandenburg	NB1, well	51	6.1	mg/l
Kühn (1997)	EUR/DE/Neubrandenburg	NB3, well	50	6.1	mg/l
Merkel (1991)	EUR/DE/Neubrandenburg	N1GT86, well	53	6.0	mg/l
Kühn (1997)	EUR/DE/Neustadt-Glewe	NG1, well	92	5.2	mg/l
Kühn (1997)	EUR/DE/Neustadt-Glewe	NG2, well	70	5.2	mg/l
Bartels (2000)	EUR/DE/Stralsund	GtSs1/85, bh	58	5.5	mg/l
Bartels (2000)	EUR/DE/Stralsund	GtSs2/85, bh	30	5.2	mg/l
Bartels (2000)	EUR/DE/Stralsund	GtSs6/89, bh	22	6.0	mg/l
Merkel (1991)	EUR/DE/Waren	GTWA1/89, well	60	6.0	mg/l
Adams (1996)	EUR/GB/AscensionIsland	#1(ws87-7), well		5.5	ppm
Adams (1996)	EUR/GB/AscensionIsland	#1(ws87-20), well		7.1	ppm
Adams (1996)	EUR/GB/AscensionIsland	#1(ws87-22), well		6.2	ppm
Adams (1996)	EUR/GB/AscensionIsland	#1(ws87-27), well		8.1	ppm
Adams (1996)	EUR/GB/AscensionIsland	#1(ws87-28), well		6.9	ppm
Traganos (1995)	EUR/GR/Langada/MygdoniaB.	12, well	38	7.0	mg/l
Traganos (1995)	EUR/GR/Langada/MygdoniaB.	14, well	37	7.0	mg/l
Traganos (1995)	EUR/GR/Langada/MygdoniaB.	16, well	33	7.2	mg/l
Traganos (1995)	EUR/GR/Langada/MygdoniaB.	17, well	34	7.1	mg/l
Grassi (1996)	EUR/GR/NeaKessani	W1, well	46	6.7	mg/l
Grassi (1996)	EUR/GR/NeaKessani	W2, well	38	6.5	mg/l
Grassi (1996)	EUR/GR/NeaKessani	W3, well	76	7.2	mg/l
Grassi (1996)	EUR/GR/NeaKessani	W4, well	64	6.7	mg/l
Grassi (1996)	EUR/GR/NeaKessani	W12, well	67	7.0	mg/l
Grassi (1996)	EUR/GR/NeaKessani	W21, well	72	6.8	mg/l
Grassi (1996)	EUR/GR/NeaKessani	W22, well	78	7.4	mg/l
Grassi (1996)	EUR/GR/NeaKessani	W23, well	39	6.6	mg/l
Grassi (1996)	EUR/GR/NeaKessani	S31, sp	44	7.0	mg/l
Grassi (1996)	EUR/GR/NeaKessani	S32, sp	54	7.0	mg/l
Grassi (1996)	EUR/GR/NeaKessani	S33, sp	53	7.0	mg/l
Grassi (1996)	EUR/GR/NeaKessani	S34, sp	39	7.0	mg/l
Grassi (1996)	EUR/GR/NeaKessani	S36, sp	42	7.0	mg/l
Grassi (1996)	EUR/GR/NeaKessani	S37, sp	42	7.1	mg/l
Grassi (1996)	EUR/GR/NeaKessani	S38, sp	40	7.1	mg/l
Grassi (1996)	EUR/GR/NeaKessani	S40, sp	42	7.0	mg/l
Grassi (1996)	EUR/GR/NeaKessani	S41, sp	37	7.0	mg/l
Grassi (1996)	EUR/GR/NeaKessani	S42, sp	39	7.0	mg/l
Grassi (1996)	EUR/GR/NeaKessani	S43, sp	40	7.0	mg/l
Grassi (1996)	EUR/GR/NeaKessani	S44, sp	41	7.0	mg/l
Grassi (1996)	EUR/GR/NeaKessani	S45, sp	30	7.0	mg/l
Grassi (1996)	EUR/GR/NeaKessani	S46, sp	35	7.0	mg/l
Grassi (1996)	EUR/GR/NeaKessani	S47, sp	42	7.2	mg/l
Grassi (1996)	EUR/GR/NeaKessani	S48, sp	52	7.1	mg/l
Szita (2000)	EUR/HU/Szarvas	gemittelt		7.6	mg/l
Arnórsson (1983)	EUR/IS/Bakki	Bakki, well1, well	134	8.4	ppm
Kristmannsdottir (1989)	EUR/IS/Bakki	h-1, well	116	8.7	ppm
Arnórsson (1983)	EUR/IS/Gjögur	Gjögur, sp	72	7.1	ppm

Na	K	Mg	Ca	Cl	SO4	HCO3	CO3	CO2	SIO2
922.00	34.00	9.50	28.70	30.00	0.45	2540.70			64.50
868.00	30.00	10.10	30.00	29.50	1.60	2556.00			70.50
128.00	3.60	0.03	1.00	37.80	38.80	207.40			86.00
850.00	1.70	0.60	39.00	1320.00	18.00			53.00	55.00
904.00	5.10	1.00	32.00	1423.00	20.00			22.00	42.00
4760.00	36.00	33.00	1435.00	9847.00	6.00			35.00	22.00
4840.00	31.00	106.00	1090.00	9750.00	14.00			29.00	15.00
2240.00	25.00	660.00	1745.00	8037.00	20.00			37.00	8.50
4460.00	27.00	70.00	755.00	8794.00	4.10			37.00	6.20
75000.00	1250.00	1300.00	6690.00	132200.00	465.00	240.00			6.00
33318.00	3264.00	442.00	7581.00	73323.00	220.00	473.00			63.98
85000.00	660.00	2520.00	17400.00	169000.00	400.00	35.00			20.00
88200.00	750.00	2710.00	18100.00	175700.00	500.00	35.00			20.00
48900.00	181.00	638.00	2080.00	80777.00	1000.00	181.17			12.72
48800.00	179.00	637.00	2150.00	81669.00	1010.00	174.46			12.25
48850.00	243.00	660.00	2080.00	80525.00	960.00	53.00			15.15
72300.00	792.00	1380.00	8840.00	131700.00	481.00	118.34			31.42
73000.00	864.00	1400.00	8612.00	131100.00	463.00	120.17			31.42
92700.00	680.00	2350.00	14350.00	175000.00	400.00	50.00			20.00
95000.00	700.00	2430.00	14700.00	179300.00	400.00	70.00			20.00
93000.00	690.00	2300.00	14450.00	175500.00	400.00	50.00			20.00
57650.00	264.00	780.00	2730.00	95615.00	900.00	40.00			16.92
18886.00	1200.00	24.50	10428.00	47100.00	411.00	60.00			242.00
2368.00	180.00	6.00	1271.00	5770.00	56.00	268.00			265.00
7261.00	497.00	17.70	3676.00	17650.00	134.00	102.00			344.00
11845.00	424.00	1419.00	465.00	20100.00	2865.00	144.00			1.80
10898.00	1089.00	131.00	2497.00	22800.00	145.00	238.00			179.00
62.07	3.13	30.14	88.00	19.50	151.77	331.00		30.80	23.30
152.65	7.04	48.13	128.80	27.30	490.90	331.01		44.00	5.30
151.73	5.47	46.68	144.00	21.27	658.01	278.75		22.00	19.00
167.83	4.69	33.74	116.80	21.27	543.70	326.66		44.00	20.80
1483.00	156.00	30.00	228.00	1675.00	229.00	1870.00			18.40
1472.00	156.00	29.00	176.00	1640.00	233.00	1712.00			26.00
1455.00	146.00	20.00	127.00	1595.00	233.00	1553.00			54.30
1455.00	146.00	16.00	134.00	1595.00	233.00	1553.00			48.60
1481.00	195.00	25.00	135.00	1702.00	226.00	1595.00			44.00
1504.00	116.00	15.00	136.00	1613.00	229.00	1577.00			49.00
1546.00	131.00	35.00	130.00	1764.00	227.00	1555.00			46.00
1380.00	135.00	48.00	228.00	1539.00	216.00	1897.00			15.00
1472.00	156.00	23.00	128.00	1613.00	248.00	1547.00			51.00
1471.00	156.00	17.00	133.00	1613.00	237.00	1551.00			52.50
1472.00	156.00	19.00	132.00	1613.00	248.00	1548.00			52.00
1471.00	156.00	16.00	132.00	1613.00	244.00	1549.00			50.50
1472.00	156.00	19.00	131.00	1613.00	241.00	1548.00			49.00
1471.00	156.00	17.00	132.00	1613.00	250.00	1550.00			50.50
1509.00	156.00	15.00	128.00	1667.00	242.00	1550.00			52.00
1472.00	156.00	17.00	148.00	1613.00	231.00	1597.00			53.00
1472.00	156.00	16.00	138.00	1613.00	231.00	1565.00			55.00
1472.00	156.00	16.00	131.00	1613.00	242.00	1567.00			53.50
1472.00	156.00	15.00	140.00	1613.00	249.00	1554.00			50.50
1472.00	156.00	18.00	128.00	1613.00	236.00	1546.00			53.00
1555.00	156.00	14.00	142.00	1684.00	255.00	1638.00			57.00
1472.00	156.00	15.00	140.00	1613.00	250.00	1554.00			52.50
1472.00	156.00	17.00	127.00	1613.00	247.00	1555.00			56.50
1472.00	156.00	16.00	138.00	1613.00	254.00	1568.00			51.00
1100.00	26.00	3.30	8.00	107.00	53.00	2684.00			70.61
387.50	19.60	0.07	67.10	658.50	122.50			6.70	133.60
386.00	19.00	0.02	74.00	634.00	121.00			8.00	133.00
715.70	17.30	3.68	759.40	2460.00	297.60			13.30	49.00

Reference	Location	Site	T	pH	unit
Arnórsson (1983)	EUR/IS/Húsatóttir	Húsatóttir, well4, well	70	8.9	ppm
Arnórsson (1983)	EUR/IS/Laugarbakkar	Laugarbakkar, well1, well	87	9.1	ppm
Kristmannsdottir (1989)	EUR/IS/Reykjanes	Rn-9 a, well			ppm
Kristmannsdottir (1989)	EUR/IS/Reykjanes	Rn-9 b, well			ppm
Arnórsson (1983)	EUR/IS/ReykjanesFsafjardard.	Reykjanes, Fsafjardard.,	84	9.1	ppm
Arnórsson (1983)	EUR/IS/Selfoss	Selfoss, well9, well	70	8.5	ppm
Kristmannsdottir (1989)	EUR/IS/Selfoss	H-10, well	84	8.6	ppm
Arnórsson (1983)	EUR/IS/Seltjarnarnes	Seltjarnarnes, well4, wel	144	8.6	ppm
Kristmannsdottir (1989)	EUR/IS/Seltjnes	Sn-6, well	117	8.4	ppm
Kristmannsdottir (1989)	EUR/IS/StadurReykjanes	H-2, well	71	7.3	ppm
Arnórsson (1983)	EUR/IS/Thjórsárdalur	Thjórsárdalur, sp	70	9.0	ppm
Bortolami (1983)	EUR/IT/AcquiTerme/AcquaTi.	Acqua Tiepida, sp	38	7.9	mg/l
Bortolami (1983)	EUR/IT/AcquiTerme/AcquaTi.	Acqua Tiepida, sp	32	7.6	mg/l
Bortolami (1983)	EUR/IT/AcquiTerme/Fornace	Fornace, well	34	8.6	mg/l
Bortolami (1983)	EUR/IT/AcquiTerme/Fornace	Fornace, well	33	8.7	mg/l
Bortolami (1983)	EUR/IT/AcquiTerme/Fornace	Fornace, well	33	8.7	mg/l
Bortolami (1983)	EUR/IT/AcquiTerme/LaBolle.	La Bollente, sp	68	8.3	mg/l
Bortolami (1983)	EUR/IT/AcquiTerme/LaBolle.	La Bollente, sp	67	8.3	mg/l
Bortolami (1983)	EUR/IT/AcquiTerme/LaBolle.	La Bollente, sp	66	8.3	mg/l
Bortolami (1983)	EUR/IT/AcquiTerme/LaBolle.	La Bollente, sp	65	8.4	mg/l
Bortolami (1983)	EUR/IT/AcquiTerme/Savoia	Savoia, well	39	8.6	mg/l
Bortolami (1983)	EUR/IT/AcquiTerme/Savoia	Savoia, well	40	8.6	mg/l
Bortolami (1983)	EUR/IT/AcquiTerme/Savoia	Savoia, well	39	8.7	mg/l
Bortolami (1983)	EUR/IT/AcquiTerme/VascaRot.	Vasca Rotonda, sp	45	8.2	mg/l
Bortolami (1983)	EUR/IT/AcquiTerme/VascaRot.	Vasca Rotonda, sp	45	8.1	mg/l
Bortolami (1983)	EUR/IT/AcquiTerme/VascaRot.	Vasca Rotonda, sp	45	8.0	mg/l
Marini (2000)	EUR/IT/AcquiTerme/Visone	1, sp	26	7.8	ppm
Marini (2000)	EUR/IT/AcquiTerme/Visone	2, sp	15	6.9	ppm
Marini (2000)	EUR/IT/AcquiTerme/Visone	3, sp	15	7.7	ppm
Marini (2000)	EUR/IT/AcquiTerme/Visone	4, sp	12	7.5	ppm
Marini (2000)	EUR/IT/AcquiTerme/Visone	5, sp	22	7.5	ppm
Marini (2000)	EUR/IT/AcquiTerme/Visone	6, sp	39	7.7	ppm
Marini (2000)	EUR/IT/AcquiTerme/Visone	7, sp	35	8.3	ppm
Marini (2000)	EUR/IT/AcquiTerme/Visone	8, sp	70	8.0	ppm
Marini (2000)	EUR/IT/AcquiTerme/Visone	9, sp	14	7.1	ppm
Marini (2000)	EUR/IT/AcquiTerme/Visone	10, sp	12	6.9	ppm
Marini (2000)	EUR/IT/AcquiTerme/Visone	11, sp	11	7.5	ppm
Marini (2000)	EUR/IT/AcquiTerme/Visone	12, sp	13	7.3	ppm
Marini (2000)	EUR/IT/AcquiTerme/Visone	13, sp	10	7.3	ppm
Marini (2000)	EUR/IT/AcquiTerme/Visone	14, sp	11	7.5	ppm
Marini (2000)	EUR/IT/AcquiTerme/Visone	15, sp	11	7.1	ppm
Marini (2000)	EUR/IT/AcquiTerme/Visone	16, sp	12	7.8	ppm
Marini (2000)	EUR/IT/AcquiTerme/Visone	17, sp	10	7.7	ppm
Marini (2000)	EUR/IT/AcquiTerme/Visone	18, sp	8	7.3	ppm
Marini (2000)	EUR/IT/AcquiTerme/Visone	19, sp	9	8.2	ppm
Marini (2000)	EUR/IT/AcquiTerme/Visone	20, sp	8	7.7	ppm
Marini (2000)	EUR/IT/AcquiTerme/Visone	21, sp	11	7.4	ppm
Marini (2000)	EUR/IT/AcquiTerme/Visone	22, sp	9	7.2	ppm
Marini (2000)	EUR/IT/AcquiTerme/Visone	24, sp	9	6.9	ppm
Marini (2000)	EUR/IT/AcquiTerme/Visone	25, sp	9	7.6	ppm
Marini (2000)	EUR/IT/AcquiTerme/Visone	26, sp	15	7.5	ppm
Marini (2000)	EUR/IT/AcquiTerme/Visone	27, sp	13	7.4	ppm
Marini (2000)	EUR/IT/AcquiTerme/Visone	28, sp	17	7.5	ppm
Marini (2000)	EUR/IT/AcquiTerme/Visone	29, sp	19	7.6	ppm
Marini (2000)	EUR/IT/AcquiTerme/Visone	30, sp	12	7.4	ppm
Marini (2000)	EUR/IT/AcquiTerme/Visone	31, sp	30	7.4	ppm
Marini (2000)	EUR/IT/AcquiTerme/Visone	32, sp	11	7.4	ppm
Marini (2000)	EUR/IT/AcquiTerme/Visone	33, sp	11	7.1	ppm
Marini (2000)	EUR/IT/AcquiTerme/Visone	34, sp	11	7.2	ppm

Na	K	Mg	Ca	Cl	SO4	HCO3	CO3	CO2	SIO2
136.10	2.88	0.05	11.30	171.70	67.60			22.00	71.80
141.40	4.20	0.07	23.70	139.50	144.90			27.40	102.30
9079.00	1388.00	0.93	1526.00	17749.00	16.20			1523.00	587.00
12564.00	1920.00	1.28	2112.00	24558.00	22.40			14.20	812.00
184.20	4.34	0.09	67.60	390.20	42.10			9.00	70.90
159.60	4.58	0.15	30.30	246.70	53.50			26.00	62.20
158.00	5.60	0.09	24.00	226.00	57.00			24.00	76.00
355.50	10.00	0.10	139.00	670.80	204.80			6.50	112.10
597.00	14.00	0.38	523.00	1617.00	305.00			10.00	123.00
11041.00	399.00	109.00	1915.00	19880.00	1535.00			40.00	69.00
115.60	3.24	0.08	52.90	49.40	302.30			5.90	58.00
770.00	11.90	5.35	130.00	1180.00	270.00	29.30			56.70
675.00	12.40	2.43	137.00	1066.00	184.00	43.30			48.80
565.00	8.80	2.43	76.10	826.00	208.00	27.50	3.60		26.20
555.00	8.80	1.95	78.50	812.00	200.00	28.10	4.20		31.70
513.00	8.10	1.33	80.90	794.00	145.00	23.20	2.40		28.70
700.00	16.30	6.10	136.00	1170.00	188.00	39.70			54.00
765.00	13.20	4.86	128.00	1180.00	292.00	39.70	1.50		48.60
755.00	14.10	2.43	135.00	1175.00	285.00	33.60	1.50		59.30
768.00	14.30	2.13	135.00	1177.00	210.00	23.20	2.40		52.60
755.00	8.60	3.40	120.00	1110.00	289.00	49.40	2.40		25.10
651.00	6.56	2.92	122.00	1060.00	255.00	19.50	1.80		29.30
700.00	7.17	2.31	126.00	1069.00	189.00	34.80	3.60		26.50
832.00	11.80	3.65	140.00	1260.00	306.00	59.80			50.90
778.00	14.30	4.38	143.00	1250.00	262.00	35.40			57.30
785.00	13.50	1.94	146.00	1213.00	196.00	25.60			53.30
435.00	6.75	4.87	35.20	609.00	76.10	257.00			25.60
19.40	4.64	65.30	623.00	26.30	1420.00	425.00			18.80
11.50	1.00	24.40	42.50	5.45	7.99	269.00			28.90
9.55	0.76	19.20	70.90	16.40	25.20	277.00			15.60
561.00	13.00	4.03	135.00	948.00	204.00	89.00			44.00
704.00	14.80	0.21	142.00	1160.00	231.00	17.00			52.30
685.00	14.70	0.54	140.00	1180.00	228.00	27.00			48.90
665.00	14.10	0.16	135.00	1130.00	230.00	23.00			52.60
12.50	3.93	63.30	132.00	8.14	222.00	474.00			37.20
12.70	3.04	27.00	175.00	30.50	73.20	502.00			19.40
4.48	1.82	35.40	59.80	3.89	19.20	342.00			12.10
8.16	1.18	37.40	93.30	8.12	88.70	378.00			18.10
3.41	0.42	33.40	68.60	5.05	35.40	331.00			14.20
4.22	0.56	27.90	77.10	5.63	67.10	297.00			20.40
229.00	4.50	12.50	102.00	390.00	122.00	233.00			24.80
6.22	1.06	68.50	32.50	6.45	74.80	370.00			41.30
4.72	1.17	34.70	63.10	7.15	32.50	322.00			10.00
2.55	0.36	1.04	18.20	3.78	2.19	56.00			8.40
2.90	0.84	5.67	52.30	8.01	17.30	168.00			8.60
6.06	0.82	2.27	26.10	4.92	8.19	87.00			15.20
5.65	0.58	24.60	95.60	4.17	95.50	311.00			26.60
5.45	1.47	20.70	109.00	6.13	70.80	361.00			12.00
4.66	1.31	17.60	12.60	5.04	12.20	119.00			19.30
5.08	0.52	9.36	85.90	5.89	31.40	260.00			9.40
19.90	4.15	42.80	58.50	10.90	65.00	364.00			15.50
5.10	1.84	26.80	110.00	5.28	103.00	365.00			12.50
35.10	4.54	11.60	87.50	7.49	73.00	315.00			9.90
40.30	3.33	37.10	49.70	4.52	46.50	391.00			17.60
11.00	6.16	21.90	119.00	28.10	82.70	341.00			14.30
188.00	5.14	40.00	46.80	47.80	273.00	476.00			22.50
3.31	0.40	27.40	91.00	3.77	135.00	255.00			24.70
27.90	2.34	49.50	143.00	36.40	176.00	469.00			24.70
7.45	2.96	29.30	132.00	12.50	65.00	440.00			16.50

Reference	Location	Site	T	pH	unit
Marini (2000)	EUR/IT/AcquiTerme/Visone	35, sp	11	7.2	ppm
Marini (2000)	EUR/IT/AcquiTerme/Visone	36, sp	9	6.9	ppm
Marini (2000)	EUR/IT/AcquiTerme/Visone	37, sp	8	7.5	ppm
Marini (2000)	EUR/IT/AcquiTerme/Visone	38, sp	10	7.5	ppm
Marini (2000)	EUR/IT/AcquiTerme/Visone	39, sp	7	7.4	ppm
Marini (2000)	EUR/IT/AcquiTerme/Visone	40, sp	14	7.6	ppm
Marini (2000)	EUR/IT/AcquiTerme/Visone	41, sp	6	7.2	ppm
Marini (2000)	EUR/IT/AcquiTerme/Visone	42, sp	9	8.2	ppm
Marini (2000)	EUR/IT/AcquiTerme/Visone	43, sp	14	7.6	ppm
Marini (2000)	EUR/IT/AcquiTerme/Visone	44, sp	32	8.7	ppm
Marini (2000)	EUR/IT/AcquiTerme/Visone	45, sp	11	7.5	ppm
Marini (2000)	EUR/IT/AcquiTerme/Visone	46, sp	32	8.6	ppm
Marini (2000)	EUR/IT/AcquiTerme/Visone	47, sp	23	7.7	ppm
Marini (2000)	EUR/IT/AcquiTerme/Visone	48, sp	43	8.0	ppm
Marini (2000)	EUR/IT/AcquiTerme/Visone	49, sp	18	7.6	ppm
Marini (2000)	EUR/IT/AcquiTerme/Visone	50, sp	15	8.6	ppm
Marini (2000)	EUR/IT/AcquiTerme/Visone	1b, sp	26	7.8	ppm
Marini (2000)	EUR/IT/AcquiTerme/Visone	5b, sp	22	7.6	ppm
Marini (2000)	EUR/IT/AcquiTerme/Visone	6b, sp	40	7.9	ppm
Marini (2000)	EUR/IT/AcquiTerme/Visone	7b, sp	36	8.3	ppm
Marini (2000)	EUR/IT/AcquiTerme/Visone	8b, sp	70	8.2	ppm
Marini (2000)	EUR/IT/AcquiTerme/Visone	15b, sp	13	7.0	ppm
Marini (2000)	EUR/IT/AcquiTerme/Visone	44b, sp	32	8.7	ppm
De Gennaro (1984)	EUR/IT/Ischia/Campania	Giardini Eden, well	27	6.4	ppm
De Gennaro (1984)	EUR/IT/Ischia/Campania	La Gondola 1, well	90	6.5	ppm
De Gennaro (1984)	EUR/IT/Ischia/Campania	Aphrodite, well	99	7.5	ppm
De Gennaro (1984)	EUR/IT/Ischia/Campania	Acque Termominerali, well	43	6.6	ppm
De Gennaro (1984)	EUR/IT/Ischia/Campania	La Gondola 2, well	93	7.5	ppm
De Gennaro (1984)	EUR/IT/Ischia/Campania	S. Michele, well	65	7.0	ppm
De Gennaro (1984)	EUR/IT/Ischia/Campania	Castiglione, well	66	7.6	ppm
De Gennaro (1984)	EUR/IT/Ischia/Campania	Romantica, well	84	8.5	ppm
De Gennaro (1984)	EUR/IT/Ischia/Campania	S. Montano, well	30	7.4	ppm
Chiodini (1988)	EUR/IT/Mofete	3/1, well		6.8	mg/l
Chiodini (1988)	EUR/IT/Mofete	3/2, well		6.6	mg/l
Chiodini (1988)	EUR/IT/Mofete	3/3, well		6.2	mg/l
Chiodini (1988)	EUR/IT/Mofete	3/4, well		6.5	mg/l
Chiodini (1988)	EUR/IT/Mofete	3/5, well		6.5	mg/l
Chiodini (1988)	EUR/IT/Mofete	3/6, well		6.3	mg/l
Chiodini (1988)	EUR/IT/Mofete	3/7, well		6.1	mg/l
Chiodini (1988)	EUR/IT/Mofete	9/1, well		6.9	mg/l
Chiodini (1988)	EUR/IT/Mofete	9/2, well		6.9	mg/l
Chiodini (1988)	EUR/IT/Mofete	9/3, well		6.6	mg/l
Dongarra (1983)	EUR/IT/PantelleriaIs/Buvira	DX, sp	23	6.3	meq/l
Dongarra (1983)	EUR/IT/PantelleriaIs/Buvira	DX, sp	19	6.9	meq/l
Dongarra (1983)	EUR/IT/PantelleriaIs/Buvira	LM, sp	21	6.5	meq/l
Dongarra (1983)	EUR/IT/PantelleriaIs/Buvira	LM, sp	21	6.9	meq/l
Dongarra (1983)	EUR/IT/PantelleriaIs/Gadir	G, sp	53	6.2	meq/l
Dongarra (1983)	EUR/IT/PantelleriaIs/Gadir	PG, sp	32	7.3	meq/l
Dongarra (1983)	EUR/IT/PantelleriaIs/Gadir	PG, sp	28	8.0	meq/l
Dongarra (1983)	EUR/IT/PantelleriaIs/Sataria	S, sp	42	6.1	meq/l
D'Amore (1987)	EUR/IT/Sardinia	Casteldoria, sp	78	7.0	ppm
D'Amore (1987)	EUR/IT/Sardinia	Fordongianus, sp	54	8.7	ppm
D'Amore (1987)	EUR/IT/Sardinia	Benetutti, sp	41	9.3	ppm
D'Amore (1987)	EUR/IT/Sardinia	Oddini, sp	32	9.6	ppm
D'Amore (1987)	EUR/IT/Sardinia	Sardara, sp	60	6.7	ppm
Kralj (2000)	EUR/SI/MurskaSobota	Sob-1, well	49	6.9	mg/l
Kralj (2000)	EUR/SI/MurskaSobota	Sob-2, well	48	7.0	mg/l
Simsek (1985)	EUR/TR/Denizli/Sarayköy-Buld	KD 1, well			ppm
Simsek (1985)	EUR/TR/Denizli/Sarayköy-Buld	KD 2, well		8.9	ppm

Na	K	Mg	Ca	Cl	SO4	HCO3	CO3	CO2	SIO2
3.74	2.03	23.90	118.00	4.76	61.80	399.00			28.20
3.25	0.28	23.30	45.80	3.53	22.30	221.00			18.00
5.93	0.17	12.40	44.10	3.76	10.20	187.00			9.70
5.18	0.45	32.00	64.80	6.00	29.10	310.00			24.00
5.29	0.30	26.00	14.40	7.88	10.50	154.00			30.30
4.17	2.13	23.90	104.00	11.10	52.00	348.00			13.90
7.64	0.27	26.20	51.80	20.90	15.90	246.00			18.00
1.80	0.18	33.20	19.90	4.22	8.10	211.00			37.90
4620.00	25.80	196.00	425.00	8895.00	4.30	85.00			13.40
501.00	8.82	0.23	78.70	767.00	171.00	26.00			28.40
105.00	8.62	25.20	60.30	71.40	105.00	348.00			20.40
142.00	1.90	0.14	7.30	80.50	78.10	127.00			20.00
561.00	12.40	4.22	133.00	928.00	196.00	101.00			42.80
695.00	14.70	0.20	143.00	1170.00	226.00	21.00			49.80
840.00	11.50	17.50	74.20	1340.00	80.10	247.00			28.30
113.00	3.20	16.00	10.00	11.00	25.10	363.00			19.40
435.00	6.43	5.09	36.30	586.00	60.00	256.00			23.90
563.00	12.40	4.12	135.00	940.00	192.00	91.00			41.90
703.00	14.60	0.21	143.00	1175.00	223.00	19.00			48.40
686.00	14.40	0.41	140.00	1180.00	222.00	29.00			46.20
661.00	14.00	0.14	136.00	1120.00	220.00	28.00			49.60
388.00	7.15	8.53	107.00	619.00	152.00	153.00			27.40
494.00	8.78	0.22	81.80	772.00	167.00	29.00			27.40
7000.00	267.00	903.00	320.00	12900.00	1414.00			125.00	36.00
5000.00	200.00	109.00	76.00	7660.00	494.00			37.00	146.00
4390.00	160.00	29.00	72.00	6700.00	446.00			24.00	158.00
3150.00	110.00	169.00	167.00	4920.00	632.00			15.00	92.00
5080.00	182.00	11.00	79.00	7656.00	479.00			53.00	133.00
4167.00	121.00	107.00	146.00	6200.00	1093.00			86.00	90.00
7000.00	238.00	599.00	238.00	11390.00	2018.00			75.00	79.00
3200.00	69.00	153.00	48.00	5015.00	600.00				206.00
12150.00	435.00	1404.00	421.00	20810.00	4028.00			42.00	6.00
12700.00	850.00	44.00	610.00	22335.00	236.00	122.00			
10500.00	710.00	42.00	540.00	17372.00	220.00	146.00			
10600.00	720.00	40.00	540.00	18436.00	190.00	146.00			
12300.00	870.00	42.00	540.00	21272.00	210.00	122.00			
12100.00	820.00	42.00	560.00	20279.00	220.00	134.00			
11900.00	810.00	40.00	320.00	19145.00	220.00	146.00			
9900.00	670.00	40.00	500.00	15900.00	200.00	134.00			
11300.00	1100.00	55.00	880.00	19500.00	114.00	152.00			
11300.00	1060.00	52.00	760.00	20000.00	112.00	159.00			
11020.00	1110.00	60.00	930.00	18900.00	116.00	137.00			
27.00	1.00	5.10	5.20	31.60	3.30				1.00
34.80	1.10	6.50	6.40	41.60	5.00				
32.00	1.10	3.80	5.60	34.60	2.90				0.80
6.80	0.60	2.60	3.50	7.30	2.50				
122.30	4.50	11.40	2.40	122.70	8.50				
115.00	4.50	6.70	2.90	109.70	8.70				3.40
120.40	5.30	4.20	3.90	110.60	13.70				
135.00	5.40	11.20	6.40	134.50	12.10				
1288.00	67.00	1.60	584.00	2929.00	87.00	24.00			66.50
225.00	1.60	0.04	13.00	296.00	46.00				43.50
165.00	2.70	0.03	11.00	203.00	33.00	41.00			41.80
173.00	2.40	0.19	11.00	237.00	37.00	44.00			44.50
977.00	46.00	4.40	25.00	511.00	83.00	1693.00			55.00
759.00	71.00	8.00	32.00	149.00	32.00	2130.00		550.00	
719.00	44.00	4.00	10.00	122.00	5.00	2020.00		411.00	
1468.00	174.00	0.72	2.70	120.00	847.00	2183.00	360.00		
1272.00	136.00	0.48	2.30	142.00	1015.00	1974.00	72.00		

Reference	Location	Site	T	pH	unit
Simsek (1985)	EUR/TR/Denizli/Sarayköy-Buld	KD 1A, well			ppm
Simsek (1985)	EUR/TR/Denizli/Sarayköy-Buld	KD 111, well			ppm
Simsek (1985)	EUR/TR/Denizli/Sarayköy-Buld	KD 6 seper., well		8.7	ppm
Simsek (1985)	EUR/TR/Denizli/Sarayköy-Buld	KD 9, well		8.0	ppm
Simsek (1985)	EUR/TR/Denizli/Sarayköy-Buld	KD 8, well			ppm
Simsek (1985)	EUR/TR/Denizli/Sarayköy-Buld	KD 7 seper., well		8.4	ppm
Simsek (1985)	EUR/TR/Denizli/Sarayköy-Buld	KD 13, well		8.5	ppm
Simsek (1985)	EUR/TR/Denizli/Sarayköy-Buld	KD 13 seper., well		8.5	ppm
Simsek (1985)	EUR/TR/Denizli/Sarayköy-Buld	KD 14 seper., well		8.5	ppm
Simsek (1985)	EUR/TR/Denizli/Sarayköy-Buld	KD 15, well		8.2	ppm
Simsek (1985)	EUR/TR/Denizli/Sarayköy-Buld	KD 16 seper., well		8.7	ppm
Simsek (2000)	EUR/TR/Germenick/Anatolia	Bozköy, sp	60	6.8	ppm
Simsek (2000)	EUR/TR/Germenick/Anatolia	Bozköy-Confined, sp	55	7.3	ppm
Simsek (2000)	EUR/TR/Germenick/Anatolia	Gümüsköy, well	39	7.0	ppm
Simsek (2000)	EUR/TR/Germenick/Anatolia	ÖB 2, well	48	6.7	ppm
Simsek (2000)	EUR/TR/Germenick/Anatolia	ÖB 9, well	89	8.2	ppm
Simsek (2000)	EUR/TR/Germenick/Anatolia	ÖB 8, well	49	7.1	ppm
Simsek (2000)	EUR/TR/Germenick/Anatolia	ÖB 1, well	30	7.2	ppm
Simsek (2000)	EUR/TR/Kizlidere/Anatolia	KD 13, well	80	7.6	ppm
Simsek (2000)	EUR/TR/Kizlidere/Anatolia	KD 6A, well	84	8.2	ppm
Simsek (2000)	EUR/TR/Kizlidere/Anatolia	R 1, well	77	8.2	ppm
Simsek (2000)	EUR/TR/Kizlidere/Anatolia	Tekkehamam, well	68	7.1	ppm
Simsek (2000)	EUR/TR/Söke/Anatolia	Tuzburgazi, sp	21	6.7	ppm
Simsek (2000)	EUR/TR/Söke/Anatolia	Karina, sp	27	7.0	ppm
Simsek (2000)	EUR/TR/Söke/Anatolia	Davutlar, well	42	6.3	ppm
Simsek (2000)	EUR/TR/Söke/Anatolia	Sazliköy, sp	27	7.6	ppm
Martinovic (2000)	EUR/YU/Belotic/Macva	BBe-1, well	34	7.1	mg/l
Martinovic (2000)	EUR/YU/Bogatic/Macva	BB-1, well	75	7.1	mg/l
Martinovic (2000)	EUR/YU/Bogatis/Macva	BB-2, well	80	7.2	mg/l
Cox (1991)	OCE/NZ/Maungamuka	Maungamuka Spring, sp	18	7.3	ppm
Cox (1991)	OCE/NZ/Mokau	Mokau Spring, east, sp		8.8	ppm
Cox (1991)	OCE/NZ/Ngawha	Ngawha Spg. Hotel well, w	58	6.6	ppm
Cox (1991)	OCE/NZ/Ngawha	L. Omapere Spring, sp	30	5.8	ppm
Cox (1991)	OCE/NZ/Ngawha	Te Pua (south) Spring, sp	14	3.4	ppm
Cox (1991)	OCE/NZ/Ngawha	Te Pua (north) Spring, sp		2.5	ppm
Cox (1991)	OCE/NZ/Ngawha	Ohaeawai Spring, sp		8.6	ppm
Sheppard (1980)	OCE/NZ/Ngawha	NG2, well		6.5	ppm
Sheppard (1980)	OCE/NZ/Ngawha	NG4, well		7.6	ppm
Sheppard (1980)	OCE/NZ/Ngawha	NG9, well		7.5	ppm
Mahon (1967)	OCE/NZ/North Island	Napier-Taupo Rd 49,Spring	49	8.4	ppm
Mahon (1967)	OCE/NZ/North Island	Baths1&2, Morere	62	6.7	ppm
Mahon (1967)	OCE/NZ/North Island	Bath, Te Puia	65	6.8	ppm
Lichti (2000)	OCE/NZ/Ohaaki	Well BR22, well		9.2	ppm
Cox (1991)	OCE/NZ/OrongaBay	Oronga Bay Spring, sp	17	8.0	ppm
Cox (1991)	OCE/NZ/Puketona	Puketona, Wait.R. Spring,		6.9	ppm
Giggenbach (1992)	OCE/NZ/Rotorua	RR619, well	64	7.7	ppm
Giggenbach (1992)	OCE/NZ/Rotorua	RR889, well	100	9.6	ppm
Giggenbach (1992)	OCE/NZ/Rotorua	RR738, well	98	9.0	ppm
Giggenbach (1992)	OCE/NZ/Rotorua	RR662, well	131		ppm
Giggenbach (1992)	OCE/NZ/Rotorua	RR280, well	100	8.9	ppm
Cox (1991)	OCE/NZ/Runaruna	Runaruna Spring, sp		7.9	ppm
Cox (1991)	OCE/NZ/Tangitu	Tangitu Spring, sp		8.5	ppm
Cox (1991)	OCE/NZ/Tangowahine	Tangowahine Spring, sp		7.3	ppm
Sunaryo (1993)	OCE/NZ/TeAroha	DTB, well	67	7.9	ppm
Sunaryo (1993)	OCE/NZ/TeAroha	DTB, well	67		ppm
Sunaryo (1993)	OCE/NZ/TeAroha	MOKENA, well	85		ppm
Sunaryo (1993)	OCE/NZ/TeAroha	MOKENA, well	85	7.9	ppm
Severne (1998)	OCE/NZ/Tokaanu-Waihi	S7 Tokaanu Hotel, well	78	7.1	ppm
Cox (1991)	OCE/NZ/Waiare	Waiare Spring, sp		7.5	ppm

Na	K	Mg	Ca	Cl	SO4	HCO3	CO3	CO2	SIO2
1251.00	132.00	0.80	0.80	123.00	742.00	1906.00	361.00		
1040.00	90.00	1.70	3.20	81.00	743.00	2043.00			
1310.00	140.00	1.24	2.60	117.00	743.00	2147.00	258.00		120.00
1120.00	49.80	10.70	6.90	107.00	813.00	1426.00	261.00		248.00
1492.00	88.00	25.00	18.00	113.00	782.00	2147.00	453.00		
1420.00	154.00	1.58	0.80	122.00	758.00	2147.00	318.00		125.00
1300.00	138.00	1.26	2.00	117.00	773.00	1880.00	360.00		120.00
1300.00	136.00	1.10	1.00	116.00	745.00	2144.00	240.00		128.00
1410.00	152.00	2.70	1.70	122.00	749.00	2388.00	180.00		128.00
1340.00	138.00	0.97	2.00	117.00	730.00	2257.00	216.00		125.00
1400.00	146.00	2.70	1.60	123.00	707.00	2385.00	198.00		360.00
1505.00	90.00	17.50	90.00	1152.12	46.30	1818.59	285.99		
1320.00	100.00	67.50	167.50	1063.50	3.66	1611.13	489.12		
705.00	62.50	30.00	152.50	957.15	37.50	729.74	56.07		
1445.00	135.00	1.00	30.00	1559.80	33.30	1419.53			
1440.00	140.00	10.00	85.00	1542.07	43.62	860.80	336.50		
1410.00	122.50	1.50	62.50	1488.90	96.02	1396.72			
1440.00	60.00	1.00	50.00	1595.25	125.85	1140.21			
1125.00	132.50	1.00	20.00	102.80	637.03	2214.48	19.56		
955.00	132.50	1.50	122.50	102.80	617.91	2055.30	32.60		
1595.00	127.50	1.50	25.00	147.12	747.95	3222.20	45.66		
905.00	92.50	4.50	45.00	88.62	988.84	820.94	218.70		
9230.00	360.00	437.50	640.00	15598.00	1577.87	267.97	39.24		
11725.00	480.00	462.50	715.00	19852.00	1858.98	119.74	22.44		
1650.00	110.00	80.00	462.50	2747.37	58.73	1185.78	112.14		
37.30	7.00	44.00	152.00	49.63	22.59	437.86	112.14		
210.00	9.00	12.00	40.00	114.00	6.00	555.00			23.00
155.00	11.00	10.00	40.00	107.00	4.00	409.00			64.00
149.00	10.00	9.00	47.00	113.00	2.00	408.00			66.00
201.00	0.60	94.00	5.90	15.00	9.00	1039.00			
434.00	60.90	80.30	8.50	73.00	9.00	1584.00			28.00
131.00	10.00	37.20	94.00	67.00	48.00	704.00			128.00
55.00	8.00	12.80	34.00	26.00	14.00	283.00			130.00
9.00	1.00	2.50	2.80	23.00	36.00	224.00			10.00
6.20	4.00	1.70	3.20	17.00	390.00				37.00
37.80	3.30	27.00	22.80	14.00	3.00	274.00			97.00
861.00	80.00	0.60	16.00	1162.00	42.00	590.00			426.00
1025.00	90.00	0.11	2.90	1475.00	27.00	298.00			464.00
1011.00	90.00	0.10	2.90	1437.00	35.00	486.00			471.00
500.00	9.00	0.10	12.00	660.00	82.00			111.00	42.00
6100.00	100.00	137.00	3900.00	16000.00	21.00			25.00	28.00
4550.00	22.00	8.00	815.00	8300.00	110.00			60.00	52.00
1007.00	217.00	0.01	1.30	1657.00	4.00	44.00			849.00
486.00	14.80	58.00	152.00	903.00	98.00	220.00			37.00
2000.00	80.70	57.00	195.00	1371.00	22.00	4471.00			62.00
185.00	14.00	0.20	10.00	159.00	33.00	228.00			157.00
1190.00	145.00	0.03	19.00	1865.00	11.00	50.00		630.00	448.00
452.00	34.00	0.12	13.00	393.00	219.00	287.00			263.00
318.00	77.00	0.13	11.60	387.00	148.00	121.00			283.00
933.00	126.00	0.04	3.10	1369.00	12.00	210.00		719.00	322.00
8474.00	238.00	181.00	37.90	9961.00	55.00	5937.00			22.00
453.00	40.80	63.00	42.70	19.00	11.00	1534.00			117.00
4365.00	305.00	162.00	99.00	5699.00	69.00	4170.00			14.00
3159.00	74.00	3.50	5.60	631.00	348.00	7700.00			108.00
3200.00	66.00	4.00	11.20	550.00	390.00	7700.00			133.00
3200.00	66.00	3.90	6.90	550.00	390.00	7500.00			135.00
3385.00	70.00	3.60	8.20	574.00	312.00	7500.00			116.00
436.00	48.00	36.60	53.00	685.00	14.00	281.00			216.00
56.00	3.50	55.50	49.80	35.00	7.00	556.00			80.00

Reference	Location	Site	T	pH	unit
Cox (1991)	OCE/NZ/Waikoura	Waikoura Spring A, sp		7.0	ppm
Giggenbach (1994)	OCE/NZ/Waiotapu	W8, well(2)		8.6	ppm
Giggenbach (1994)	OCE/NZ/Waiotapu	BR, well		6.5	ppm
Giggenbach (1994)	OCE/NZ/Waiotapu	WK, well		8.3	ppm
Mahon (1967)	OCE/NZ/Wairakei	27		8.5	ppm
Mahon (1967)	OCE/NZ/Wairakei	44		8.6	ppm
Mahon (1967)	OCE/NZ/Wairakei	9		6.9	ppm
Mroczek (1999)	OCE/NZ/Wairakei	FP10, well		8.6	mg/l
Wood (1997)	OCE/NZ/Wairakei	WK-307, well	93	7.8	mg/l
Wood (1997)	OCE/NZ/Wairakei	WK-230, well	175	6.5	mg/l
Reyes (1999)	OCE/NZ/Wairakei	WK-620, well	129	6.6	ppm
A.R.W.B. (1980)	OCE/NZ/Waiwera/NorthIsland	1, well	47	8.7	mg/l
A.R.W.B. (1980)	OCE/NZ/Waiwera/NorthIsland	8, well	49	8.7	mg/l
A.R.W.B. (1980)	OCE/NZ/Waiwera/NorthIsland	11, well	45	8.6	mg/l
A.R.W.B. (1980)	OCE/NZ/Waiwera/NorthIsland	27, well	36	8.4	mg/l
A.R.W.B. (1980)	OCE/NZ/Waiwera/NorthIsland	12, well	49	8.7	mg/l
A.R.W.B. (1980)	OCE/NZ/Waiwera/NorthIsland	18, well	40	8.8	mg/l
A.R.W.B. (1980)	OCE/NZ/Waiwera/NorthIsland	31, well	41	8.4	mg/l
A.R.W.B. (1980)	OCE/NZ/Waiwera/NorthIsland	34, well	42	8.8	mg/l
A.R.W.B. (1980)	OCE/NZ/Waiwera/NorthIsland	29, well	47	8.5	mg/l
A.R.W.B. (1980)	OCE/NZ/Waiwera/NorthIsland	33, well	47	8.5	mg/l
A.R.W.B. (1980)	OCE/NZ/Waiwera/NorthIsland	40, well	20	7.0	mg/l
A.R.W.B. (1980)	OCE/NZ/Waiwera/NorthIsland	22, well	43	8.9	mg/l
A.R.W.B. (1980)	OCE/NZ/Waiwera/NorthIsland	21, well	35	8.9	mg/l
A.R.W.B. (1986)	OCE/NZ/Waiwera/NorthIsland	12, well		8.7	mg/l
A.R.W.B. (1994)	OCE/NZ/Waiwera/NorthIsland	22, well		8.8	mg/l
A.R.W.B. (1994)	OCE/NZ/Waiwera/NorthIsland	29, well		8.3	mg/l
A.R.W.B. (1994)	OCE/NZ/Waiwera/NorthIsland	25, well		8.6	mg/l
A.R.W.B. (1994)	OCE/NZ/Waiwera/NorthIsland	12, well		8.6	mg/l
A.R.W.B. (1997)	OCE/NZ/Waiwera/NorthIsland	22, well		8.9	mg/l
A.R.W.B. (1997)	OCE/NZ/Waiwera/NorthIsland	29, well		8.4	mg/l
A.R.W.B. (1997)	OCE/NZ/Waiwera/NorthIsland	25, well		8.6	mg/l
A.R.W.B. (1997)	OCE/NZ/Waiwera/NorthIsland	12, well		8.7	mg/l
A.R.W.B. (1997)	OCE/NZ/Waiwera/NorthIsland	31, well		8.6	mg/l
A.R.W.B. (1986)	OCE/NZ/Waiwera/NorthIsland	32, well	45	8.5	mg/l
A.R.W.B. (1980)	OCE/NZ/Waiwera/NorthIsland	31, well		8.4	mg/l
A.R.W.B. (1980)	OCE/NZ/Waiwera/NorthIsland	33, well		8.3	mg/l
A.R.W.B. (1980)	OCE/NZ/Waiwera/NorthIsland	35, well		8.3	mg/l
A.R.W.B. (1986)	OCE/NZ/Waiwera/NorthIsland	21, well		8.9	mg/l
Cox (1991)	OCE/NZ/Wekaweka	Wekaweka Spring, sp		6.7	ppm

Na	K	Mg	Ca	Cl	SO4	HCO3	CO3	CO2	SIO2
740.00	18.80	78.00	465.00	186.00	31.00	3200.00			26.00
457.00	40.00	0.01	3.40	591.00	77.00	128.00			423.00
886.00	172.00	0.00	0.60	1360.00	10.00	438.00			812.00
1290.00	220.00	0.01	18.50	2250.00	34.00	0.50			620.00
1200.00	200.00	0.05	17.50	2156.00	25.00	23.00			660.00
1320.00	225.00	0.04	17.50	2260.00	36.00	19.00			650.00
1190.00	75.00	0.06	35.00	1950.00	16.00	22.00			320.00
1279.00	190.00	0.01	18.70	2068.00	29.00				533.00
123.00	3.10	0.52	9.10	11.00	10.40	339.00			128.00
118.00	12.10	0.06	9.50	96.00	53.00	165.00			255.00
174.00	11.80	0.88	45.00	185.00	41.00	300.00			240.00
730.00	6.90	0.17	41.25	1160.00	14.40	1.11	16.80		32.30
715.00	8.30	0.17	42.45	1140.00	4.40	0.20	16.90		33.70
1080.00	10.70	0.71	105.73	1770.00	68.80	0.60	14.40		32.10
980.00	9.30	1.68	100.92	1600.00	73.20	10.50	9.20		30.90
790.00	9.30	0.17	52.46	1260.00	19.20	0.60	16.40		33.10
715.00	4.85	0.56	36.84	1132.00	3.60	2.00	17.80		28.60
780.00	7.80	0.68	62.88	1244.00	40.60	13.80	6.70		29.60
665.00	2.58	0.19	38.45	1040.00	14.00	9.60	11.30		26.70
975.00	10.80	4.57	113.74	1610.00	76.20	12.20	12.50		31.10
935.00	9.70	0.75	90.51	1520.00	71.60	6.70	11.30		30.20
152.00	10.50	38.39	125.35	166.00	79.40	558.00			25.60
590.00	3.64	0.12	28.03	930.00	2.20	7.70	15.00		26.20
490.00	1.58	1.24	26.43	770.00	6.60	17.60	12.60		22.10
690.00	9.20	0.10	42.80	1140.00	11.10	18.90			30.20
550.00	3.12	0.20	26.40	911.80	1.40	32.70			52.90
740.00	7.66	1.70	59.20	1266.00	16.10	41.70			63.83
750.00	5.16	0.30	51.20	1290.00	6.70	28.70			62.55
650.00	8.30	0.10	46.40	1130.00	3.50	29.70			67.19
548.00	3.08	0.10	25.30	847.00	1.40	33.10			54.83
700.00	8.25	3.70	54.50	1186.00	15.70	42.00			67.90
764.00	5.86	0.50	48.00	1336.00	15.70	23.10			68.97
668.00	8.75	0.40	45.00	1180.00	2.40	29.80			70.25
676.00	6.50	0.40	43.00	1144.00	10.50	27.40			63.83
916.00	9.40	0.90	71.20	1470.00	43.00	21.80			28.60
650.00	5.59	0.40	45.60	1150.00	10.00	29.30			62.54
713.00	6.48	0.50	56.00	1280.00	9.60	28.70			64.90
770.00	6.74	1.10	54.40	1270.00	10.70	31.50			64.90
498.00	1.80	0.15	26.20	790.00	2.10	33.10			42.19
3536.00	51.90	11.70	1120.00	7540.00	60.00	71.00			35.00

Lecture Notes in Earth Sciences

For information about Vols. 1–29
please contact your bookseller or Springer-Verlag

Vol. 30: E. G. Kauffmann, O. H. Walliser (Eds.), Extinction Events in Earth History. VI. 432 pages. 1990.

Vol. 31: K.-R. Koch, Bayesian Inference with Geodetic Applications. IX, 198 pages. 1990.

Vol. 32: B. Lehmann, Metallogeny of Tin. VIII, 211 pages. 1990.

Vol. 33: B. Allard, H. Borén, A. Grimvall (Eds.), Humic Substances in the Aquatic and Terrestrial Environment. VIII, 514 pages. 1991.

Vol. 34: R. Stein, Accumulation of Organic Carbon in Marine Sediments. XIII, 217 pages. 1991.

Vol. 35: L. Håkanson, Ecometric and Dynamic Modelling. VI, 158 pages. 1991.

Vol. 36: D. Shangguan, Cellular Growth of Crystals. XV, 209 pages. 1991.

Vol. 37: A. Armanini, G. Di Silvio (Eds.), Fluvial Hydraulics of Mountain Regions. X, 468 pages. 1991.

Vol. 38: W. Smykatz-Kloss, S. St. J. Warne, Thermal Analysis in the Geosciences. XII, 379 pages. 1991.

Vol. 39: S.-E. Hjelt, Pragmatic Inversion of Geophysical Data. IX, 262 pages. 1992.

Vol. 40: S. W. Petters, Regional Geology of Africa. XXIII, 722 pages. 1991.

Vol. 41: R. Pflug, J. W. Harbaugh (Eds.), Computer Graphics in Geology. XVII, 298 pages. 1992.

Vol. 42: A. Cendrero, G. Lüttig, F. Chr. Wolff (Eds.), Planning the Use of the Earth's Surface. IX, 556 pages. 1992.

Vol. 43: N. Clauer, S. Chaudhuri (Eds.), Isotopic Signatures and Sedimentary Records. VIII, 529 pages. 1992.

Vol. 44: D. A. Edwards, Turbidity Currents: Dynamics, Deposits and Reversals. XIII, 175 pages. 1993.

Vol. 45: A. G. Herrmann, B. Knipping, Waste Disposal and Evaporites. XII, 193 pages. 1993.

Vol. 46: G. Galli, Temporal and Spatial Patterns in Carbonate Platforms. IX, 325 pages. 1993.

Vol. 47: R. L. Littke, Deposition, Diagenesis and Weathering of Organic Matter-Rich Sediments. IX, 216 pages. 1993.

Vol. 48: B. R. Roberts, Water Management in Desert Environments. XVII, 337 pages. 1993.

Vol. 49: J. F. W. Negendank, B. Zolitschka (Eds.), Paleolimnology of European Maar Lakes. IX, 513 pages. 1993.

Vol. 50: R. Rummel, F. Sansò (Eds.), Satellite Altimetry in Geodesy and Oceanography. XII, 479 pages. 1993.

Vol. 51: W. Ricken, Sedimentation as a Three-Component System. XII, 211 pages. 1993.

Vol. 52: P. Ergenzinger, K.-H. Schmidt (Eds.), Dynamics and Geomorphology of Mountain Rivers. VIII, 326 pages. 1994.

Vol. 53: F. Scherbaum, Basic Concepts in Digital Signal Processing for Seismologists. X, 158 pages. 1994.

Vol. 54: J. J. P. Zijlstra, The Sedimentology of Chalk. IX, 194 pages. 1995.

Vol. 55: J. A. Scales, Theory of Seismic Imaging. XV, 291 pages. 1995.

Vol. 56: D. Müller, D. I. Groves, Potassic Igneous Rocks and Associated Gold-Copper Mineralization. 2nd updated and enlarged Edition. XIII, 238 pages. 1997.

Vol. 57: E. Lallier-Vergès, N.-P. Tribovillard, P. Bertrand (Eds.), Organic Matter Accumulation. VIII, 187 pages. 1995.

Vol. 58: G. Sarwar, G. M. Friedman, Post-Devonian Sediment Cover over New York State. VIII, 113 pages. 1995.

Vol. 59: A. C. Kibblewhite, C. Y. Wu,Wave Interactions As a Seismo-acoustic Source. XIX, 313 pages. 1996.

Vol. 60: A. Kleusberg, P. J. G. Teunissen (Eds.), GPS for Geodesy. VII, 407 pages. 1996.

Vol. 61: M. Breunig, Integration of Spatial Information for Geo-Information Systems. XI, 171 pages. 1996.

Vol. 62: H. V. Lyatsky, Continental-Crust Structures on the Continental Margin of Western North America. XIX, 352 pages. 1996.

Vol. 63: B. H. Jacobsen, K. Mosegaard, P. Sibani (Eds.), Inverse Methods. XVI, 341 pages, 1996.

Vol. 64: A. Armanini, M. Michiue (Eds.), Recent Developments on Debris Flows. X, 226 pages. 1997.

Vol. 65: F. Sansò, R. Rummel (Eds.), Geodetic Boundary Value Problems in View of the One Centimeter Geoid. XIX, 592 pages. 1997.

Vol. 66: H. Wilhelm, W. Zürn, H.-G. Wenzel (Eds.), Tidal Phenomena. VII, 398 pages. 1997.

Vol. 67: S. L. Webb, Silicate Melts. VIII. 74 pages. 1997.

Vol. 68: P. Stille, G. Shields, Radiogenetic Isotope Geochemistry of Sedimentary and Aquatic Systems. XI, 217 pages. 1997.

Vol. 69: S. P. Singal (Ed.), Acoustic Remote Sensing Applications. XIII, 585 pages. 1997.

Vol. 70: R. H. Charlier, C. P. De Meyer, Coastal Erosion – Response and Management. XVI, 343 pages. 1998.

Vol. 71: T. M. Will, Phase Equilibria in Metamorphic Rocks. XIV, 315 pages. 1998.

Vol. 72: J. C. Wasserman, E. V. Silva-Filho, R. Villas-Boas (Eds.), Environmental Geochemistry in the Tropics. XIV, 305 pages. 1998.

Vol. 73: Z. Martinec, Boundary-Value Problems for Gravimetric Determination of a Precise Geoid. XII, 223 pages. 1998.

Vol. 74: M. Beniston, J. L. Innes (Eds.), The Impacts of Climate Variability on Forests. XIV, 329 pages. 1998.

Vol. 75: H. Westphal, Carbonate Platform Slopes – A Record of Changing Conditions. XI, 197 pages. 1998.

Vol. 76: J. Trappe, Phanerozoic Phosphorite Depositional Systems. XII, 316 pages. 1998.

Vol. 77: C. Goltz, Fractal and Chaotic Properties of Earthquakes. XIII, 178 pages. 1998.

Vol. 78: S. Hergarten, H. J. Neugebauer (Eds.), Process Modelling and Landform Evolution. X, 305 pages. 1999.

Vol. 79: G. H. Dutton, A Hierarchical Coordinate System for Geoprocessing and Cartography. XVIII, 231 pages. 1999.

Vol. 80: S. A. Shapiro, P. Hubral, Elastic Waves in Random Media. XIV, 191 pages. 1999.

Vol. 81: Y. Song, G. Müller, Sediment-Water Interactions in Anoxic Freshwater Sediments. VI, 111 pages. 1999.

Vol. 82: T. M. Løseth, Submarine Massflow Sedimentation. IX, 156 pages. 1999.

Vol. 83: K. K. Roy, S. K. Verma, K. Mallick (Eds.), Deep Electromagnetic Exploration. X, 652 pages. 1999.

Vol. 84: H. V. Lyatsky, G. M. Friedman, V. B. Lyatsky. Principles of Practical Tectonic Analysis of Cratonic Regions. XX, 369 pages. 1999.

Vol. 85: C. Clauser, Thermal Signatures of Heat Transfer Processes in the Earth's Crust. X, 111 pages. 1999.

Vol. 86: H. V. Lyatsky, V. B. Lyatsky, The Cordilleran Miogeosyncline in North America. XX, 384 pages. 1999.

Vol. 87: M. Tiefelsdorf, Modelling Spatial Processes. XVIII, 167 pages. 2000.

Vol. 88: S. Rodrigues-Filho, G. Müller, A Holocene Sedimentary Record from Lake Silvana, SE Brazil. XII, 96 pages. 1999.

Vol. 89: M. Makhous, The Formation of Hydrocarbon Deposits in the North African Basins. XII, 329 pages. 2000.

Vol. 90: R. Klees, R. Haagmans (Eds.), Wavelets in the Geosciences. XVIII, 241 pages. 2000.

Vol. 91: I. Gilmour, C. Koeberl (Eds.), Impacts and the Early Earth. XVIII, 445 pages. 2000.

Vol. 92: P. C. Hansen, B. H. Jacobsen, K. Mosegaard (Eds.), Methods and Applications of Inversion. X, 304 pages. 2000.

Vol. 93: A. Montanari, Ch. Koeberl, Impact Stratigraphy. XIII, 364 pages. 2000.

Vol. 94: M. Breunig, On the Way to Component-Based 3D/4D Geoinformation Systems. XI, 199 pages. 2001.

Vol. 95: A. Dermanis, A. Grün, F. Sansò (Eds.), Geomatic Methods for the Analysis of Data in the Earth Sciences. XII, 265 pages. 2000.

Vol. 96: J. Cobbing, The Geology and Mapping of Granite Batholiths. XII, 141 pages. 2000.

Vol. 97: H. J. Neugebauer, C. Simmer, Dynamics of Multiscale Earth Systems. XIII, 357 pages. 2003.

Vol. 98: T. Takanami, G. Kitagawa, Methods and Applications of Signal Processing in Seismic Network Operations. XIV, 266 pages. 2003.

Vol. 99: M. S. Rosenbaum, A. K. Turner (Eds.), New Paradigms in Subsurface Prediction. XXVI, 397 pages. 2003.

Vol. 100: H. Burchard, Applied Turbulence Modelling in Marine Waters. XIV, 215 pages. 2002.

Vol. 101: A. Lang, K. Hennrich, R. Dikau (Eds.), Long Term Hillslope and Fluvial System Modelling. 246 pages. 2003.

Vol. 102: W. Smykatz-Kloss, P. Felix-Henningsen (Eds.), Paleoecology of Quaternary Drylands. VI, 250 pages. 2003.

Vol. 103: M. Kühn, Reactive Flow Modeling of Hydrothermal Systems. XI, 261 pages. 2003.

Batch number: 10703972

Printed by Printforce, the Netherlands